CHEMISTRY IN THE LABORATORY

Charles W. J. Scaife
Union College

O. T. Beachley, Jr.
State University of New York — Buffalo

Saunders Golden Sunburst Series
SAUNDERS COLLEGE PUBLISHING
Philadelphia Fort Worth Chicago
San Francisco Montreal Toronto
London Sydney Tokyo

Chemistry in the Laboratory

ISBN # 0-03-058353-5

9   066   9876543

Saunders College Publishing
Holt, Rinehart and Winston
The Dryden Press

The cover of this book shows an artist's rendition of an experiment illustrating both the dropping of a liquid from a buret and a chemical reaction. It is designed to stimulate interest and to illustrate dramatic and esthetic effects. As such it violates several safety precautions and would not be accomplished this way in the laboratory. In particular, one would at least place the reacting sample within a container such as a watch glass or a beaker.

PREFACE

This book is designed to provide experiments for both a general chemistry course and for an inorganic chemistry course as covered either within general chemistry or as a separate course. This book has been written through the collaboration of two authors, one at a small, private, liberal arts college, and the other at a large, public, state university. They have taught general chemistry to a wide spectrum of college students ranging from nonscience majors in humanities and social sciences through science majors, engineers, and advanced placement chemistry majors. In the past, they have written laboratory experiments relating to a two-week laboratory experience for middle school students, a college laboratory course for nonscience majors, a laboratory manual for general chemistry, and both sophomore-level and senior-level inorganic chemistry courses. Thus, they brought extensive experience into their writing.

The experiments were selected for a variety of reasons, and a good balance of experiments is provided. Several are designed primarily to teach laboratory techniques and skills that will be required in later experiments or courses. A couple involve the determination of physical properties. A few involve the use of instruments other than analytical balances. Some give practice in the synthesis and characterization of species. Some provoke thoughtful appreciation of the experimental basis of chemistry. Some illustrate fundamental chemical principles. Some require gravimetric or volumetric analyses characteristic of classical quantitative analysis courses. Several involve separations and identifications characteristic of classical qualitative analysis courses. A number relate to the recent increasing emphasis in inorganic reaction chemistry and weave the reaction chemistry of common elements into experiments involving the handling of many important materials. Most of the experiments can be utilized with either small or large groups, but several are inappropriate for a large class of general chemistry students at a public university. Many of the experiments involve the excitement and frustration of unknown samples.

The length of many experiments is such that they can be completed in a single laboratory period. An instructor often has considerable time flexibility, even within a single period, by designating only well defined parts of experiments to be accomplished, by having different students work on different parts and then share data, or by specifying a lesser number of runs than the experiment suggests to check reproducibility. Some experiments are specifically designed as multi-week experiments, with the longest ones requiring up to four weeks. However, an instructor again has the flexibility of assigning only a part of these experiments as a one-week effort.

The format of the experiments is designed to ensure maximum teaching efficiency and benefit for the students. The format of most experiments is the same so that students can quickly develop study and work patterns and can understand the expectations of the instructor early in the term. Most experiments have an **INTRODUCTION** followed by some or all of the divisions called **REFERENCES**, **EXPERIMENTAL PROCEDURE**, **PRE-LABORATORY QUESTIONS**, **DATA**, **CALCULATIONS**, and **POST-LABORATORY QUESTIONS**. The introductions provide a general background including most calculations required and state clearly the purpose of the experiment. **REFERENCES** give sources in which students can find additional background in more detail than that given in the introduction. **EXPERIMENTAL PROCEDURE** is usually an explicit and detailed segment, although students are occasionally asked to devise their own methods and then get approval for using them from their instructors. Techniques that are used frequently are discussed in detail and illustrated clearly in figures in an extensive section on **LABORATORY METHODS** that precedes the experiments. A reference is given in **EXPERIMENTAL PROCEDURE** the first time each Laboratory Method is required in a given experiment. Laboratory safety is stressed in a section on **SAFETY PRECAUTIONS** that

Preface

precedes the experiments, and possible hazards are pointed out in each experiment by **CAUTION:** followed by words of warning that are underlined and boldfaced. Chemicals were chosen very carefully for their safety and their low cost. In addition, the amounts of chemicals used were kept small to maximize safety and to minimize costs and strains on hood space and ventilation equipment. Equipment needs are minimal, and alternative pieces of equipment are often suggested. **PRE-LABORATORY QUESTIONS** are designed to get students to study **EXPERIMENTAL PROCEDURES** carefully <u>before</u> the laboratory and to understand what procedures they are going to perform, why these procedures are necessary, and what calculations are required for the laboratory report. Instructors may require some or all of these questions to be turned in, check during laboratory to see that assigned questions were answered, or give a quiz covering one or more of these questions at the beginning of the laboratory period. The **DATA** section provides space for recording numerical data for quantitative experiments or observations for qualitative experiments. The **DATA** section also has questions relating to the qualitative observations that students must answer and turn in as part of their report. The **CALCULATIONS** section provides tables for recording results calculated from the data as well as space for sample calculations. **POST-LABORATORY QUESTIONS** are designed to get students to think more deeply about illustrated concepts after the experiment has been completed, to consider what effect certain procedural errors might have had on their calculated results or conclusions, or to apply what they have learned to new but related situations. Instructors may require one or more of these to be turned in, or they may give a quiz on one or more of these at the next laboratory period. Enough questions are provided to give flexibility in assigning different questions for different laboratory sections or during different terms. Some questions are intentionally written in a flexible fashion so that instructors can change them each term by designating specific words or formulas. Pages can easily be removed to hand in as parts of reports.

We wish to express our appreciation for the assistance we have received from many sources. Some of our colleagues on the staffs at Union College and the State University of New York at Buffalo have made valuable suggestions. In particular, we acknowledge the suggestions and critical appraisals of Professor Charles F. Weick of Union College and the extensive contribution of ideas and supervision of testing by Priscilla B. Clarke of SUNY, Buffalo. In addition, Laurie LeTarte and David Shinberg tested some of the experimental procedures at Union College, and Rebecca Scaife and Priscilla Scaife did much of the word processing and make-up of the manuscript. We also appreciate the efforts and suggestions of countless students who endured early drafts of some of these experiments. Our thanks also go to John Vondeling and the other members of the staff of Saunders College Publishing who encouraged the writing of this book and saw it through production. Finally, we are grateful for the patience, encouragement, and assistance of our wives and children during the preparation of the manuscript.

Charles W. J. Scaife

O. T. Beachley, Jr.

CONTENTS

Contents

EXPERIMENTS (continued)

GENERAL INSTRUCTIONS

The following general instructions should be studied carefully before any laboratory experiments are performed.

The science of chemistry rests solidly on observations and quantitative data obtained from careful and critical experimentation by people who are curious to know how matter behaves. Some first-hand experience with apparatus and experimental methods is essential for a real understanding of the factual knowledge and basic principles of chemistry. You gain this experience by performing certain of the experiments in this book. Other of the experiments expose you to new concepts and ideas, sometimes even before they are discussed in lecture. This is appropriate because chemistry frequently advances from experimental results and interpretations to concepts and ideas, not the other way around. Other experiments give you experience in scientific inquiry and decision making. You will often be forced to make decisions on what seems like insufficient evidence, just as you frequently have to do in everyday nonscientific decisions. All experiments are written in such a way that you can begin with no prior knowledge providing one or more of the references are studied, but are also designed to build on any sound previous experience you may have had in chemistry. Those students who go on to more advanced courses in chemistry will acquire additional skills in synthetic methods and quantitative experimental techniques. Laboratory experiments are both fun and frustrating, and you should be prepared to experience both of these emotions.

A. General Laboratory Facilities

The laboratory is equipped with chemical workdesks that provide space on which to work and contain individual equipment drawers or lockers. The desks are also equipped with sources of water and fuel gas, with aspirators, and sometimes with fume hoods. Sinks are available for washing apparatus and for disposing nontoxic aqueous solutions. Solid waste containers are provided, with a specific one being designated for GLASS. Specifically labeled containers will also be available on the reagent bench for disposing of designated solids or liquids used in particular experiments. If you do not have a fume hood at your desk, there are ventilated hoods nearby in which you can safely perform experiments that liberate noxious gases or fumes. Beam balances or electronic top-loading balances on equipment benches or shelves are used for rough weighings. More sensitive analytical balances are used for more precise weighings. Additional shelves or benches provide space for special equipment or chemicals required for each day's experiment. Other special equipment as well as replacements for common apparatus or expendable supplies are signed out from the stockroom near the laboratory. All items of special equipment should be returned either to the equipment bench or to the stockroom before the end of each laboratory period. Much of the special equipment in the laboratory is both expensive and easily stolen, thus adding considerably to laboratory fees. If you see people (especially those younger than college age) around the building who apparently don't belong there, or if you spot any apparently suspicious activities, notify a faculty or staff member of the Chemistry Department immediately.

Do not place personal items like coats, knapsacks, or books on the laboratory desks because they take up valuable work space and may deteriorate as a result of a chemical spill that was not properly cleaned up. Cloakrooms or lockers are normally provided for your personal items. Do not leave valuables in the cloakrooms.

EQUIPMENT
(alphabetical order)

Beaker

Wide-mouth
Bottle

Buchner Funnel

Leveling
Bulb

Bunsen Burner

Burner Tip

Flame Spreader

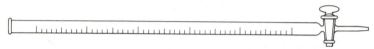
Buret with Glass Stopcock

Crucible and
Cover

Crucible Tongs

Erlenmeyer Flask

Evaporating Dish

Drying
Tube

File

Filter Flask

Florence Flask

Funnel

Graduated Cylinder

Gooch Crucible
and Adapter

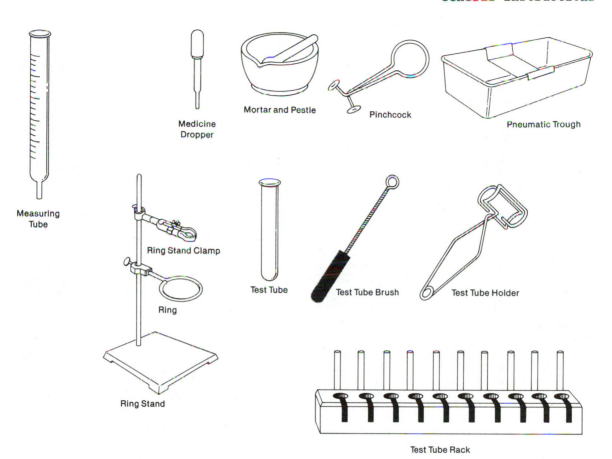

Measuring
Tube

Medicine
Dropper

Mortar and Pestle

Pinchcock

Pneumatic Trough

Ring Stand Clamp

Ring

Ring Stand

Test Tube

Test Tube Brush

Test Tube Holder

Test Tube Rack

Thermometer

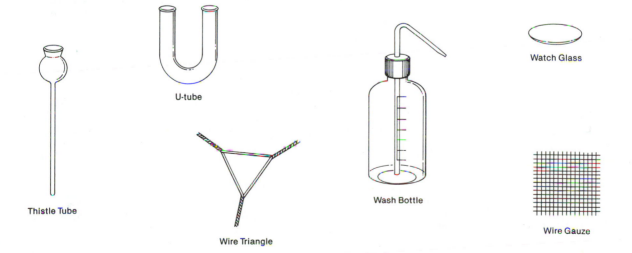

Thistle Tube

U-tube

Wire Triangle

Wash Bottle

Watch Glass

Wire Gauze

3

General Instructions

B. Individual Locker Equipment

You are assigned a drawer or locker containing equipment that you will use and be responsible for during the term. The contents of your drawer or locker should be inspected and compared to the list provided during the first laboratory period to determine that nothing is missing and that all pieces are in working condition. An alphabetical chart of equipment on pages 2 and 3 will help you identify pieces on the check-in sheet provided by your instructor. However, not all pieces of equipment in your drawer or locker are shown on the chart. The chart may also help you identify special pieces that are not in your individual drawer or locker and that must be obtained from the equipment bench or stockroom for particular experiments. Return all special equipment to the equipment bench or stockroom at the end of each laboratory period.

Be sure to lock your drawer or locker before leaving the laboratory at the end of each period and to bring your key or combination to the next period. During or at the end of the term you will be required to replace all equipment that has been used up, lost, or broken. If you withdraw from the course, arrange with your instructor to check out your apparatus immediately. The Department of Chemistry reserves the right to open your drawer or locker in an emergency (for example, to search for or repair a leak in the plumbing).

C. Reagents

The reagents for experiments assigned to your section for each period will be found on shelves or benches at the sides of the laboratory. Use only the reagents assigned to your section. **Read the labels on bottles very carefully to make certain that you are using the correct reagents.** So that there can be no confusion, reagent bottles should be labeled with a chemical name, a concentration if the reagent is not a pure liquid or a solid, and a chemical formula. **Never carry reagent bottles away from the reagent bench.** Bring test tubes, beakers, flasks, or graduated cylinders with you to obtain liquid reagents. Use beakers or sheets of paper that are provided near the balances to obtain solid reagents. Take only the quantity of reagent specified in the experiment. **Measure the quantities as carefully as significant figures given in the experiment require.** Correct data may not be obtained, and accidents may result if specified amounts are not used. Never dip a stirring rod or medicine dropper into a reagent bottle containing liquid. Instead, transfer a small quantity of the reagent to a test tube or beaker, and dip the stirring rod into that. Handle stoppers so that they remain clean, and be sure to replace them into the correct bottles. Never return reagents that you have been using to the original reagent bottles because dangerous contamination may result. **If you spill reagents, clean them up immediately,** checking first with your instructor if you have any doubts about how to do that safely. An absorbent mixture is normally available for spills of acid solutions and some other liquid reagents.

D. Care of the Laboratory

Good housekeeping is a prerequisite for safe and accurate experimentation. Keep your desk and the reagent bench clean and tidy at all times. A dustpan and brush are available to clean up any pieces of broken glass or other solids. At the end of each period, clean the apparatus you have used, and place it in order in your drawer or locker. Wipe off the top of your desk, and wash off any material you may have spilled on the front of the desk.

Dispose of solid wastes by placing them in one of the containers provided for that purpose; **never put any solid materials in the sinks.** A specific solid waste container is labeled for GLASS. Pour waste liquids into the designated containers.

E. Experimental Work

Experiments will be assigned at least one week in advance and perhaps even in a syllabus given to you at the beginning of the term. Study carefully the **INTRODUCTION**, **EXPERIMENTAL PROCEDURE**, specifically referenced **LABORATORY METHODS**, and **PRE-LABORATORY QUESTIONS** before coming to the laboratory. One purpose of the **PRE-LABORATORY QUESTIONS** is to help you understand what procedures you are going to perform and why they are necessary. Your instructor may require you to turn in the **PRE-LABORATORY QUESTIONS** or may quiz you on them at the beginning of the laboratory. Refer to your lecture notes or to appropriate sections of your textbook in preparing for the laboratory. Following prior careful preparation, you will be able to work efficiently in the laboratory, and you will derive more benefit from the experiments. Ask questions before you start the experiment if there is a procedure or technique that you do not understand.

Follow directions rigorously. Observe closely what happens; don't let even minor pieces of evidence slip by you. Record data and behavior carefully in the **DATA** section of your experiment. Be sure you understand what is happening at each stage. Your instructor may ask you questions during your experimentation to prod you to think more deeply about what you are doing and what you hope to learn.

Attendance is obviously required, and a passing grade in the course requires completion of the laboratory work. Make-up laboratories may be permitted at some colleges and universities, but only for valid reasons and only by approval of both instructors involved in writing.

F. Constructing Apparatus

Diagrams of the more complicated setups of apparatus are given in each experiment, and use of model setups may be demonstrated by your instructor. Be sure to study the diagram before you attempt to construct a setup. Always use clean glassware. Support all of the heavier parts firmly, either by clamping them in position or by resting them on the top of the desk. Do not support equipment with books or notebooks.

G. Laboratory Record

Always fill in your name, student ID number, section, date, and instructor on the **DATA** page as you begin an experiment. Also write down the number or letter of your unknown, if appropriate. Record numerical data in the tables that are designed to help you organize your data. Reminders given in **EXPERIMENTAL PROCEDURE** refer you to specific tables when you need to record numerical data. Your instructor may require you to hand in a carbon copy of your numerical data before you leave the laboratory, and will describe a method for accomplishing this. Record observations clearly and concisely in the spaces provided. Reminders given in **EXPERIMENTAL PROCEDURE** refer you to specific tables when you need to record observations. Each such space in a table is indicated by **Observation** as a reminder. No observation is too minor to record. It is easy to decide later that something is unimportant, but it is impossible to base a conclusion on evidence that was not recorded and cannot be recalled. Specimen records may be posted on a bulletin board in the laboratory. Use these as models for writing your own notes. Your numerical data and observations must be recorded in the laboratory as you perform the experiment. Above all, **write legibly.** Numerous errors are made in laboratory reports because students misread their own numerical data or observations.

General Instructions

H. Laboratory Reports

Lengthy, extensive reports are not required for most experiments. As a minimum, your instructor will require you to turn in sections of each experiment marked **DATA** and **CALCULATIONS** for quantitative experiments and just the section marked **DATA** for qualitative or descriptive experiments. As a maximum, your instructor may also require any or all **POST-LABORATORY QUESTIONS** or other questions provided. In qualitative experiments **QUESTIONS** relating to **Observations** made in the laboratory must be answered. In either type of experiment you should strive to finish the **CALCULATIONS** or **QUESTIONS** before leaving the laboratory. You work more rapidly and efficiently while the procedures and evidence are fresh in your mind, the immediate effort solidifies and helps you to remember more of the concepts, your instructor is readily available for help particularly if you have any difficulty with **CALCULATIONS** or with interpreting evidence to answer **QUESTIONS**, and you don't have to face the trauma of completing a laboratory report the night or early morning before it is due at a time when you have long since forgotten the details of what it was about. Appropriate pages can be removed or torn out easily when you are ready to turn them in. Most instructors have a penalty for late reports, and the penalty frequently gets stiffer after graded reports are already turned back to other students.

Plagiarism in the preparation of laboratory reports must be scrupulously avoided. The definition of plagiarism and the penalties for this practice are described in student handbooks and specific pamphlets published by your college or university and are usually available in the office of the Dean. Several amplifying statements relate specifically to this course. Except for several experiments in which you work with a partner, **you are expected to collect laboratory data independently.** Even in partner experiments your report should be prepared independently. It is often advantageous and a useful learning experience to consult with fellow students concerning both concepts and calculations required for laboratory reports; however, all critical thinking and final calculations and writing must be your own.

I. Evaluation of Your Work

Various instructors will weigh distinct factors differently, but normally instructors will consider most or all of the following criteria in evaluating your performance: (1) your ability to set up apparatus properly; (2) your dexterity and skill in performing experiments; (3) your knack for recording numerical data and qualitative observations effectively; (4) your proficiency in performing calculations correctly (wrong calculations receive little credit!); (5) the accuracy and precision of your results; (6) your adroitness in interpreting experimental results to draw reasonable conclusions; and (7) your understanding of concepts and ideas as illustrated through oral discussions in the laboratory, written **POST-LABORATORY QUESTIONS**, or written questions on quizzes. Finally, all of the above criteria can be learned over the course of a term; thus, your instructor will also weigh improvement rather heavily.

SAFETY PRECAUTIONS

Many chemicals and combinations of chemicals are potentially dangerous if handled carelessly. Likewise, improper manipulation of certain pieces of equipment leads to potentially hazardous situations. However, with proper precautions, accidents can be avoided. The following safety precautions are by no means complete, but should increase your awareness of how to "experiment defensively" just as you must learn to "drive defensively."

A. Use of this Book

The experiments in this book were carefully selected and written so as to minimize chances of accident or injury. Therefore, it is important for you to **follow the EXPERIMENTAL PROCEDURES carefully and rigorously.** Moreover, **do not perform any experiments** (from books or of your own design) **that are not assigned unless you are specifically authorized to do so by your instructor.** Failure to cooperate may result in your being expelled from the laboratory because of the hazard to yourself and to others.

Some areas of minimal hazard arise even in the **EXPERIMENTAL PROCEDURES** of this book. These are pointed out clearly by **CAUTION:** followed by appropriate words of warning that are both boldfaced and underlined (for example, **CAUTION: <u>Do not heat the tube so hot that it melts or sags</u>.**). Simpler minor words of warning are just boldfaced and underlined and do not have **CAUTION:** preceding them (for example, **<u>gently</u>**). Words or phrases that are just underlined or just boldfaced are not safety precautions, but are emphasized for other reasons.

B. Common Sense

Your own common sense is one of your best sources of safety precautions. Common sense often dictates actions earlier than might be required by written guidelines and frequently prescribes more stringent actions than might be required by written guidelines. One of the purposes of department safety training in the form of lectures, films, and hand-outs is to increase your awareness of safety and correct procedures so that your common sense can better prompt you <u>before</u> emergency situations arise. Use your common sense liberally. Think before you do something in the laboratory.

C. Eye Protection

Safety goggles must be worn in the laboratory at all times and should be purchased at the bookstore before coming to your first laboratory. They should be put on as soon as you open your laboratory drawer or locker and not taken off until you are ready to lock your drawer or locker and leave the laboratory. Even when you are not actually doing experiments, you must wear your safety goggles because your neighbor may accidentally splash something into your face. The only exception to this is that your instructor may allow you in the laboratory without safety goggles during pre-laboratory lectures, quizzes, or calculation periods when <u>nobody</u> is doing any experiments.

Prescription glasses do <u>not</u> provide adequate protection against splashes from above, below, or the sides, and are therefore inadequate by themselves. Prescription glasses can, however, be worn safely and comfortably under several styles of safety goggles. Contact lenses <u>cannot</u> be worn in chemistry laboratories; you must have prescription glasses if eye correction is required. Some contact lenses are degraded by vapors of organic solvents or inorganic acids or bases. Soft contact lenses are permeable to vapors and allow them to reach and be trapped at the cornea and slowly cause damage there. Both hard and soft contact lenses are very difficult for your instructor or someone else to remove from your eye in case of emergency.

Safety Precautions

Learn the location of the nearest eyewash station and how to use it and also the nearest sink. If a chemical is splashed into your eye, immediately rinse your eye thoroughly with large amounts of water to dilute and remove the chemical.

D. Strong Acids and Bases

Strong acids and bases are the most common source of accidents in general chemistry laboratories. They cause holes in your clothing and burns and/or discoloration of your skin if spills or spatters occur. Therefore, a number of precautions are necessary: (1) wear rubber gloves when directed for transferring strong acids or bases between reagent bottles and other containers; (2) never wear halter tops, shorts, or open sandals in the laboratory because they fail to provide protection against spills and spatters for your chest, legs, and feet; (3) tie back long hair, and do not wear scarves, neckties, and loose blouses, shirts, and sweaters that might drape into containers of acids or bases; (4) when diluting concentrated acids or bases, add the concentrated acid or base slowly to a large volume of water (not vice versa) so that excessive heat will be absorbed by the water and not cause spattering; (5) never point a test tube either at yourself or toward your neighbors or look directly down into a beaker, crucible, flask, or test tube that is being heated; (6) if strong acids or bases are spilled on your skin or clothing, immediately rinse the affected area with large amounts of water to minimize damage by diluting the acid or base and decreasing the rate of reaction; and (7) if strong acids or bases are spilled on your desk or the floor, cover the spill with a neutralizing absorbent provided by your instructor until any reaction ceases, sweep up the absorbent and place it in a labeled container in the hood, and wash the affected area with a sponge or towel.

E. Fires

Fires are uncommon in general chemistry laboratories because flammable solvents are not used very frequently. Nevertheless, you should adhere to the following simple precautions: (1) smoking in the laboratory is prohibited at all times; (2) check to see if people around you are using flammable solvents such as methanol, ethanol, acetone, or others before striking a match to light your burner; (3) never reach over or around a burner; (4) tie back long hair so that there is no chance of it draping over a burner and catching fire; (5) do not wear loose scarves, neckties, and blouses, shirts, or sweaters that might drape over a burner and catch fire; (6) avoid wearing synthetic fabrics because of their flammability; and (7) learn the location of the nearest fire blanket, safety shower, and fire extinguisher and how to use them.

F. Miscellaneous Precautions

Some precautions relate to but do not fit into the above categories: (1) never work in the laboratory unless an instructor is present; (2) no horseplay or practical joking will be tolerated because it involves too much danger to yourself and your neighbors; (3) report immediately to your instructor or the nearest instructor if you have an accident; (4) if you are directed to check the odor of a chemical, be cautious about smelling it, fanning the vapors gently toward your nose with your hand; (5) never eat food or drink fluids in the laboratory because of the danger of accidentally ingesting toxic chemicals; (6) wear old clothing in the laboratory or even a laboratory apron to protect your clothes; and (7) follow carefully the directions under **GENERAL INSTRUCTIONS** and **LABORATORY METHODS** because many of these relate directly to safety.

Your instructor may require you to locate items of safety equipment in the laboratory and to answer questions about safety precautions before beginning experiments the first day in the laboratory.

LABORATORY METHODS

When you are learning how to play a musical instrument such as a tuba, it is important to learn some basic techniques such as how to hold the tuba, which fingers to use on the valves, how to press your mouth to the mouthpiece for playing high notes versus low notes, and how to tune the tuba to other instruments. It is also imperative that you learn the capabilities and limitations of the tuba, the highest note and lowest note - for example - so that you do not waste your time trying to play notes that are impossible to reach on the tuba. Learning basic techniques as well as capabilities and limitations of these techniques is also a prerequisite for effective work in a chemistry laboratory.

The basic equipment and the experimental methods described here will be used frequently in your work in the laboratory. Attention is directed especially to methods for making quantitative measurements and to the uncertainties in such measurements. Study these sections carefully, particularly as they are assigned in and apply to individual experiments. Learn how to perform these techniques with both precision and speed. Form the habit of analyzing critically the possible sources of error that may be involved in your measurements and how you can minimize these errors. Finally, consider the extent to which your data are uncertain, and recognize the absurdity of carrying computations to too many digits. See APPENDICES **A** and **B** for a discussion of significant figures, errors in measurements, and propagation of errors in calculations.

A. Measuring Length

1. Centimeter ruler. A device that you will use frequently to measure short distances is shown in FIGURE A.1. It is often called a centimeter ruler because it is numbered in centimeters. The metric units are also compared with inches in FIGURE A.1.

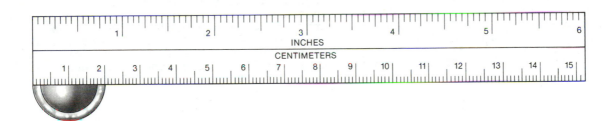

FIGURE A.1. Using a centimeter ruler.

Suppose that you are measuring the outside diameter of a test tube shown under the centimeter ruler in FIGURE A.1. The diameter is recorded as 2.00 cm, the first two digits, 2.0, being read with certainty, and the third digit, 0, being doubtful or estimated with an uncertainty of perhaps 2 in that digit or ±0.02 cm overall. Only three digits are recorded because any more digits to the right would not be valid significant figures.

2. Meter stick. A meter stick is frequently used to measure longer distances or to determine the height of mercury columns when measuring pressures. Uncertainties in the heights of mercury columns are typically larger (±0.4 mm to ±1 mm) because the mercury meniscus is at least the thickness of the meter stick away from the scale and is more difficult to align, tubing containing mercury easily forms a scum on its surface making it harder to see the meniscus, and occasional leaks cause minor fluctuations in heights of mercury columns.

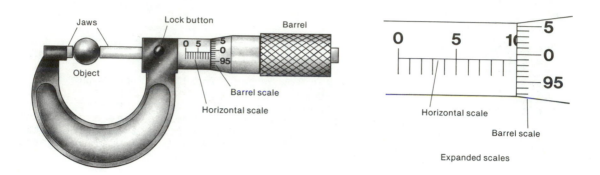

FIGURE A.2. Using a micrometer.

3. **Micrometer.** More precise measurements of short distances are often made using a micrometer like the one shown in FIGURE A.2. A micrometer is a sensitive and expensive instrument that must be handled very carefully. It is utilized as follows. Hold the micrometer with its horseshoe shape to your left with the jaws open end upward. The markings on the horizontal scale should then be facing you. Open the jaws of the micrometer far enough to receive the solid object by rotating the top of the right end of the barrel toward you with your right hand. Place the solid object between the jaws of the micrometer with your left hand, and rotate the top of the right end of the barrel away from you so as to close the jaws just snugly, <u>but not tightly</u>, against the object. Then read the distance from the horizontal scale and barrel scale as described below. Some micrometers have a lock button that can be pushed in to hold the barrel from turning while you are reading the barrel scale. The lock button is released by pushing it toward you from the back side.

Many micrometers are calibrated in millimeters (mm) and can be read to the nearest 0.01 mm. You read the length to the nearest 1 mm from the horizontal scale and then add the tenth and hundredth digits from the barrel scale as shown on the expanded scale of FIGURE A.2. The expanded scale indicates 9 mm (4 divisions beyond 5 mm on the horizontal scale) plus 0.99 mm (between 95 and 0 on the barrel scale). Thus, the ball bearing in the jaws of the micrometer has a diameter of 9.99 mm.

B. **Measuring Volume**

Volumes of liquids or gases are measured by using glass or plastic apparatus that have been calibrated to read in units of volume as shown in FIGURE B.1. Certain apparatus such as graduated cylinders, volumetric flasks, and some pipets are designed to <u>contain</u> a specific volume of liquid when filled to the appropriate calibration mark. Other apparatus such as burets and some pipets are designed to <u>deliver</u> a specific volume of liquid into another container.

The most important prerequisite for using volumetric apparatus is that it always be <u>cleaned thoroughly</u> before use by scrubbing with a brush and soap or detergent solution and rinsing carefully first with tap water and then with distilled water. The apparatus should then be allowed to dry or should be rinsed several times with a few milliliters of the liquid to be used, draining each portion completely between rinses. If the apparatus is clean, liquids will form a continuous film on the apparatus and will drain completely from the apparatus. However, if the apparatus is dirty, liquids will not wet the surface uniformly and will not drain completely from the apparatus. Droplets adhering to dirty surfaces can cause considerable errors in your volume measurements.

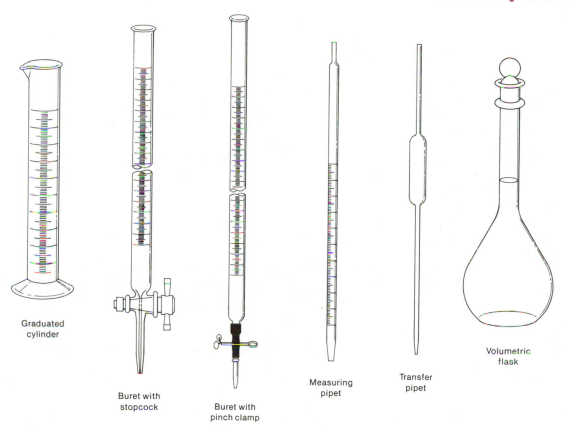

FIGURE B.1. Volumetric apparatus.

    1. Graduated cylinder. Graduated cylinders are used to contain
approximate volumes of liquids. They typically have uncertainties of about
±2%. Thus, you might measure 18.2 ± 0.4 mL in a 25-mL graduated cylinder or
67 ± 1 mL in a 100-mL graduated cylinder. Some 10-mL graduated cylinders
are calibrated to the nearest 0.1 mL and have uncertainties of about ±1%.
Thus, you might measure 6.37 ± 0.06 mL in such a 10-mL graduated cylinder.

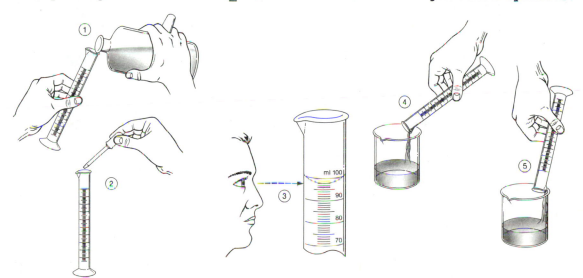

FIGURE B.2. Using a graduated cylinder.

Laboratory Methods

The steps for using a graduated cylinder are illustrated in FIGURE B.2. The required number of milliliters of a liquid is obtained by pouring or dropping it into the graduated cylinder ① until the lowest point of the downward curved surface of the liquid (called the <u>meniscus</u>) is at the desired mark. This is easier to accomplish if a medicine dropper is used to add the last few drops ②. When you read the volume of the liquid, which reads 98 mL, **make certain that your eye is at the same level as the meniscus** ③. This avoids a reading error due to parallax. A good light source behind the cylinder also makes reading easier.

To transfer this measured volume of liquid quantitatively to another container, pour by inverting the cylinder, and hold it in an inclined position for 15 seconds to allow it to drain ④. Remove the last drop from the lip of the cylinder by touching the lip to the inner wall of the receiving container ⑤.

2. Buret. Burets are long, narrow, finely graduated tubes as shown in FIGURE B.1. They are designed to <u>deliver</u> precise and variable volumes of liquids into other containers. A stopcock or pinch clamp at the base of the buret provides for release of volumes between about 0.05 mL and the total volume of the buret. Burets are especially useful for adding solutions stepwise in small increments. Burets typically have uncertainties of about ±0.1%. Thus, you might measure 21.33 ± 0.02 mL using a buret.

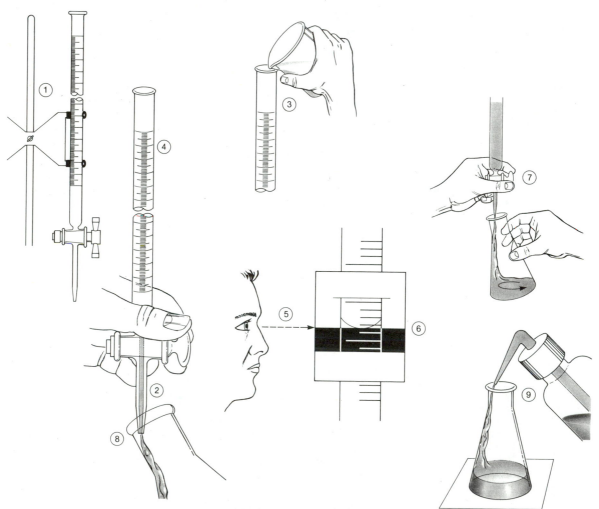

FIGURE B.3. Using a buret.

The steps for using a buret are illustrated in FIGURE B.3. First, secure the buret in a buret clamp attached to a ringstand ①. If you have never used a buret or have only used it a few times, it is wise to practice manipulation of the stopcock or pinch clamp to adjust the liquid flow using distilled water. Partially fill a clean buret with distilled water. Then adjust the stopcock or pinch clamp a number of times until you have the feel of the degrees of turn or pressure required to release liquid one drop at a time or in a steady flow. The stopcock should be handled with the thumb and two fingers ② along with a slight inward pressure on the plug to prevent leakage. The pinch clamp should be handled with the thumb and third finger. Do not use the buret for an actual experiment until you are comfortable carrying out these operations with relaxed muscles.

Fill a clean buret that has already been rinsed with your reagent solution with a few more milliliters of solution than you need for the task at hand ③. Then open the stopcock or pinch clamp long enough to fill the buret tip below the stopcock or pinch clamp and to remove any air bubbles from the buret tip ②. Since the volume of solution delivered by the buret is always determined by taking the difference between an initial volume and final volume, it is not necessary or even worth the effort to adjust the initial reading to the zero calibration mark. Therefore, record the initial volume wherever it is by observing the position on the graduated scale of the lowest portion of the meniscus ④. **Make certain that your eye is at the same level as the meniscus** ⑤. A dark background placed behind the buret and at or just below the meniscus makes it easier to read ⑥.

Place the receiving flask under the buret and over white paper which enhances visibility at the endpoint. **Make certain that you have added the required drops of an indicator solution if you are performing an acid-base titration or an oxidation-reduction titration.** Open the stopcock or pinch clamp carefully to adjust the liquid flow from dropwise to a rapid flow as desired ⑦. When as much solution as is needed has been delivered, close the stopcock or pinch clamp, and touch the inner wall of the receiving container to the buret tip to remove any hanging drop ⑧. If you are using the buret for titration to an endpoint, then rinse the wall of the receiving flask with distilled water ⑨, and record the final volume by observing the new position of the meniscus. If you are using the buret not for titration to an endpoint, but simply to deliver a carefully measured volume of a liquid, do <u>not</u> rinse the wall of the receiving vessel with distilled water before observing and recording the final volume.

3. **Volumetric flask.** Volumetric flasks are used to <u>contain</u> precise volumes of liquids at a specified temperature. They are usually pear shaped and flat bottomed and have a long narrow neck with a single calibration mark as shown in FIGURE B.1. They are always used when standard solutions of precise concentration are being prepared. They typically have uncertainties of about $\pm 0.1\%$. Thus, you might measure 10.00 ± 0.01 mL in a 10-mL volumetric flask or 100.0 ± 0.1 mL in a 100-mL volumetric flask.

The steps for using a volumetric flask for preparing a standard solution are illustrated in FIGURE B.4. A precisely measured volume of liquid or weighed sample of solid is transferred to the volumetric flask ①. Sufficient solvent is added to fill the flask about one-quarter full ②, and the flask is then stoppered and swirled to dissolve the solute ③. Sufficient solvent is then added to fill the flask about three-quarters full, and the stoppering and swirling is repeated to insure thorough mixing. Sufficient solvent is then added to bring the meniscus just to the calibration mark **when viewed with your eye at the same level as the meniscus.** The last couple milliliters should be added drop by drop with a medicine dropper ④. Finally, the flask is stoppered securely, and with the fingers of one hand holding the stopper in place, the flask is inverted and swirled ⑤ and then placed right side up through a few cycles in order to get thorough mixing. Remember that liquids diffuse slowly, and insufficient mixing will result in concentration gradients in the flask.

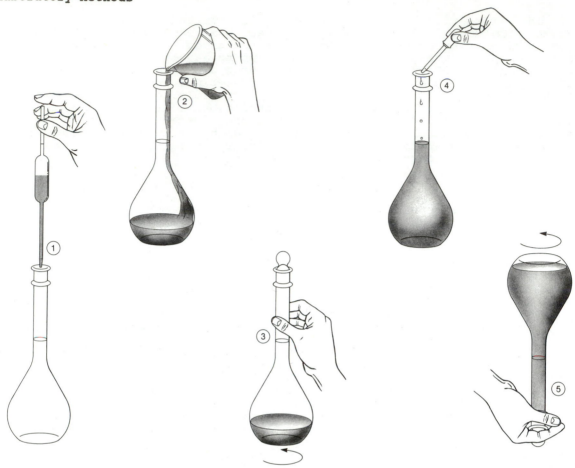

FIGURE B.4. Using a volumetric flask.

    4. Pipet. Pipets are used to measure precise volumes of liquids into other containers. Pipets are of two types as shown in FIGURE B.1. One type has graduation marks along its length and is called a <u>measuring pipet</u>. Another type has only a single calibration mark for a designated volume and is called a <u>transfer pipet</u>. Pipets typically have uncertainties in the range of ± 0.1-0.5%. For some models the uncertainty is printed on the pipet. Thus, you might measure 0.800 ± 0.004 mL using a measuring pipet or 25.00 ± 0.05 mL using a transfer pipet.

    The steps for using a transfer pipet are illustrated in FIGURE B.5. Use a rubber bulb to fill the pipet by suction to a point well above the calibration mark ①. **CAUTION:** <u>**Never use suction by mouth to fill a pipet**</u>. Be careful not to allow liquid to enter the rubber bulb because the liquid may react with and eventually decompose the rubber bulb. Then remove the rubber bulb, and simultaneously cover the top of the pipet with your index finger ②. After wiping the lower stem and tip of the pipet with a towel or tissue ③, release the pressure from your index finger slightly, and drain solution into a waste beaker until the meniscus drops slowly to the calibration mark ④. Then release your index finger, and allow the pipet to drain into the desired container ⑤, touching the pipet to the side of the container to drain all of the calibrated volume ⑥. If you are using a measuring pipet, you only allow it to drain to the desired mark in step ⑤. For <u>to deliver</u> pipets you do <u>not</u> blow out the remaining small volume of liquid in the tip because these pipets have been calibrated assuming that small volume is retained. For <u>to contain</u> pipets you blow out the entire volume of liquid. Pipets are usually labeled as to which kind they are.

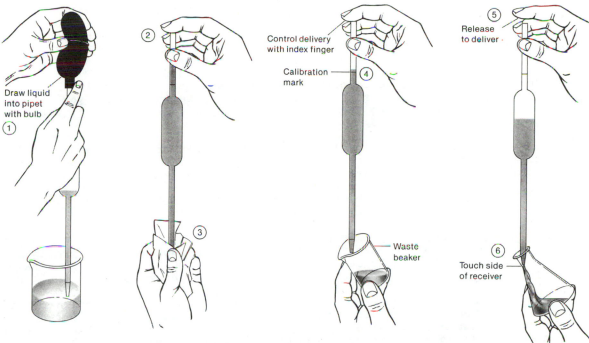

FIGURE B.5. Using a pipet.

Fancier safety pipet fillers such as the one illustrated in FIGURE B.6 may be available for your use in place of a rubber bulb. Safety pipet fillers are rubber or plastic devices designed to fill pipets with liquids and to deliver liquids from pipets with speed and precision, but without the **hazardous practice of sucking liquids into pipets by mouth**. They allow safe

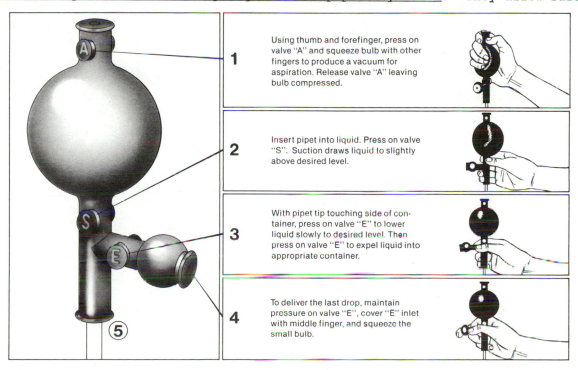

1 Using thumb and forefinger, press on valve "A" and squeeze bulb with other fingers to produce a vacuum for aspiration. Release valve "A" leaving bulb compressed.

2 Insert pipet into liquid. Press on valve "S". Suction draws liquid to slightly above desired level.

3 With pipet tip touching side of container, press on valve "E" to lower liquid slowly to desired level. Then press on valve "E" to expel liquid into appropriate container.

4 To deliver the last drop, maintain pressure on valve "E", cover "E" inlet with middle finger, and squeeze the small bulb.

FIGURE B.6. Using a safety pipet filler.

transfer of corrosive, toxic, infectious, odoriferous, radioactive, or sterile liquids. Most safety pipet fillers have a specially designed throat ⑤ that permits their use with all standard laboratory, industrial, or serological pipets as well as most lambda and micropipets. Safety pipet fillers are easy to operate with one hand, leaving your other hand free to hold the pipet or other equipment. Steps are outlined in FIGURE B.6 (modi-fied from **Fisher Scientific Company advertisement and instruction sheet**). CAUTION: <u>Follow the procedures rigorously and carefully so that you do not draw corrosive liquid into the rubber bulb and so that you do not acciden-tally spill corrosive liquid.</u>

Two other precautions are also pertinent. If any reagents accidentally get into any part of the safety pipet filler, squeeze the reagents out immediately, flush with water, and squeeze the water out completely. Moreover, the large bulb should never be left compressed after usage, but should be allowed to fill by pressing valve "A". Storage in a compressed state weakens the bulb so that it cannot be used again.

5. Repipet dispenser. A repipet dispenser is a device that is used to <u>deliver</u> pre-set approximate volumes of liquids both rapidly and precisely. Repipet dispensers typically deliver volumes up to 100. mL with uncertain-ties of about ±1%. Thus, you might measure 4.00 ± 0.04 mL or 40.0 ± 0.4 mL.

The repipet dispenser in FIGURE B.7 is easy to operate (**modified from Labindustries Repipet Dispenser Instructions**). It must first be primed as follows if that has not already been done. Remove the outlet tip closure. Pump the plunger up and down <u>very slowly</u> with short strokes to remove all air bubbles and to fill the Teflon inlet tube, the glass inlet tube, and the Teflon outlet tip with reagent solution contained in the bottle. Tipping the dispenser slightly in a counterclockwise direction and tapping the dispenser barrel with your finger will assist you in removing bubbles by forcing them into the Teflon outer tip and eventually out of the dispenser. In this manner priming can be accomplished without losing reagent solution, particularly if short strokes of the plunger are used after the reagent solution has reached the level of the Teflon outer tip.

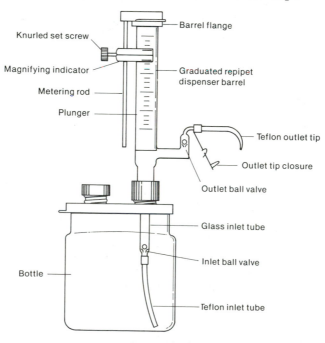

FIGURE B.7. Using a repipet dispenser.

The magnifying indicator used for precise volume adjustments is set as follows. Make certain that the plunger is fully inserted into the dispenser barrel. Then slide the magnifying indicator up or down on the metering rod so as to align visually the two red lines on the magnifying indicator with the desired black line for the volume you require on the scale of the dispenser barrel. Rotating the magnifying indicator farther from the dispenser barrel increases magnification to a maximum of about ten times. When the magnifying indicator is set at the required volume, lock it in place by turning the knurled set screw until it is finger tight. Do <u>not</u> use pliers or other tools to tighten the knurled set screw.

The repipet dispenser can be made to deliver the pre-set volume as follows. Lift the plunger <u>very slowly</u> until it is stopped by the magnifying indicator impinging against the barrel flange. Lifting the plunger gently eliminates errors caused by ball valves bounced from their seats to produce a pre-dispense droplet. If you lose a droplet while the plunger is still in the raised position, lower the plunger slightly to bring the air-reagent meniscus back to the outlet tip, and then raise the plunger gently to its uppermost position. Finally, after waiting a second for the inlet ball valve to return to its seat, press the plunger down gently all the way to deliver the pre-set volume.

Do <u>not</u> replace the outlet tip closure. Your instructor will do that at the end of the period to prevent recession of the reagent solution back into the reservoir and to prevent reagent evaporation. However, the Teflon outlet tip and the inside of the outlet tip closure must both be dry before the outlet tip closure is placed over the Teflon outlet tip so that the outlet tip closure can be removed easily for future use.

C. Measuring Mass

The mass of a portion of matter is conveniently determined by balancing it against standard masses in a process called weighing. The choice of a balance to accomplish this depends upon the total mass of the sample to be weighed and the precision desired.

1. **Triple-beam balance.** Large approximate masses up to about 600 g or even higher can be determined with an uncertainty of about ± 0.2 g using a triple-beam balance (sometimes called a platform balance) shown in FIGURE C.1. This balance is also frequently used for quick approximate weighings after which a sample will be weighed on a balance having greater precision.

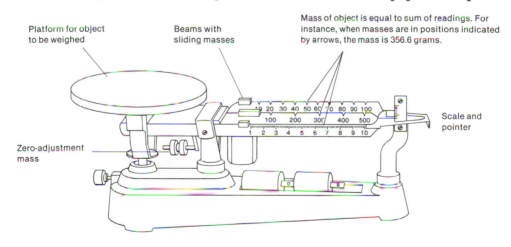

FIGURE C.1. Triple-beam balance.

Laboratory Methods

A triple-beam balance is utilized as follows. First check the balance with nothing on the clean platform and the sliding masses in their zero positions on the beams. The pointer should swing about an equal number of divisions to either side of the zero point. (Turning the zero-adjustment mass will effect this.) Place the container to be weighed on the platform, and adjust the sliding-beam masses until the pointer swings in the same way with respect to the zero point as it did when the platform was empty. This is accomplished easily by moving the 100-g mass on the center beam one notch too far to the right and then moving it back left one notch, moving the 10-g mass on the back beam one notch too far to the right and then moving it back left one notch, and then adjusting the 1-g mass on the front beam to achieve balance. For masses over 600. g many models require hanging one or more masses lying in the base of the balance on a hook at the far right end of the beams. The mass of the object is equal to the sum of the readings on the beams. Weigh samples of solid or liquid chemicals by first weighing a clean, dry container or a sheet of paper. Then add the sample, and obtain a new mass from which the sample mass can be determined by difference. **CAUTION:** **Never place chemicals directly on the platform. If you spill any chemical on the balance or table, clean it up immediately**.

2. **Quadruple-beam balance.** Masses up to about 300 g can be determined with an uncertainty of about ± 0.02 g using a quadruple-beam balance shown in FIGURE C.2.

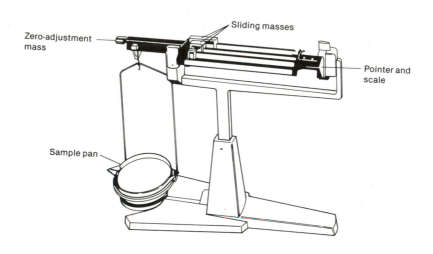

FIGURE C.2. Quadruple-beam balance.

A quadruple-beam balance is utilized as follows. First check the balance with nothing on the hanging sample pan at the left of the balance and with the sliding masses in their zero positions on the beams at the upper center of the balance. The magnetically damped pointer at the right of the balance should point to the zero point at the center of its scale. (Turning the zero-adjustment mass will effect this.) Then place the container to be weighed on the sample pan. Adjust the sliding masses on the beams until the pointer at the right of the balance again points to the zero point at the center of its scale. This is accomplished easily by moving the 100-g mass on the back beam one notch too far to the right and then moving it back left one notch, moving the 10-g mass one notch too far to the right on its beam and then moving it back left one notch, moving the 1-g mass one notch too far to the right on its beam and then moving it back left one notch, and then adjusting the 0.1-g mass on the front beam to achieve balance. The mass of the object is equal to the sum of the readings on the four beams. **CAUTION:** **Never place chemicals directly on the sample pan,** but instead first weigh a clean, dry container or sheet of paper and add the sample to obtain a new mass from which the sample mass can be determined by difference. Make certain that the balance is clean when you remove your paper or container and leave the balance for the next person to use.

   3. Electronic top-loading balance. There are a great variety of electronic top-loading balances available having ranges from about 100 g to 12,000 g and having uncertainties from +0.001 g to +1 g, depending on the range. One model is shown in FIGURE C.3. These balances may be very expensive and require tender loving care in their operation. Such balances with uncertainties of +0.01 g or larger are often used in place of beam balances described in sections C.1 and C.2 because of their speed in weighing. Thus, when the term "beam balance" is used in experiments, your instructor may designate that an electronic top-loading balance be used instead. On the other hand, electronic top-loading balances with +0.001-g uncertainty may be used as analytical balances whenever higher sensitivity is not required. Thus, when the term "analytical balance" is used in experiments, your instructor may designate either an electronic top-loading balance with +0.001-g uncertainty or an electronic single-pan balance described in section C.4.

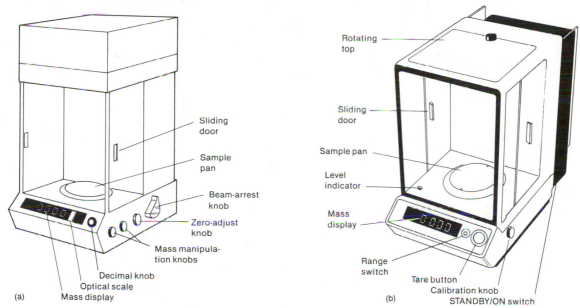

FIGURE C.3. Electronic top-loading balance.

   Electronic top-loading balances are easy to operate because no masses are actually handled. They function according to the following principles. When an object is placed on the balance pan, the object forces a mass of metal downward into the field of a permanent magnet. This causes a restoring force in the form of a current applied to an electromagnet. This electromagnet repels the mass of metal upward until the balance attains a null point. The current that is required to accomplish this is converted first to a digital signal and then to a mass on a digital display. Temperature and humidity affect the linearity of the field strength of the permanent magnet and thus limit the accuracy of the balances.

   Your instructor will have set up the balance, plugged in the power cord, set the integration time switch to the NORMAL position, and depressed the switch on the back of the balance to ON at the beginning of the laboratory. Please leave this switch ON throughout the period because it is better for the electronic components not to switch the balance OFF and ON any more than necessary. Moreover, the position of the integration time switch must <u>not</u> be changed for any reason unless the balance is first switched OFF. In addition, the draft protector must be left surrounding the pan and/or over the balance to help eliminate the effects of air currents. Your instructor will switch the balance OFF and unplug it at the end of the laboratory.

Most electronic top-loading balances are operated as follows. Depress the tare button to display zero on the digital readout. Place a piece of weighing paper or a container into which sample is to be weighed **gently** onto the weighing pan.   The tare mass of the paper or container will appear on the bright, glare-free, digital readout within about 2 seconds required for stabilization.   On most models a stability indicator, g, will appear to the right of the mass on the digital readout when the reading stabilizes. Record the tare mass of the paper or container. Then add sample **carefully without spilling** to the paper or container, and read and record the gross mass in the same fashion. **CAUTION: Never place chemicals directly on the sample pan.**   If the weighing range has been exceeded, an "E" will appear on the digital readout of most models. The net mass of sample is just the difference between the gross mass and the tare mass. Make certain that the balance is clean when you remove your paper or container and leave the balance for the next person to use.

If the tare mass is not required for later calculations, you can depress the tare switch at the right front of most balances after placing your paper or container on the pan. The balance will zero instantly and show zero on the digital readout. Then add sample carefully as before, and read the net mass of sample directly from the digital readout.   Consecutive multi-step taring is also possible.

4. Single-pan analytical balance. There are a great variety of single-pan analytical balances, but most have a capacity of 80.-150. g with an uncertainty of ±0.0001 g. Since the effects of chemical fumes are very critical at this level of uncertainty, these balances are usually kept in a balance room separate from the laboratory.   These balances are very expensive and require tender loving care in their operation.   Two basic types are shown in FIGURE C.4.a and C.4.b.

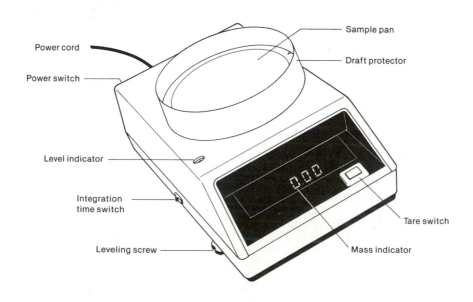

FIGURE C.4.   Single-pan analytical balances.

The first type, shown in FIGURE C.4.a, operates on the principle of removing precise internal masses to compensate for the mass of a sample until equilibrium is restored. It is easy to operate because no masses are actually handled. An electrical system controlled by knobs adjusts the internal masses very rapidly, and an optical lever mechanism registers the total mass directly on a readout panel. The balance has a beam with a knife-edge on the underside near its center that acts as a fulcrum. On the rear end of the beam is a fixed counterweight, which is exactly equal in mass to two components on the front end of the beam, a hanger holding precise internal masses, and the pan supported by a knife-edge. Thus, when there is no object on the pan, the beam is at equilibrium, and the balance registers zero mass. When a sample is placed on the pan, the total mass on the front of the beam is then greater than that of the fixed counterweight at the rear of the beam. In order to restore equilibrium, mass-manipulation knobs must be turned to remove internal masses from the hanger exactly equal to the mass of the sample. The sum of the masses removed registers on a readout panel.

The components of the balance and their proper functions must be thoroughly understood in order to obtain reproducible and reliable weighings and to avoid damaging the balance. CAUTION: Specific directions for the use of the particular single-pan analytical balances in your laboratory will be presented by your instructor and will be posted in your balance room. However, the various models of this balance, which differ primarily in the location of knobs and in the registering of masses, require the following basic steps.

The first step involves adjusting the zero point. This must be performed each time the balance is used. Make certain that the pan is empty and clean, that the windows are closed (air currents must be avoided), and that all masses on the readout panels are set to zero. Locate the beam-arrest knob, which appropriately positions the beam between and during weighings so as to prevent damage to the knife-edges from excessive wear or sudden impact. Normally the beam-arrest knob is kept in the arrest position, in which no knife-edges are in contact because the beam is completely supported by lift mechanisms. Slowly rotate the beam-arrest knob to the full-release position, in which the beam and optical lever mechanisms are supported only on the knife-edges and thus swinging freely. After the optical scale comes to rest, turn the zero-adjust knob until the optical scale indicates a reading of zero. Then slowly return the beam-arrest knob to the arrest position. If you cannot make this adjustment, ask your instructor for help.

Weighing of a sample is accomplished as follows. Make certain that the beam-arrest knob is in the arrest position. Open the sliding door of the balance chamber, place the sample to be weighed on the pan, and close the sliding door of the balance chamber. Use tongs to handle the sample to avoid absorption of moisture or oil from your hands. CAUTION: Never place chemicals directly on the pan; instead, add them to a suitable weighed container or piece of weighing paper. CAUTION: Never place a hot object on the pan, since convection currents from the rising of warm air push up the pan and reduce the apparent mass. Slowly rotate the beam-arrest knob to the semirelease position, in which the beam has some freedom of movement but in which contact with the knife-edges is restricted. Turn the 10-g mass-manipulation knob to increasing masses until the optical scale makes a pronounced jump (or shows a minus sign or "remove weight" sign on several models); then turn the knob back one stop. Repeat the same procedure with the 1-g and 0.1-g mass-manipulation knobs. Slowly return the beam-arrest knob to the arrest position. Then gently turn the beam-arrest knob to the full-release position. When the optical scale has come to equilibrium, adjust the decimal knob (or micrometer knob or digital control knob on several models) as appropriate on your particular balance. The mass of the sample is read from left to right, observing the decimal point. Record the mass. Slowly return the beam-arrest knob to the <u>arrest position</u>. Open the sliding door, remove the sample, close the sliding door, and set all the mass knobs to zero. Clean the balance carefully with a brush as necessary.

The second type of single-pan analytical balance, shown in FIGURE C.4.b, operates on the same principles as the electronic top-loading balance described in section **C.3**. **CAUTION:** <u>Specific directions for the use of the particular balances in your laboratory will be presented by your instructor and will be posted in your balance room</u>. However, most models of this balance, which differ primarily in the location of knobs, require the following basic steps. Depress the STANDBY/ON switch, if necessary. Select either a <u>course range</u> or a <u>fine range</u> using the range switch. Depress the tare button to display 0.000 g or 0.0000 g. **CAUTION:** <u>Never place chemicals directly on the pan, and never place a hot object on the pan</u>. Place the object to be weighed <u>gently</u> onto the pan either through the sliding doors or a rotating top; and read the mass from the digital display. Open the sliding door, remove the object, and close the sliding door. Taring can also be accomplished as described in section **C.3**.

One specific example of the second type of single-pan analytical balance is the Mettler AE 100 balance. Your instructor will have already set up the balance by plugging it in, calibrating it, setting the integration time, and setting the stability dectector. The display should read 0.0000. To weigh an object, open the sliding door, place the object on the pan, and close the door. **CAUTION:** <u>Never place chemicals directly on the pan, and never place a hot object on the pan</u>. When the green pilot light to the left of the display goes off, the balance has achieved stability, and you may read the display. To tare, briefly press the control bar. The display is first blanked out, and then will read 0.0000. The container mass is tared out, and the weighing range is available for weighing sample. When sample is being added quickly, the last two decimal places are automatically blanked out, thereby allowing for better following of the mass increase. These decimal places will return as the balance stablizes. If the weighing range is exceeded, the upper portions of all digits appear: "--------". Consecutive multi-step taring is also possible.

D. Heating Materials

A number of methods are used to supply heat in the laboratory depending upon the intensity of heat required, the duration of heat required, the stability of temperature required, and whether or not a flame can be used because of the absence or presence of flammable solvents.

1. Hot tap water. If only gentle warming of a sample is required, it is often sufficient to hold the container under hot tap water or to immerse the container in a beaker of hot tap water. If constant temperature is required over a period of time, a water bath held at constant temperature by a resistance heater with a thermostat is appropriate. More frequently, stronger heating is necessary, and a gas burner of some kind is used.

2. Bunsen burner. A Bunsen burner, shown in FIGURE D.1.a, is designed so that the fuel gas may be mixed with the proper amount of air to give complete combustion and produce the maximum amount of heat. In order to use this burner properly, you must understand its construction and adjustment. Before you use this burner in the laboratory for the first time, examine its construction, and connect it to a gas line and light it. When lighting a Bunsen burner, light a match first, and then turn on the gas. This prevents a minor explosion of an already accumulated gas-air mixture. For easy lighting, always bring the lighted match in just at the top of the burner barrel and not some distance above the barrel. Try closing the air inlet collar to obtain a cool, yellow, <u>luminous flame</u> shown in FIGURE D.1.b. This results because the gas is not mixing with sufficient air to effect complete combustion, and the carbon particles that are formed become incandescent. Use this flame for mild heating.

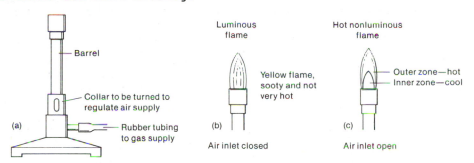

FIGURE D.1. Bunsen burner and types of flames.

A hot, pale blue, but nearly invisible, <u>nonluminous flame</u> is obtained by turning the collar at the base of the burner so that air can enter and mix with the gas. The supply of gas should be regulated so that the flame is about 10. cm high, and the air openings should be adjusted so that all luminosity is gone and two zones appear in the flame as shown in FIGURE D.1.c. Too much air will cool the flame or may blow it out. Use this flame for strong heating up to about 1000°C.

If a flame of intermediate heat intensity is required, the flame can be adjusted between the extremes of the luminous flame and the nonluminous flame by proper rotation of the collar regulating the air supply. Practice these adjustments before using the Bunsen burner in an experiment.

Containers holding samples must be supported carefully when being heated to eliminate the danger of their tipping over and perhaps scalding you or your neighbor. Thus, as shown in FIGURE D.2, beakers, flasks, evaporating dishes, and casseroles are supported on a wire gauze over an iron ring attached to a ring stand; test tubes are held in test tube holders; and crucibles are supported in a clay or wire triangle over an iron ring attached to a ring stand. Any hot containers should be handled with care using gloves, a towel, a "hot hand," or tongs as appropriate. Finally, it should be noted that a boiling water bath is often used when a relatively low, constant temperature must be maintained because the boiling water bath cannot exceed 100°C.

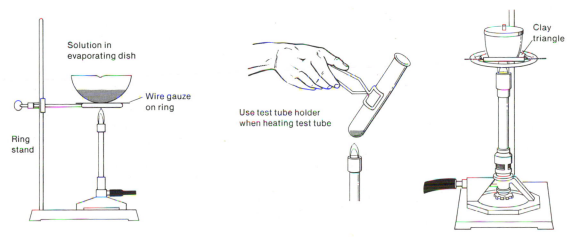

FIGURE D.2. Supporting containers being heated.

3. **Fisher burner.** A Fisher burner (or Meker burner), shown in FIGURE D.3.a, attains slightly higher temperatures (about 1200°C) than a Bunsen burner and can provide up to twice the heat output of a Bunsen burner. A Fisher burner has a stainless steel grid at the top that channels the flame through separate openings and minimizes the temperature variation throughout the flame. Maximum temperature and heat output is obtained by opening the needle valve to provide maximum gas flow and then adjusting the air inlet collar until there is a continuous, flat layer of pale blue inner cones rising 1-2 mm above the grid as shown in FIGURE D.3.b. The hottest part of the flame is just above this layer of pale blue cones. As with the Bunsen burner, too much air will cool the flame and may blow it out.

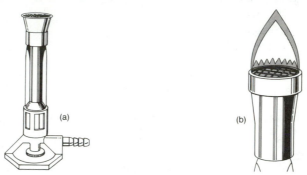

FIGURE D.3. Fisher burner and hottest flame.

4. **Gas-oxygen burner.** A variety of gas-oxygen burners are available. They are designed either to be held in your hand or to sit on a desk. They can easily achieve temperatures well above 1300°C that are required to work or blow borosilicate glass. All models must be connected to sources of both gas and oxygen and have valves to control the flows of both gas and oxygen to the burner tip. Your instructor will provide specific directions for operating the particular gas-oxygen burner in your laboratory if you need to use one.

5. **Hotplate.** A hotplate is considerably more expensive than a burner and is used primarily in situations in which an open flame is not allowed for safety reasons because of the presence of a flammable solvent. Clearly this type must be electrically heated. A hotplate is also used frequently for convenience when a number of containers must be held at the same temperature for a period of time. This type may be heated electrically, or with gas if no flammable solvent is present.

E. **Measuring Temperature**

Temperature is most commonly measured using a thermometer constructed of glass. The thermometer has a terminal bulb surmounted by a capillary tube. The bulb contains mercury or some other suitable liquid, which is free to expand up the capillary as the temperature rises. The capillary tube is calibrated to read in degrees of temperature (usually on the Celsius scale for work in chemistry). Typical thermometers that you are likely to use in this course may be calibrated to the nearest degree and have an uncertainty of ±0.2-0.5°C or may be calibrated to the nearest 2°C and have an uncertainty of ±1°C. CAUTION: __Because of the long, slender construction of a thermometer it is very easy to break by bumping it into other pieces of apparatus or the desktop; therefore, be very careful in handling it.__ Notify your instructor immediately if a thermometer should break so that proper clean-up of the mercury can be implemented.

To measure the temperature of a sample, the bulb of the thermometer is totally immersed in the sample (or the thermometer is submerged to an immersion mark) and allowed to stand until the level of liquid in the capillary becomes stationary. The position of the meniscus then is read as the temperature.   Each thermometer is designed for use over a specific range of temperature, say -10° to 110°C, which is marked on the stem.   Be sure to note this range.   **CAUTION:**   <u>**Never heat the thermometer above the upper limit; if the liquid expands too much, it will break the thermometer**</u>.

F. Inserting Glass Tubing into Rubber Stoppers

The most dangerous aspect of manipulating glass tubing in terms of the potential for serious cuts and bleeding involves the insertion of glass tubing into rubber stoppers or the removal of glass tubing from rubber stoppers.   Therefore, <u>**you must take every precaution to follow the pre- scribed procedures with great care**</u>.

When you insert glass tubing into rubber stoppers or rubber connectors, first slide a 10-cm piece of (1/4-in. ID x 3/8-in. OD x 1/16-in. wall) Tygon tubing, if available, over the glass tubing (**C. W. J. Scaife, J. Chem. Educ., 1984, 61, 838**). Then moisten the glass tubing and the holes in rubber stoppers or rubbers connectors with glycerol or water. Wrap the glass tubing and the rubber stopper with a towel, and grasp the tubing <u>close</u> to the end you wish to introduce into the rubber stopper or connector as shown in FIGURE F.1. The Tygon tubing and the towel will protect your hands should the glass break.   **CAUTION:** <u>**Do not push on the tubing too strenu- ously, but twist it slowly and cautiously into place.   Your hand grasping the glass tubing should always be within 1 inch of the rubber stopper. Never try to force the glass tubing into place**</u>. If the glass tubing does not go into the rubber stopper without forcing it, enlarge the hole in the rubber stopper with a series of rotations of your triangular file.   Then lubricate the rubber stopper with glycerol, and try again.

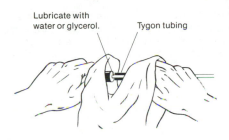

Lubricate with
water or glycerol.   Tygon tubing

Grasp tubing with a towel
close to the stopper. Twist
and apply pressure cautiously.

FIGURE F.1.   Inserting glass tubing into a rubber stopper.

When you remove glass tubing from a rubber stopper, first slide a piece of Tygon tubing, if available, over the end of the glass tubing that you will grab, and then moisten the glass tubing with glycerol next to the rubber stopper on the side toward which you will slide it. Then wrap the glass tubing and the rubber stopper with a towel, grasp the tubing close to the rubber stopper on the end you wish to pull from the rubber stopper, and twist slowly while **cautiously** pulling the glass tubing from the rubber stopper. Your hand grasping the glass tubing should always be within 1 inch of the rubber stopper. **CAUTION:** <u>**Never jerk the glass tubing from the rubber stopper**</u>. Glass tubing can always be removed from rubber stoppers more easily immediately following an experiment than some weeks later.

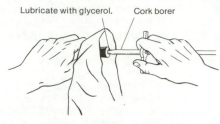

Lubricate with glycerol.   Cork borer

Grasp tubing and stopper with a towel.

FIGURE F.2. Removing glass tubing from a rubber stopper.

The procedure illustrated in FIGURE F.2 (**A. Bosch, J. Chem. Educ., 1973, 50, 113**) is useful for glass tubing that simply cannot be removed from a rubber stopper by the normal technique, particularly if the glass tubing has broken inside the rubber stopper. Slide a hand cork borer that is slightly larger than the glass tubing over the glass tubing, and lubricate the lower outside portion of the cork borer with glycerol. Then force the cork borer between the glass tubing and the rubber stopper to break the seal between the glass and the rubber. This can be accomplished easily because the cork borer is designed for both pressure and twisting. The glass tubing will then slip out easily. Removal of the cork borer leaves a reusable rubber stopper.

G. Transferring Liquids from Bottles

A first very important step is to **read the label on the bottle carefully and be certain that you have chosen the correct bottle.** So that there can be no confusion, reagent bottles should be labeled with a chemical name, a concentration if the reagent is not a pure liquid or a solid, and a chemical formula.

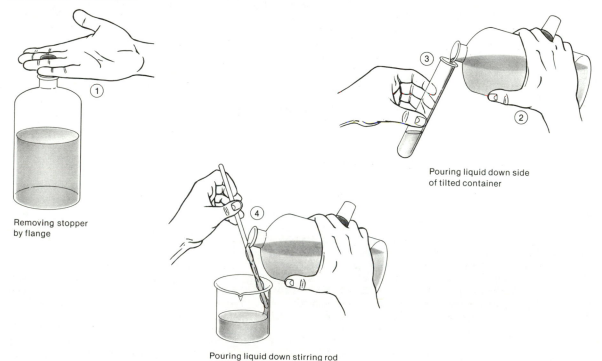

Removing stopper
by flange

Pouring liquid down side
of tilted container

Pouring liquid down stirring rod

FIGURE G.1. Procedures for pouring liquid from a bottle.

26

When obtaining liquids from bottles, a primary concern is to avoid contamination of the reagent solution by using procedures described here and illustrated in FIGURE G.1. Two precautions assure that this objective is met. First, **never insert another piece of apparatus such as a pipet into the reagent bottle.** You should instead pour slightly more than the required volume of liquid into a graduated cylinder or beaker and then pipet the liquid from the new container. Second, **never allow the part of a stopper that fits into a bottle to touch other surfaces.** If the stopper has a flat top or screw cap, place it upside down on your desktop before pouring from the bottle. If the stopper has a vertical flange, grasp this between the fingers ①, remove the stopper, and then lift the bottle with the same hand ②, leaving the other hand free.

A second concern is to prevent spillage while pouring liquids from bottles. This is accomplished by bringing the neck of the bottle into contact with the rim of the tilted receiving vessel such as a test tube or beaker ③, and pouring down the inside of this vessel so that no spattering occurs and the liquid does not spill back down the outside of the bottle. Pouring the liquid down a stirring rod ④ also works well. If liquid is spilled down the outside of the bottle, be sure to rinse it and wipe it clean before returning it to the shelf.

The two concerns above are now frequently eliminated through the use of plastic squeeze bottles rather than glass bottles. The plastic bottles are not prone to breakage and have jets that allow convenient delivery even into narrow containers such as 10-mL graduated cylinders. Liquids are delivered from the plastic bottles by placing the receiving vessel under the tip and squeezing the bottle gently as shown in FIGURE G.2. Be careful to release the pressure of your squeeze before you have the required amount of liquid. The primary disadvantages of plastic bottles are that the jets often drip on standing and thus require a tray under them; they develop cracks, especially when containing organic solvents for extended periods; and they leak around screw seals or molded seals. In the latter two instances, besides the leaks, squeezing no longer causes delivery of liquid, and they are of little use except as containers from which you can pour.

Squeeze plastic bottle gently to deliver liquid.

FIGURE G.2. Procedure for delivering liquid from a squeeze bottle.

H. Decanting Liquids

Decanting is useful for separating a liquid from a solid that has already settled to the bottom of a container. A crude form of decanting is used when you clean a fish tank. After removing the fish, you carefully pour the water from the tank, leaving behind the sand, shells, and plants. Decanting is a gentle pouring off of a liquid without disturbing the solid sediment. For example, a liquid from one beaker may be poured, without dribbling, down the side of a second beaker by holding a stirring rod against both beakers as shown in FIGURE H.1.

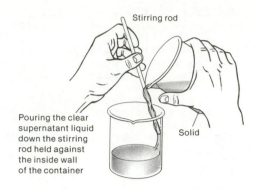

Stirring rod

Pouring the clear
supernatant liquid
down the stirring
rod held against
the inside wall
of the container

Solid

FIGURE H.1.   Procedure for decanting.

I.   Filtering and Washing Solids

Filtering is an effective method for separating considerable amounts of a finely divided solid from a liquid in which it is suspended. Two types of filtration are commonly used in general chemistry laboratories, depending on whether aspirators are available. Gravity filtration must be used in the absence of aspirators. It is slow because liquid is pulled through the filter paper only by gravity. Vacuum filtration is faster than gravity filtration because liquid is pulled through the filter paper by a partial vacuum provided by an aspirator. Your instructor will designate which method you are to use.

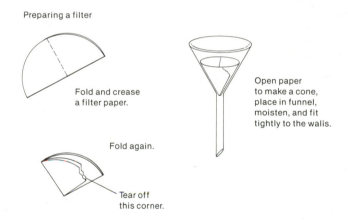

Preparing a filter

Fold and crease
a filter paper.

Fold again.

Tear off
this corner.

Open paper
to make a cone,
place in funnel,
moisten, and fit
tightly to the walls.

FIGURE I.1.   Preparing for gravity filtration.

1.   Gravity Filtration. Prepare for gravity filtration using the procedures illustrated in FIGURE I.1. Fold and crease a disk of filter paper on a diameter. Then fold the paper again to form a quadrant, but do not crease the paper this time. Tear off one corner of the paper. Open the paper so that three layers including the tear are on one side and one layer on the other, and place it in a funnel. Press the paper against the sides of the funnel, and moisten it with water or another solvent so that it will adhere to the funnel. The paper should not reach to the top of the funnel.

Procedures for gravity filtration are illustrated in FIGURE I.2. Support the funnel on an iron ring attached to a ring stand, and place a beaker under the funnel so that the stem of the funnel touches the side of the beaker. Pour the suspension onto the filter paper. It is best to pour down a glass rod held in a vertical position against the lip of the beaker with its lower end close to the filter. Never fill the filter with liquid up to the top of the paper. Use a stirring rod or rubber policeman, if available, to transfer as much as possible of your solid to the filter paper. If no further washing is specified, use a wash bottle containing distilled water or whatever solvent is appropriate, and rinse any remaining material onto the filter. The liquid that passes through the filter is termed the **filtrate**, and the solid left on the filter paper is called the **residue**. If solid appears in the filtrate, the filter paper may have been torn or was not seated properly, or you may have to boil the suspension gently for a few minutes to flocculate or form larger particles of the precipitate and/or use finer filter paper, and repeat the gravity filtration.

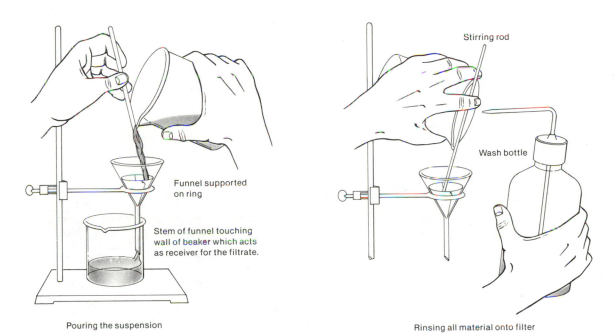

Funnel supported on ring

Stem of funnel touching wall of beaker which acts as receiver for the filtrate.

Stirring rod

Wash bottle

Pouring the suspension

Rinsing all material onto filter

FIGURE I.2.   Procedures for gravity filtration.

If you are directed to wash impurities from the residue on the filter paper, add the washing solvent first to the original beaker and then onto the filter paper so as to remove last traces of solid from the beaker. Swirl the filter assembly to cause intimate contact with the residue, and allow each portion of the washing solvent to drain before adding the next. Remove the filter paper and residue after first lifting one edge of the filter paper with a spatula.

2. **Vacuum Filtration.** Procedures for vacuum filtration are shown in FIGURE I.3. Clamp the filter flask to a ringstand, and insert the rubber stopper surrounding the neck of the Buchner funnel tightly into the mouth of the flask. Place filter paper of the appropriate diameter into the Buchner funnel, and moisten the paper with distilled water or whatever solvent is being used. Connect the aspirator tubing to the sidearm of the flask, and turn on the aspirator fully by opening the appropriate faucet. The filter paper should be centered in and seated tightly to the base of the Buchner funnel. Pour the suspension to be filtered slowly onto the filter paper. It is best to pour down a glass rod held in a vertical position against the lip of the beaker with its lower end close to the desired section of the filter paper. Use a stirring rod or rubber policeman, if available, to transfer as much as possible of your solid to the filter paper. If no further washing is specified, use a wash bottle containing distilled water or whatever solvent is appropriate, and rinse any remaining material onto the filter. Remove by suction as much solvent as possible from the solid on the filter paper. The liquid that passes through the filter is called the **filtrate**, and the solid remaining on the filter paper is called the **residue.**

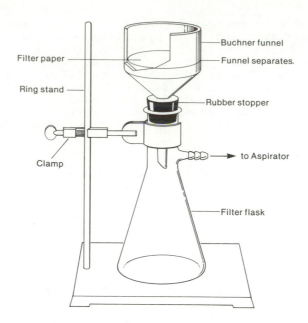

Filter paper —
Ring stand —
Clamp —

Buchner funnel —
Funnel separates. —
Rubber stopper —
to Aspirator —
Filter flask —

FIGURE I.3.   Procedures for vacuum filtration.

If solid appears in the filtrate, the filter paper may have been torn or was not seated properly, or you may have to boil the suspension gently for a few minutes to flocculate the precipitate and/or use finer filter paper, and repeat the vacuum filtration. **Disconnect the aspirator tubing from the sidearm of the flask**, and turn off the aspirator. Failure to disconnect the tubing first may result in water backing from the aspirator into your flask, at least diluting and possibly contaminating your filtrate. Analogous procedures are used for vacuum filtration through a Gooch crucible and adapter that replace the Buchner funnel and rubber stopper in FIGURE I.3.

If you are directed to wash impurities from the residue on the filter paper, make certain that the aspirator is turned off. Otherwise, washing solvent will be pulled through so rapidly that little intimate contact and washing will occur. Add the washing solvent first to the original beaker and then onto the filter paper so as to remove last traces of solid from the beaker. Carefully mix the solvent with the residue either by swirling the filter assembly or using a rubber policeman, but taking care not to disturb the filter paper. Then turn on the aspirator weakly, and allow each portion of washing solvent to drain before turning off the aspirator and adding the next portion. Finally, remove as much solvent from the residue as possible with the aspirator on fully. Disconnect the aspirator tubing from the flask, and turn off the aspirator. The top portion of some Buchner funnels separates from the funnel part, and can be placed flat on the desktop to prevent tipping and spilling. Remove the filter paper and residue after first lifting one edge of the filter paper with a spatula.

J. Centrifuging Solids

Centrifuging is another method for separating a finely divided solid from a liquid in which it is suspended. Unlike filtering, which can be used to separate large amounts of solids from considerable volumes of liquids, centrifuging is only useful for separating small amounts of solids suspended in up to a few mL of liquids. The device used for centrifuging is called a centrifuge and is shown in a cut-away drawing in FIGURE J.1. A centrifuge has a rotor head mounted on a shaft that can be rotated at high speeds by an electric motor, thus developing considerable centrifugal force. The centrifugal force causes finely divided solids in suspension in the test tube to settle very rapidly and completely to the bottom of the test tube.

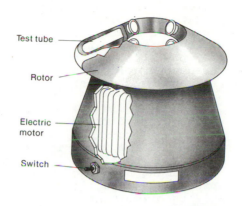

Test tube

Rotor

Electric
motor

Switch

FIGURE J.1. A centrifuge.

A centrifuge is used as follows. Obtain a 10 x 75-mm or slightly larger test tube, and make certain that it fits properly into the centrifuge available for your use. **CAUTION:** <u>Check the test tube to be absolutely certain that it has no tiny cracks</u>. Pour the suspension into the test tube, or carry out the chemical reaction to produce the suspension right in the test tube. Place the test tube in one of the sockets of the rotor so that it is in an inclined position with its mouth toward the center. **CAUTION:** <u>It is important to balance each tube by mounting, diametrically opposite to it, a second tube similarly loaded with the same volume of suspension or of water</u>. Obviously test tubes must be labeled or numbered if several differ-ent suspensions are placed in the centrifuge at the same time. Spin the rotor once by hand to make sure it is free to spin. Turn the switch ON, and let the rotor spin for about 2 minutes. Then turn the switch OFF, and let the rotor slow to a stop. **CAUTION:** <u>Do not touch the rotor while it is spinning, no matter how slowly</u>. Remove the test tubes. The clear liquid can be decanted carefully from the lower solid layer, or the liquid can be drawn off with a medicine dropper. You can wash any adhering solution from the solid by adding distilled water or other appropriate solvent, stirring to suspend the solid again, and centrifuging again. If a poor separation is achieved the first time, it may help to heat the test tube in a boiling water bath for a few minutes to flocculate the precipitate, and then centri-fuge again, perhaps for a longer time.

K. Preparing Solutions

The question that students ask most frequently when preparing solutions is, "Is this dissolved?" That question must be answered by visual inspec-tion that you can perform easier than your instructor.   The key is to determine whether two distinct phases are present, in which case the solute is <u>not</u> completely dissolved in the solvent, or whether a single, homogeneous solution is present, in which case the solute is dissolved in the solvent, and a solution has been prepared. Therefore, if you are trying to dissolve a solid in a liquid solvent, and few tiny solid particles are still visible when you hold the container up to a good light source, the solid sample is not yet completely dissolved.

A gas can be dissolved in a liquid solvent in which it is soluble by bubbling the gas through the liquid. Normally stirring is not necessary because the bubbling provides sufficient mixing. Thus, red-brown nitrogen dioxide, NO_2, gas can be bubbled into water and does not escape from the water because it dissolves. Conversely, colorless hydrogen, H_2, gas bubbles into and out of water because H_2 is insoluble in water.

Laboratory Methods

One liquid can be dissolved in another liquid with which it is miscible if the two liquids are poured together and stirred or shaken. Thus, ethanol, CH_3CH_2OH, and water mix in all proportions to form a solution whereas cyclohexane, C_6H_{12}, forms a layer above water after the two are stirred in a test tube. It is important to stir aqueous solutions when performing chemical reactions because aqueous solutions do not mix readily, especially if they have quite different densities and are in narrow test tubes.

A solid can be dissolved in a liquid solvent in which it is soluble by mixing the solid with the liquid by stirring or shaking. A solid dissolves more rapidly if it is first ground to a fine powder to increase surface area using a mortar and pestle. If the solid increases in solubility as the temperature is raised, heating the mixture, as shown in FIGURE D.2, will facilitate dissolving, by increasing both the rate of dissolving and the total solubility. The approximate solubility of a solid can be determined by adding small measured amounts of solvent successively with stirring to a weighed sample of the solid until it dissolves completely.

L. Evaporating Liquids

A solution of a nonvolatile solid in a volatile solvent can be separated into its components by removing the liquid solvent by evaporation. The solution is poured into an evaporating dish which has a shape that provides a large surface area for both heating and evaporation. The porcelain dish is supported on a wire gauze attached to a ring stand and is heated as shown in FIGURE D.2. A small porcelain casserole is used in place of the evaporating dish when very small volumes of liquid are evaporated. **Never evaporate from a test tube.** The volatile solvent evaporates, and the nonvolatile solid remains in the dish. **Avoid spattering** by controlling the rate of heating and using a cool, luminous flame, shown in FIGURE D.1.b, once boiling begins. When the solution becomes sufficiently concentrated, the solid will begin to crystallize. If the heating is continued until all of the solvent has evaporated, all of the solid can be recovered. If the liquid is a flammable organic solvent such as methanol, CH_3OH, ethanol, CH_3CH_2OH, or diethyl ether, $CH_3CH_2OCH_2CH_3$, do not heat over an open flame; instead, use a heat lamp shown in FIGURE M.1.b, an electrical hotplate, or a steam bath. Any hot containers should be handled with care using gloves, a towel, a "hot hand," or tongs as appropriate.

M. Drying Solids

In experiments where a precipitate has been formed, collected on a filter paper, and washed free of solution, the sample must be dried before it is weighed for the purpose of determining a yield or performing any kind of analysis. Three common devices for drying solids in general chemistry are an oven, a steam bath, and a heat lamp.

An oven is the most expensive of the three but also requires the least attention and can provide a fairly stable temperature except when the door is opened frequently. If an oven is available for use in your laboratory, your instructor will have already preset the temperature, and you should not alter the temperature setting. <u>Be very careful when opening and closing the door and when removing your sample so that you do not spill someone else's sample in the oven or on the floor</u>. Small pieces of equipment such as crucibles should be placed on watch glasses or in beakers so that they don't tip over while sitting on the oven racks. All watch glasses and beakers should have a label with your name on it. Any hot containers should be handled with care using gloves, a towel, a "hot hand," or tongs as appropriate. Don't try to slide containers onto flat surfaces such as a book from which the containers can easily slip off onto the floor.

A simple steam bath is made by setting up the assembly shown in FIGURE M.1.a, half-filling the beaker with water, and placing a watch glass, concave side up, on the beaker. Steam, generated by boiling the water gently, heats the watch glass. If the filter paper containing the sample is spread on the watch glass, the filter paper and the sample can be dried without being heated above the boiling point of water.

A very convenient way of drying a sample is to spread it on a watch glass and place the watch glass under a heat lamp as shown in FIGURE M.1.b. The lamp should be suspended high enough (about 30.-35 cm) above the sample so that the sample is not overheated and the filter paper is not charred.

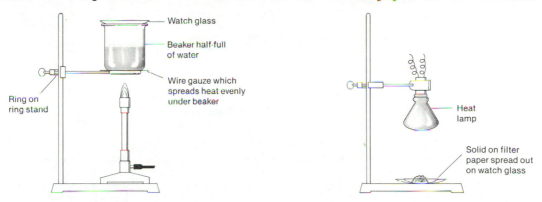

FIGURE M.1. Drying with a steam bath and a heat lamp.

N. Bausch and Lomb Spectronic 20 Spectrophotometer

A spectrophotometer is used to determine the percent transmittance or the absorbance of a sample (usually a solution) as a function of wavelength. The Bausch and Lomb Spectronic 20 Spectrophotometer sacrifices some photometric accuracy in percent transmittance or absorbance in return for simplicity of operation and relatively low cost (about $950). It allows continuous selection of wavelength throughout the 340-600 nm range accurate to ±2.5 nm, and reading of absorbance accurate to 1-3% over the range from 0 to 1.5.

In this instrument light intensity from a tungsten lamp as the source is detected by a photocell. The photocell then converts the light intensity into a proportional electric current. The electric current is then measured and displayed on an ammeter calibrated directly in percent transmittance and absorbance. We will read absorbance because we are interested in measuring concentration, and absorbance is directly proportional to concentration whereas percent transmittance is not.

The Bausch and Lomb Spectronic 20 Spectrophotometer, shown in FIGURE N.1, is calibrated as follows (**modified from the Bausch and Lomb Spectronic 20 Spectrophotometer Operator's Manual**). If it has not already been done, plug in the power cord, turn the instrument ON by rotating the amplifier control (lower left front) clockwise past a click, and let the instrument warm up for about 20. minutes. With nothing in the sample holder (top left front) and the door closed, rotate the amplifier control clockwise or counterclockwise to align the meter needle with "infinity" on the absorbance scale. The door of the sample holder must be closed when calibrating the instrument or when taking readings so that no stray light from the room enters the instrument. The next step is to set the desired wavelength by rotating the wavelength control (top right front) until the desired wavelength on the wavelength scale appears under the hairline.   Then half fill a clean cuvette of specialty glass and of uniform wall thickness (or a 13 x 100-mm test tube if designated by your instructor) with your solvent. Make certain that the solvent is free of bubbles and that the exterior surface of the cuvette or test tube is wiped clean with a tissue. Then open the sample holder, and insert the cuvette or test tube gently into the

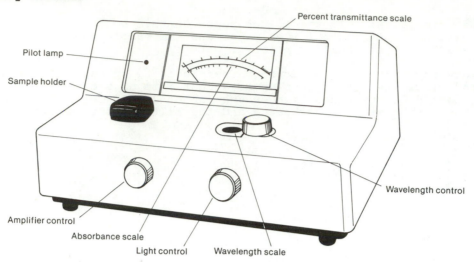

FIGURE N.1. Bausch and Lomb Spectronic 20 Spectrophotometer.

sample holder until it stops. If you are using a cuvette, skip the next seven bracketed sentences. [Rotate the light control (lower right front) clockwise or counterclockwise to bring the meter needle to an absorbance of about 0.1. Find a point where small rotations of the test tube cause no change in the absorbance reading when the sample holder is closed. Then use a pen or grease pencil to place an index mark on the test tube opposite the index mark on the plastic sample holder. This indexed test tube should always be inserted in the sample holder in this same configuration. This indexed test tube will later be emptied and used to contain your various samples. Repeat the same procedure with a second clean, 13 x 100-mm test tube half-filled with solvent. This indexed test tube will be your reference test tube and will contain solvent throughout your readings.] With the cuvette or the indexed reference test tube in the sample holder, rotate the light control clockwise to align the meter needle with "zero" on the absorbance scale. Remove the cuvette or the reference test tube containing solvent, and the spectrophotometer is now ready to measure the absorbance of a sample solution.

The "zero" absorbance adjustment, made with the cuvette or the indexed reference test tube in the sample holder, must be checked each time the wavelength is changed. In addition, during operation at a fixed wavelength for extended periods of time, both the "infinite" absorbance adjustment, made with nothing in the sample holder, and the "zero" absorbance adjustment, made with the cuvette or the indexed reference test tube in the sample holder, must be checked periodically.

An absorbance measurement of a sample is obtained as follows. Rinse a clean cuvette or the indexed sample test tube with distilled water and then with several milliliters of sample solution. Half fill the cuvette or the test tube with sample solution, and make certain that the solution is free of bubbles and that the exterior surface of the cuvette or the test tube is wiped clean with a tissue. Then insert the cuvette or the test tube carefully into the sample holder, and close the door. Read the absorbance of that solution directly from the meter. Then remove the cuvette or the test tube from the sample holder, and rinse the cuvette or the test tube with distilled water so that it is ready to receive the next sample. Repeat the above procedure until all absorbance readings at a given wavelength have been obtained.

If nobody else will use the instrument during your laboratory period, turn OFF the instrument by rotating the amplifier control counterclockwise past a click, and unplug the power cord.

34

O. Preparing Cold Baths

A cold bath or a slush bath is composed of an organic solvent cooled to its melting point by liquid nitrogen. The slush bath is a solid-liquid equilibrium mixture of the solvent and maintains a temperature at the melting point of the solvent as long as both phases are present. The solvent is chosen because it has a melting point at the temperature desired. Moreover, the solvent must also have a low vapor pressure and toxicity and should not present a fire hazard. **CAUTION**: <u>Liquid nitrogen is very cold</u> (-196°C) <u>and can cause severe burns if it splashes onto your hands. Therefore, wear the rubber gloves provided whenever handling liquid nitrogen in order to protect your hands. Do not use a "hot hand" when handling liquid nitrogen. If liquid nitrogen is trapped inside the "hot hand", it will burn your hand.</u>

The Dewar flasks that you will use to contain slush baths or liquid nitrogen are double-walled vessels constructed from borosilicate (Pyrex) glass. A vacuum between the walls provides excellent insulation. Borosilicate glass is appropriate for use over wide temperature ranges because of its low coefficient of thermal expansion and its high resistance to thermal shock. However, borosilicate glass is not very resistant to mechanical shock. Therefore, **be careful not to bump or drop the Dewar flasks or to jam the wooden stirring rods into them because they can be broken easily**. The Dewar flasks are frequently covered with a protective plastic mesh or tape to prevent injury from broken glass in case of implosion.

Using the following procedure, prepare carefully <u>in a hood</u> a slush bath composed of an appropriate organic solvent and liquid nitrogen. Fill a 250-mL Dewar flask about two-thirds full of the solvent. Pour liquid nitrogen, contained in a 1-L Dewar flask, in several-mL increments very slowly and carefully into the solvent, stirring the solvent constantly with a wooden stirring rod. If any chunks of solid solvent form, be careful to stir thoroughly until they are melted or form a slush. Continue the addition of liquid nitrogen and stirring until the consistency of the slush bath is that of a very thick milkshake or of applesauce depending on the solvent. If much of the slush melts before you are finished using the slush bath, repeat the slow addition of liquid nitrogen with stirring until the original consistency of the slush bath is obtained.

P. Using Test Papers

The primary concern when using test papers is not to contaminate the solution being tested. All test papers are impregnated with some kind of chemical that takes part in the chemical reaction that is responsible for the formation of the species exhibiting the color of the positive test. For example, starch-potassium iodide test paper is impregnated with starch and potassium iodide, KI. Therefore, **never dip the test paper into the solution being tested; always bring the solution out to the test paper**. This is most easily accomplished by placing the test paper in the concave dip of a <u>clean</u> watch glass, and bringing a drop of solution out to the test paper using a <u>clean</u> stirring rod or medicine dropper. Unlike test papers, indicators, which change the color of an entire solution being titrated, must be added directly to the solution.

A number of gas tests are based on reactions of the gases with one or more <u>moistened</u> test papers. Several different moist test papers can be placed in the concave dip of a watch glass which can then be inverted over a test tube from which gas is evolving. This procedure allows you to get ready for the gas tests before the reaction evolving gas is initiated and also permits you to use several different test papers simultaneously. To conserve test papers in experiments where they are used repeatedly, it may be necessary to tear test papers into small pieces.

Laboratory Methods

Q. Measuring pH

   1. pH Indicators. The pH of a solution can be measured qualitatively
by using pH indicator papers or pH indicator solutions. The pH indicator
papers are impregnated with one or more acid-base indicators.   The pH
indicator solutions have one or more acid-base indicators dissolved in them.

   Acid-base indicators are usually weak acids that ionize as follows.

   Indicator molecules(aq) H^+(aq) + Indicator ions⁻(aq)

The indicator molecules have quite different colors from the indicator ions,
and the color of the indicator will depend upon the position of the equilibrium and thus the pH.   This allows you to check the acidity or alkalinity
of a solution by observing the color of an indicator. Many indicator papers
or indicator solutions contain a single indicator molecule that undergoes a
color change over a fairly narrow range of about 1 pH unit. Thus, you can
only tell whether you are below that pH range, above that pH range, or in
that pH range. For example, an indicator molecule might be yellow, its ion
might be blue, and various shades of green would be observed in the intermediate range. Some indicator papers or indicator solutions contain a
mixture of indicators, and may give a continuous variation of colors over a
wide pH range. By comparing the color observed experimentally with a color
chart for that mixture of indicators, you will be able to determine the pH
of a solution easily to within 1 pH unit. **If you are using indicator paper,
be sure to use a fresh place on the paper for each test.**

   2. pH Meter. A pH meter is an instrument for determining the pH of a
small volume of solution simply, quickly, and reproducibly, without changing
the composition of the solution. A pH meter is a sensitive electrometer and
is comprised of two electrodes and a voltmeter. When the two electrodes are
immersed in a sample solution, the circuit is complete.   The half-cell
potential of one electrode is affected by pH changes whereas the half-cell
potential of the other electrode, called the reference electrode is not.
The voltmeter senses the potential difference between the electrodes and
registers that difference on a scale calibrated directly in pH units.

   There are a great variety of pH meters available covering a range of
sensitivities and ease of operation. Most pH meters are very expensive and
require tender loving care in their operation. Many pH meters have the
general appearance shown in FIGURE Q.1 and differ primarily in the

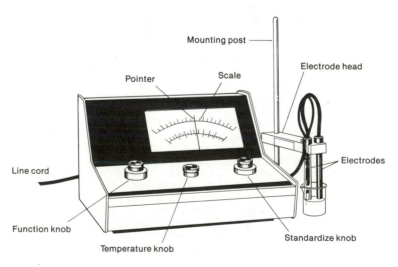

FIGURE Q.1 A pH meter.

36

placement of knobs and the type of electrode mounting device. The pH meters usually require the following basic steps in their operation.

The first step in the operation of a pH meter is to standardize it. This is accomplished by immersing the electrodes in a buffer solution of known pH and adjusting the meter reading to the pH of that solution. For precise results the pH of the buffer solution should be close to that of the sample solutions being measured, and the temperature of the buffer solution and the sample solutions should be the same.

Standardize your pH meter using the following procedure. If it is not already done, turn the function knob to the STANDBY position, plug in the line cord, and let the pH meter warm up for at least 10. minutes. Remove the protective cap or vial from the electrodes, if necessary. Using a wash bottle, rinse the electrodes carefully with distilled water, collecting the waste washings in a beaker. **Always rinse the electrodes with distilled water immediately after removing them from any solution to avoid carry-over and eventual contamination of one solution by another.** Carefully raise and lower the electrodes on the mounting post in order to immerse the electrodes into the buffer solution contained in a small beaker. **CAUTION: The glass electrodes are exceedingly fragile so that great care must be exercised when adjusting the position of the electrodes. Never let the electrodes touch the bottom of a beaker.** Set the temperature knob to 25°C (or another appropriate temperature), and turn the function knob from STANDBY to pH or, on some pH meters, to READ. Then adjust the standardize knob until the pointer indicates the exact pH of the buffer solution. Wait a few seconds to be certain that the reading remains constant. The standardize knob should not be turned again until you perform another standardization. Finally, turn the function knob back to STANDBY, carefully raise the electrodes from the buffer solution, and rinse the electrodes with distilled water, collecting the waste washings in a beaker.

After the pH meter is standardized, the second step in its operation is to determine the pH of a sample solution using the following procedure. Carefully raise and lower the electrodes on the mounting post in order to immerse the electrodes in the sample solution, but do not let the electrodes touch the bottom of the beaker. Swirl the solution **gently** and **carefully**. Then turn the function knob from STANDBY to pH or, on some pH meters, to READ. Read the pH of the sample solution from the scale, and record it. Finally, turn the function knob back to STANDBY, carefully raise the electrodes from the sample solution, and rinse the electrodes with distilled water, collecting the waste washings in a beaker.

If the pH of other sample solutions is to be measured, repeat the procedure in the previous paragraph. If more than a couple minutes will elapse before the next measurement, immerse the electrodes in distilled water. At the end of the laboratory period, be sure that the function knob is on STANDBY, disconnect the line cord, rinse the electrodes thoroughly with distilled water, and replace the protective cap or vial containing fresh distilled water.

Manipulating Glass Tubing

PURPOSE OF EXPERIMENT: Learn to cut, fire-polish, bend, and draw out small glass tubing.

In many chemical experiments it is necessary to use glass tubing to construct or connect various pieces of apparatus. Glass tubing or rod supplied by manufacturers in 1.2-meter lengths can be adapted by various operations to meet your needs.

Two types of glass tubing will be used in this experiment: (1) soft glass, also called soda lime glass or flint glass, and (2) borosilicate glass, frequently called by the tradenames Pyrex (Corning Glass Works) or Kimax (Kimble Glass Company).

Glass has a long history and has been produced to achieve a variety of properties.   A few objects made from glass date back as far as 3000 years. Basic recipes for soda lime glass are approximately 2000 years old, whereas borosilicate glass was first produced by Corning Glass Works in 1912. The primary advantage of borosilicate glass is its remarkable resistance to thermal shock. Borosilicate glass will withstand large temperature changes that would shatter soda lime glass because it expands only about one-third as much as soda lime glass upon equivalent heating. In addition, boro-silicate glass is relatively inert to most chemicals.   It has excellent resistance to acids except for hydrofluoric acid, HF, which etches its sur-face, and it is attacked by hot basic solutions only on prolonged contact. There is also a price difference between soft glass and borosilicate glass depending on the object. The 1.2-meter lengths of small-diameter soft glass and borosilicate glass tubing can be purchased in bulk quantities for the same price [about $.30 per piece for 4-mm outside diameter (OD) tubing]. As diameter increases borosilicate glass becomes progressively more expensive (about $2.00 per piece versus $1.00 per piece for 18-mm OD tubing). The 2:1 price differential is a reasonable approximation for many larger items of glassware.

For purposes of this experiment soft glass and borosilicate glass differ primarily in their thermal characteristics as shown below.

	Soft Glass	Borosilicate Glass
Annealing temperature (^{O}C)	480	550
Softening or sagging temperature (^{O}C)	675	820
Working or blowing temperature (^{O}C)	1010	1245

A Bunsen burner reaches about 1000°C and potentially can be used for all of the above operations except working borosilicate glass. A Fisher burner reaches slightly higher temperatures and is more appropriate for fire-polishing and bending small-diameter borosilicate glass tubing, but a gas-oxygen burner is required for bending large-diameter borosilicate glass tubing and for most working of borosilicate glass. Each of these burners may be used in this experiment (Laboratory Methods D).

In this experiment you will cut, fire-polish, bend, and draw out small glass tubing.   Although glass rod will not be used in this experiment, the same procedures can be used to cut, fire-polish, bend, and draw out small glass rod.   At the same time, you may make a number of pieces that will be used in apparatus for later experiments.

EXPERIMENTAL PROCEDURE

(Study this section and the PRE-LABORATORY QUESTIONS before coming to the laboratory. **Wear safety goggles when performing this experiment.**)

Obtain a 1.2-m length of 8-mm OD soft glass tubing and a 1.2-m length of 6-mm OD borosilicate glass tubing from your instructor.

A. Cutting Small Glass Tubing

Both soft glass and borosilicate glass tubing can be cut to any desired length by the following procedure. **CAUTION:** <u>Utilizing this procedure rigorously and carefully will assure a planar cut or flat break of the glass tubing rather than a jagged break that could cause a serious cut in your fingers or hands</u>.

Lay the glass tubing flat on your desk top, and measure whatever length is desired. Moisten the edge of a triangular file, and with that edge bear down firmly on the glass tubing where you want to make the break. Make a <u>single</u> fairly deep scratch across the tubing as shown in FIGURE 1.1. Note that older files are often sharper and make deeper scratches with less pressure when used near either end rather than near the middle of the file. Remember also that the file is not a saw and that it scratches effectively only when moving away from you.

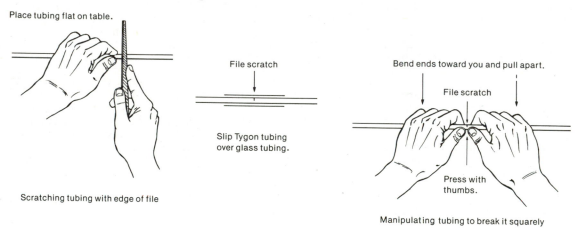

FIGURE 1.1. Cutting small glass tubing.

Before breaking the glass tubing moisten it with water, and slip a 10-cm piece of Tygon tubing over the glass tubing so that the Tygon tubing extends equally on either side of the scratch as shown in FIGURE 1.1. Use 1/4" ID x 3/8" OD x 1/16" wall Tygon tubing for 6-mm OD glass tubing, and use 3/8" ID x 1/2" OD x 1/16" wall Tygon tubing for 8-mm OD glass tubing. The Tygon tubing protects your fingers and prevents cuts in case of an irregular jagged break of the glass tubing (**C. W. J. Scaife, J. Chem. Educ., 1984, 61, 838**).

To break the tubing grasp it in both hands with your thumbs close together and the scratch between them but on the opposite side of the tubing as shown in FIGURE 1.1. Bend the glass tubing gently but firmly, pressing away from you with your thumbs and pulling toward you with your hands, at the same time pulling your hands apart. The tubing will break at the point where the scratch was made. Then remove the two pieces of glass tubing from the Tygon tubing. If a planar break was made, the ends of the glass tubing are ready for fire-polishing. If a jagged break was made, repeat the process until you have a planar break.

B. Fire-polishing the Ends of Small Glass Tubing

The ends of freshly cut glass tubing have sharp edges that can easily cause cuts, and the ends must be fire-polished or rounded smoothly by the following procedure.

The sharp end of soft glass tubing may be fire-polished by holding it in the edge of the upper zone of a hot, nonluminous Bunsen flame and slowly rolling it back and forth between your thumb and fingers as shown in FIGURE 1.2. Remove the tubing from the flame as soon as the sharp edges have become rounded; if you heat the glass tubing too long, it may close at the end. Hold the fire-polished glass tubing in your hand for a couple minutes, or place it on your heat-proof ceramic board to cool. Don't grab the glass again too soon. CAUTION: <u>Remember that hot glass looks exactly like cool glass</u>.

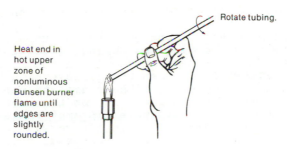

Heat end in hot upper zone of nonluminous Bunsen burner flame until edges are slightly rounded.

Rotate tubing.

FIGURE 1.2. Fire-polishing soft glass tubing.

The same procedure is used for fire-polishing borosilicate glass tubing except that it is rolled backward and forward at the edge of the hot, upper, darker blue flame of a Fisher burner as shown in FIGURE 1.3.

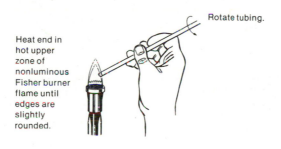

Heat end in hot upper zone of nonluminous Fisher burner flame until edges are slightly rounded.

Rotate tubing.

FIGURE 1.3. Fire-polishing borosilicate glass tubing.

CAUTION: <u>Always fire-polish the ends of pieces of glass tubing or rod before they are used</u>. Make certain that the ends of glass tubing are cool before making bends or inserting the tubing into rubber stoppers.

C. Bending Small Glass Tubing

Soft glass tubing can be bent to any desired angle using the following procedure. Draw the desired angle in pencil on a heat-proof ceramic board. Place a wing tip on your Bunsen burner. This will help you obtain a broad flame for heating a portion of glass tubing lengthwise so that it can be bent smoothly over roughly a 3-cm length. Otherwise a very sharp bend and collapsed tubing would result. Close the air inlets of your Bunsen burner, and light the gas. Regulate the flow of gas to give a flame about 5 cm high, and open the air inlets until the luminosity of the flame just disappears and two zones are present as shown in FIGURE 1.4.

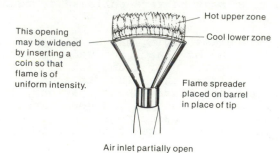

This opening may be widened by inserting a coin so that flame is of uniform intensity.

Hot upper zone

Cool lower zone

Flame spreader placed on barrel in place of tip

Air inlet partially open

FIGURE 1.4. Flame for bending soft glass tubing.

Heat the glass tubing of appropriate length with the flame centered where you want the bend, supporting the tubing lengthwise in the upper zone of the flame. Be careful to rotate it slowly in the flame to insure uniform heating as shown in FIGURE 1.5. When the glass tubing has softened so that it will bend with ease, remove it from the flame and bend it to the desired angle, simultaneously pressing the bend flat on the heat-proof ceramic board and allowing it to cool. **CAUTION: <u>Never attempt to heat very short pieces of glass tubing held in your hands</u>;** it is a sure way to burn your hands. Use two pieces of soft glass tubing to make bends having a 90° angle and a 60° angle. Try to make symmetrical bends as shown in FIGURE 1.5 and to avoid problems leading to the unsatisfactory bends in FIGURE 1.5.

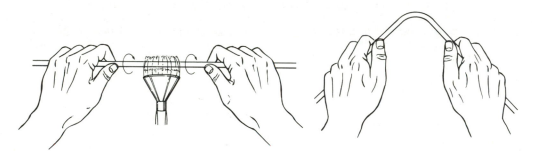

Be sure to use flame spreader and rotate tube lengthwise in hot zone until glass is pliable.

Remove from flame and quickly bend. Press bend flat on ceramic board and leave it there until cool.

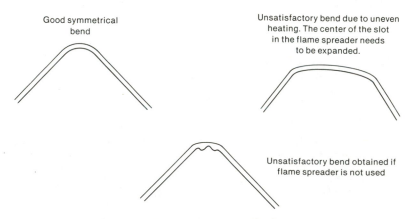

Good symmetrical bend

Unsatisfactory bend due to uneven heating. The center of the slot in the flame spreader needs to be expanded.

Unsatisfactory bend obtained if flame spreader is not used

FIGURE 1.5. Bending soft glass tubing.

The same procedure is used for bending borosilicate glass tubing except that heating is done in the upper hot zone of a Fisher burner flame as shown in FIGURE 1.6. This provides the additional heating required to soften borosilicate glass easily. The Fisher burner usually heats uniformly over enough breadth to allow for a symmetrical bend that is not too sharp. Use two pieces of borosilicate glass tubing to make a bend having a 90° angle and to make a U-tube. You may have to heat over a longer length than just the width of the Fisher burner flame when making the U-tube.

Heating the tube while rotating it

FIGURE 1.6.   Bending borosilicate glass tubing.

D.   Drawing out Glass Tubing and Making a Glass Jet

Soft glass tubing can be drawn out to make a glass jet by the following procedure. Grasp a piece of soft glass tubing of appropriate length lightly between the thumb and fingers of each hand, the knuckles of the left hand being up and those of the right hand being down. Hold the tubing horizontally in the upper part of a nonluminous Bunsen burner flame, and rotate the tubing slowly and constantly by rolling it backward and forward between your thumb and fingers as shown in FIGURE 1.7. Whenever the glass tubing is being heated in the flame, the tubing should be kept in motion, but this motion need not be rapid. While rotating the tubing during the heating, push your hands gently toward each other so that the walls of the tubing become somewhat thicker. Be careful not to twist the tubing. When the tubing is quite soft, remove it from the flame, and gently but firmly pull the two ends apart, keeping it taut until it becomes rigid. The tubing should be constricted to a diameter of about 1 mm as shown in FIGURE 1.7.

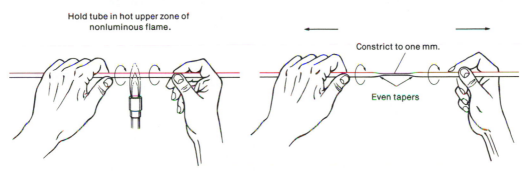

Hold tube in hot upper zone of nonluminous flame.

Constrict to one mm.

Even tapers

Heating the tube while rotating it

Remove from flame, continue rotation, and draw out.

FIGURE 1.7.   Drawing out soft glass tubing.

Experiment 1

Cut the tubing that has been constricted at a point where one end tapers down to about 1 mm as shown in FIGURE 1.8. Fire-polish both ends, being especially careful not to close the small tip. Repeat the process with the other piece so as to make two soft glass jets.

Cut and fire polish.

FIGURE 1.8. Making soft glass jets.

The same procedure is used for drawing out borosilicate glass tubing except that a gas-oxygen burner is used to provide the additional heat required for working borosilicate glass. Your instructor will show you how to adjust a gas-oxygen burner properly if one is available. If a gas-oxygen burner is not available, try making two glass jets from borosilicate glass using the upper part of a nonluminous Fisher burner flame for heating. Note the long time required to get the borosilicate glass to the point where it can be drawn out at all.

E. Checking for Stress in Glass

If either soft glass or borosilicate glass is heated, softened, and worked, and then allowed to cool rapidly and unevenly so that all regions are not simultaneously at the same temperature, patterns of internal stress are developed in the glass. If the internal stress is great enough, the glass may crack or shatter during cooling. Internal stress in a piece of glass is not apparent to the naked eye. However, internal stress can be observed easily through the use of a simple polariscope because glass alters the plane of polarized light depending on the extent of the internal stress. Unstrained glass appears uniformly clear in the polariscope whereas patterns of internal stress appear as bands of various colors and shading depending on whether portions of a glass piece are under tension, under compression, or unstressed.

A Bethlehem polariscope (or various other models) may be used to check for stress in glass as shown in FIGURE 1.9 (**modified from Bethlehem Model P6 Portable Polariscope instruction sheet**). Turn on the flashlight switch to provide a light source. Rotate the polaroid on the flashlight until the light is polarized as indicated by a deep purple color when looking at the light source through the viewer. Then hold the glass to be tested between the viewer and the polaroid on the flashlight, and note any colors or shading that is present.

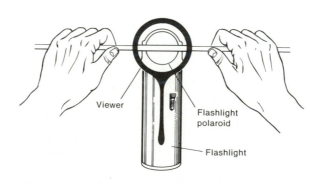

FIGURE 1.9. Checking stress in glass tubing.

Observe a short piece of 6-mm borosilicate glass tubing (held in one hand only) between the viewer and the light source. Then grasp the glass tubing with both hands positioned exactly as they were for cutting glass tubing in FIGURE 1.1. Exert a weak bending force on the glass tubing - **not enough to break it**. Record any differences that you observe from the glass tubing held in one hand. Record also what colors are observed for portions of the glass under tension and what colors are observed for portions of the glass under compression. Check one or more of your bends for internal stress.

Internal stress in glass can be removed by a process called annealing. The glass is heated slowly to the appropriate temperature at which molecules can flow over each other to release the patterns of stress. Then the glass is cooled slowly and uniformly. Annealing is particularly important for glass that will be exposed later to thermal or mechanical shock, not a requirement for the pieces that you may make for later experiments in this course.

F. **Making Pieces of Glass Tubing for Later Experiments**

Practice each of the procedures described previously with both 8-mm soft glass tubing and 6-mm borosilicate glass tubing. Then use 6-mm borosilicate glass tubing to make any pieces designated by your instructor for use in later experiments. Note that 6-mm borosilicate glass tubing is required in order that it will later fit into rubber stoppers or rubber tubing. Have your instructor check your pieces. Store your pieces carefully in your desk drawer so that they do not get broken.

Student ID number _____

Section _____ Date _____

Instructor _____   **PRE-LABORATORY QUESTIONS**

1. How do you adjust a Bunsen burner to obtain a relatively cool luminous
    flame?

2. How do you adjust a Bunsen burner to obtain a relatively hot nonluminous
    flame?

3. How do you adjust a Fisher burner to obtain a very hot nonluminous
    flame?

Experiment 1

4. Describe clearly the type of burner and the kind of flame you would use for:

    a. Fire-polishing small-diameter soft glass tubing.

    b. Fire-polishing small-diameter borosilicate glass tubing.

    c. Bending small-diameter soft glass tubing.

    d. Bending small-diameter borosilicate glass tubing.

    e. Drawing out small-diameter soft glass tubing.

    f. Drawing out small-diameter borosilicate glass tubing.

Student ID number _____

Section _____ Date _____

Instructor _____   **POST-LABORATORY QUESTIONS**

1. a. What difference do you observe for glass tubing held in one hand in the polariscope versus glass tubing with a weak bending force exerted on it?

    b. What color is glass tubing that is stressed by being under tension?

    c. What color is glass tubing that is stressed by being under compression?

2. What is likely to happen if a piece of soft glass tubing and a piece of borosilicate glass tubing are heated, softened, fused together, and allowed to cool? Why?

3. If you were going to construct a sink trap for the laboratory, would you use soft glass or borosilicate glass? Why?

4. If you were going to construct a glass lid for a casserole used for baking, would you use soft glass or borosilicate glass? Why?

5. If you were going to construct drinking glasses, would you use soft glass or borosilicate glass? Why?

6. If you were going to construct a glass coffee pot, would you use soft glass or borosilicate glass? Why?

7. If you were going to construct a thermos bottle to contain hot coffee or iced lemonade, would you use soft glass or borosilicate glass? Why?

Density of a Solid

PURPOSE OF EXPERIMENT: Determine the density of an unknown solid.

A substance, whether solid, liquid, or gas, has a number of characteristic properties that either alone or in combination can be determined experimentally in order to identify the substance. One characteristic property is the density of a substance.

Density is defined as the mass per unit volume of a sample. The density of a given sample can be determined experimentally by measuring the mass by weighing and the volume by any number of methods and then dividing the mass by the volume. Thus, if a piece of glass rod is found to weigh 6.21 grams and to displace a volume of 2.22 cubic centimeters, the density can be calculated as

$$\text{density} = \frac{\text{mass}}{\text{volume}} = \frac{6.21 \text{ g}}{2.22 \text{ cm}^3} = 2.80 \text{ g/cm}^3$$

Neither mass nor volume are characteristic properties of a substance because both depend on the size of the sample. However, the ratio of mass/volume is a characteristic property because mass increases as volume increases. Volumes of solids are often expressed in cubic centimeters or centimeters cubed (cm^3) so that densities of solids are then expressed in grams per cubic centimeter (g/cm^3). Volumes of liquids are frequently expressed in milliliters (mL) so that densities of liquids are usually expressed in grams per milliliter (g/mL). Since 1 mL = 1 cm^3 exactly, the two sets of units are equivalent and can be used interchangeably. Volumes of gases are frequently expressed in liters (L) so that densities of gases are usually expressed in grams per liter (g/L).

Densities of liquids, excepting water below 4°C, decrease slightly with increasing temperature and, neglecting liquid mercury, cover a rather small range. In contrast, densities of solids are essentially independent of temperature and cover a rather wide range as shown in TABLE 2.1.

TABLE 2.1. Densities of some solids.

Substance	Density (g/cm^3)	Substance	Density (g/cm^3)
Balsa wood	0.11-0.14	Rubber	1.1-1.2
Cork	0.22-0.26	Bone	1.7-2.0
Seasoned oak	0.60-0.90	Aluminum	2.70
Cardboard	0.69	Cement, set	2.7-3.0
Leather, dry	0.86	Iron	7.86
Butter	0.86-0.87	Lead	11.3
Ice	0.917	Osmium	22.5

In this experiment you will determine the density of an unknown solid sample provided by your instructor. You should also become proficient at measuring masses and volumes using a variety of methods and should gain a good understanding of precision and accuracy in experimental measurements.

REFERENCES

(1) Kotz, J. C., and Purcell, K. F., <u>Chemistry and Chemical Reactivity</u>, Saunders College Publishing, Philadelphia, 1987, sections 1.4 and 1.5.

(2) Masterton, W. L., Slowinski, E. J., and Stanitski, C. L., <u>Chemical Principles</u>, 6th ed., Saunders College Publishing, Philadelphia, 1985, sections 1.2 and 1.5.

EXPERIMENTAL PROCEDURE

(Study this section and the PRE-LABORATORY QUESTIONS before coming to the laboratory. **Wear safety goggles when performing this experiment.**)

Obtain an unknown solid sample from your instructor. Record in TABLE 2.2 the color and shape of your sample and/or the letter stamped on your sample.

A. Measurement of mass

If your unknown solid sample is not thoroughly dry, wipe it carefully with a towel or tissue, and let it air dry for a few minutes.

Your instructor will tell you what type of balance to use for this experiment and will demonstrate the operation of that balance (Laboratory Methods **C**). Weigh your unknown solid sample on the balance, making three independent measurements. Record the masses in TABLE 2.2A. Be careful to record all data obtained in this experiment to the appropriate number of significant figures (APPENDIX **A**).

B. Measurement of volume

Several different methods are used to measure the volume of your unknown solid sample. These methods and the results calculated from them are designed to clarify the meaning of precision and accuracy (APPENDIX **B**) in experimental measurements.

1. Measuring dimensions. The volume of a regularly shaped solid can be calculated from the measurements of appropriate dimensions. These dimensions can be measured with a centimeter ruler or with a more complex device like a micrometer.

If your unknown solid sample is a regularly shaped solid, measure the appropriate dimensions required to determine its volume using both a centimeter ruler and a micrometer, if both are available to you. [Your instructor will demonstrate the use of a micrometer (Laboratory Methods **A**).] Make three independent measurements of each required dimension with each device. Record your data in TABLE 2.2**B1**.

2. Measuring volume of water displaced. An easy way to measure the volume of either a regularly shaped or an irregularly shaped solid is to measure the volume of a liquid that it displaces. Obviously the liquid used must meet several requirements in order for this method to work: (1) the liquid must not dissolve the solid, (2) the liquid must not react with the solid, and (3) the liquid must have a lower density than the solid so that the solid will sink completely in the liquid. Water will serve as an appropriate liquid in this experiment since all of the unknown solid samples are insoluble in and unreactive toward water, and all of the samples have a greater density than water.

Clean the <u>largest</u> graduated cylinder in your desk (25-mL, 50-mL, or 100-mL). Fill it exactly four-fifths full with water, using a medicine dropper to make final adjustments (Laboratory Methods **B**). Record this initial volume in TABLE 2.2**B2**. Tilt the graduated cylinder to about a 30° angle from your benchtop, and slide your solid unknown sample carefully down the side of the graduated cylinder. This procedure prevents water from splashing out of the graduated cylinder and prevents the solid sample from crashing against the bottom of the graduated cylinder and breaking it. With the graduated cylinder vertical again, tap the graduated cylinder gently to remove any air bubbles adhering to the solid sample. Record in TABLE 2.2**B2** the final volume of water in the graduated cylinder. Pour the water and solid sample carefully from the graduated cylinder. Make three independent volume measurements altogether by this method.

3. Measuring mass of water displaced. Another way to determine the volume of either a regularly shaped or an irregularly shaped solid is to measure the mass of a liquid that the solid displaces and then to calculate the volume of the liquid displaced from its density. Water will serve as an appropriate liquid in this method.

The following calculation involving an unknown irregularly shaped solid illustrates this method. Suppose that an empty 10-mL graduated cylinder weighs 36.776 grams. After an unknown sample is added to the cylinder, it weighs 44.343 grams. Moreover, when enough water at 20° is added to give a total volume of 4.00 mL, the cylinder weighs 47.158 grams. Calculate the density of the unknown sample.

Calculate the mass of water.

```
 47.158 g   graduated cylinder, sample, and water
-44.343 g   graduated cylinder and sample
  2.815 g   water in cylinder
```

The density of water at 20.°C from APPENDIX **C** is 0.9982 g/mL.

Calculate the volume of water.

$$\frac{2.815 \text{ g}}{0.9982 \text{ g/mL}} = 2.820 \text{ mL water in cylinder}$$

Calculate the volume of sample.

```
 4.00 mL    water and sample
-2.820 mL   water in cylinder
 1.18 mL    sample   or   1.18 cm³   sample
```

Calculate the mass of sample.

```
 44.343 g   graduated cylinder and sample
-36.776 g   graduated cylinder
  7.567 g   sample
```

Calculate the density of sample.

$$\frac{7.567 \text{ g}}{1.18 \text{ cm}^3} = 6.41 \text{ g/cm}^3 \text{ of sample}$$

Clean the <u>smallest</u> graduated cylinder in your desk into which your unknown solid sample will fit. Dry the graduated cylinder carefully with a towel or by drawing air from it through a medicine dropper connected to an aspirator (Laboratory Methods **I**). Then weigh the graduated cylinder, and record the mass in TABLE 2.2**B3**. Tilt the graduated cylinder to about a 30°

angle from your benchtop, and slide your unknown solid sample carefully down the side of the graduated cylinder. Weigh the graduated cylinder and solid sample, and record the mass in TABLE 2.2B3. Leave the solid sample in the graduated cylinder, and add water exactly to the nearest 1-mL mark above the point where water completely covers the solid sample. Tap the graduated cylinder gently to remove any air bubbles adhering to the solid sample. Use a medicine dropper to make final volume adjustments, and record the final volume in TABLE 2.2B3. Weigh the graduated cylinder, solid sample, and water, and record the mass in TABLE 2.2B3. Determine the temperature of the water, and record it in TABLE 2.2B3. Then pour the water and solid sample carefully from the graduated cylinder. Make three independent sets of measurements altogether by this method.

Dry the unknown solid sample, and return it to your instructor.

Perform the calculations in TABLE 2.3 including sample calculations for one run on the back of TABLE 2.3.

Section _____ Date _____

Instructor _____   **PRE-LABORATORY QUESTIONS**

1. Why is it necessary that the unknown sample be dry when it is weighed in part **A**?

2. What is the meaning of the term <u>significant figures</u> when referring to experimental measurements?

3. If the stabilized mass on the dial of a digital top-loading balance reads 4.200 g, which digits should you record on your data page? How many significant figures does your recorded number have? Why?

4. Given the liquid level shown in a 10-mL graduated cylinder, what volume would you record on your data page? Why?

5. Define clearly what you understand by

    a. <u>Precision</u> in experimental measurements.

    b. <u>Accuracy</u> in experimental measurements.

Experiment 2

6. Assume that you have an unknown solid for which the actual mass is 6.816 g. Which of the following sets of experimental mass measurements has the higher <u>precision</u>? Which set yields the more <u>accurate</u> result? Indicate clearly your reasoning.

6.832 g	6.817 g
6.828 g	6.800 g
6.827 g	6.808 g
6.829 g	6.821 g

7. Complete the following table for each of the geometric shapes.

<u>Geometric shape</u>	Formula for <u>calculating volume</u>	Dimension(s) that <u>must be measured</u>
Cube		
Rectangular solid		
Cylinder		
Sphere		

Empirical Formula of a Compound

PURPOSE OF EXPERIMENT: Determine the empirical formula of lead iodide.

The empirical formula of a compound tells us the types of atoms present in a compound as well as the ratio of different types of atoms. The empirical formula does <u>not</u> tell us the actual number of atoms in the molecule. For example, the empirical formula, BCl_2, represents a compound that has two atoms of chlorine for every boron atom. The subscripts of 1 (unwritten) and 2 are the smallest whole numbers that can be used to express the atomic ratio. The molecular formula indicates the actual number of atoms of each element in one molecule. The molecular formula, B_2Cl_4, tells us that one molecule contains two boron atoms and four chlorine atoms. The simplest ratio of atoms in this B_2Cl_4 molecule is BCl_2, the empirical formula.

An ionic compound is an aggregrate of ions, these being present in a definite ionic ratio that is characteristic of the compound. Since there are no discrete molecules in the infinite lattice of an ionic substance, only an empirical formula can be used to express its composition. Thus, for nickel(II) chloride, which is composed of a collection of Ni^{2+} and Cl^- ions in the ratio of one to two, the empirical formula is $NiCl_2$. There is no molecular formula for nickel(II) chloride.

If you have 100.0 g of calcium chloride, you have 36.1 g of calcium and 63.9 g of chlorine. Another way of expressing the composition is to say that calcium chloride is 36.1% Ca and 63.9% Cl by mass. If there are 36.1 g of calcium in 100.0 g of compound, you have 0.900 mole of calcium. If there are 63.9 g of chlorine in 100.0 g of compound, you have 1.80 mole of chlorine.

In this experiment you will determine the empirical formula of a compound composed of lead and iodine. A weighed quantity of lead is reacted with nitric acid, HNO_3, solution. The resulting lead nitrate solution is then reacted with sodium iodide, NaI, solution to form insoluble lead iodide, which is filtered, dried, and weighed. From your experimental data you can calculate the percentage composition and the ratio of moles of lead to moles of iodine in the compound, and then write the empirical formula.

REFERENCES

(1) Kotz, J. C., and Purcell, K. F., <u>Chemistry and Chemical Reactivity</u>, Saunders College Publishing, Philadelphia, 1987, sections 2.5 and 2.11a.

(2) Masterton, W. L., Slowinski, E. J., and Stanitski, C. L., <u>Chemical Principles</u>, 6th ed., Saunders College Publishing, Philadelphia, 1985, sections 2.5 and 3.1-3.3.

EXPERIMENTAL PROCEDURE

(Study this section and the PRE-LABORATORY QUESTIONS before coming to the laboratory. <u>Wear safety goggles when performing this experiment</u>.)

This experiment will be run in duplicate to prove that the quantitative results are reproducible. The two runs should be performed simultaneously in order to save time. Record all experimental data in TABLE 3.1. Care in weighing is very important in this experiment. Therefore, review Laboratory Methods C describing the use of balances very carefully.

Using a beam balance, weigh two samples of granulated lead, Pb, of 0.13 - 0.15 g each onto pieces of weighing paper. Mark two beakers of 250-mL or more capacity as numbers 1 and 2. Using an analytical balance, weigh each sample of lead on its weighing paper, transfer each lead sample to the respective beaker by carefully tapping the paper, and then weigh each paper. Record your masses in TABLE 3.1 under the appropriate run.

Obtain 20. mL of 3 $\underline{M}$ nitric acid, HNO_3, solution in your graduated cylinder. **CAUTION: <u>HNO_3 of this concentration is corrosive and causes burns on your skin or holes in your clothing. If any spills or spatters occur onto your skin or clothing, rinse the affected area thoroughly with water</u>**. (When nitric acid solution touches skin, the skin turns a deep yellow color.) Pour half of the 3 $\underline{M}$ HNO_3 onto each sample of lead in beakers 1 and 2.

Cover beaker 1 with a clean watch glass, and heat it **gently** over a cool, luminous Bunsen burner flame (Laboratory Methods D) until steam exits from under the watch glass. **CAUTION: <u>Do not boil</u>**. Reheat periodically if gas evolution ceases, but do not boil. After all the lead has reacted, add 20. mL of distilled water. Heat the solution again until it steams, but **<u>do not boil</u>**. Remove the beaker from the ring stand, and let it stand for several minutes. Weigh approximately 0.8 g of sodium iodide, NaI, onto weighing paper on a beam balance. Transfer the NaI to a 100-mL or larger beaker, and add 40. mL of distilled water. Heat this solution until it steams, allow the solution to cool slightly, and then add half of it slowly with stirring to beaker 1. (Save the other half for beaker 2.) Allow the mixture to cool to room temperature while you repeat the procedure with beaker 2. Some of the lead iodide precipitate may be colloidal at this point. Therefore, grow larger aggregated crystals by heating the precipitate **gently** with constant stirring for 5-10. minutes, but **<u>do not boil</u>** the solution. Then allow the mixture to cool.

Using an analytical balance, weigh a piece of filter paper for either a long-stem funnel or a Buchner funnel as designated by your instructor. Use this filter paper to filter the lead iodide precipitated in run 1 either by gravity filtration or by vacuum filtration (Laboratory Methods I). Use additional distilled water and a rubber policeman, if available, to help transfer the precipitate completely from the beaker to the filter paper. Remove the filter paper from the funnel. Dry the paper and its contents by heating on a steam bath, with a heat lamp, or in an oven (Laboratory Methods M) as designated by your instructor. While the precipitate for run 1 is drying, repeat the procedure of this paragraph to filter the solid for run 2, and wash and dry it. When the filter paper and solid for each run are completely dry (heat for 10. minutes after the solid starts to crack), allow them to cool to room temperature. Weigh the filter paper and solid for each run on an analytical balance, and record the masses in TABLE 3.1. Repeat the drying procedure for 5 minutes, and reweigh. Your change in mass between dryings must be no more than 0.001 g, or the drying procedure should be repeated a third time. Place your filter papers and solids in a beaker on the reagent bench marked LEAD IODIDE RESIDUE.

Perform the calculations in TABLE 3.2 including sample calculations for one run on the back of TABLE 3.2. The percentage composition is the percent of lead and the percent of iodine by mass in lead iodide. For example, the percent of lead in lead iodide gives the number of grams of lead in 100.0 g of lead iodide. Calculate the number of moles of lead and of iodine in 100.0 g of lead iodide, making use of your percentage values for the average of your two runs. The number of moles of iodine per mole of lead will tell you the ratio of atoms of iodine to lead in the compound. Writing these numbers as subscripts for the chemical symbols of lead and iodine will give you the empirical formula.

Analysis of an Oxygen-containing
Compound or Mixture

PURPOSE OF EXPERIMENT: Determine the percent composition by mass of metal in an unknown oxygen-containing compound or mixture by decomposing and/or reducing the compound or mixture to the metal.

The metals in the periodic table combine with oxygen in a variety of ratios. Although no one metal, M, forms all of them, at least the following empirical formulas for metal oxides are known: M_2O, MO, M_2O_3, MO_2, M_2O_5, MO_3, M_2O_7, and MO_4.

Many of the oxides of the less active metals, which include many of the transition metals, are easily reduced by heating with a chemical reducing agent. Such reactions are important in the commercial preparation of metals from their oxides. For example, iron(III) oxide, Fe_2O_3, is reduced by heating it with hydrogen gas, H_2, the metal and water being formed as products.

$$Fe_2O_3(s) + 3\ H_2(g) \xrightarrow{\text{heat}} 2\ Fe(s) + 3\ H_2O(g)$$

A number of other chemical reducing agents such as coke (carbon) or an active metal like aluminum are also used to prepare metals in industry.

$$NiO(s) + C(s) \xrightarrow{\text{heat}} Ni(s) + CO(g)$$

$$3\ Mn_3O_4(s) + 8\ Al(s) \xrightarrow{\text{heat}} 9\ Mn(s) + 4\ Al_2O_3(s)$$

However, natural gas will be used as the reducing agent in this experiment because it is somewhat safer to work with than hydrogen gas, it is readily available from the gas jets in the laboratory, and it is less expensive than active-metal reducing agents. Natural gas is commonly recovered from petroleum deposits and is a complex mixture of hydrocarbons, particularly methane, CH_4, and ethane, C_2H_6. However, it contains small amounts of other hydrocarbons at least up to C_6 as well as traces of carbon dioxide, CO_2, helium, He, hydrogen sulfide, H_2S, nitrogen, N_2, and water. When natural gas reduces a heated metal oxide, the metal is produced along with water vapor and a mixture of carbon monoxide, CO, and carbon dioxide, CO_2. The following equation describing the reduction of copper(I) oxide represents one possible combination of reactants and products.

$$3\ Cu_2O(s) + CH_4(g) \xrightarrow{\text{heat}} 6\ Cu(s) + 2\ H_2O(g) + CO(g)$$

The metals in the periodic table also form other oxygen-containing compounds. These compounds usually involve oxyanions such as hydroxide, OH^-; carbonate, CO_3^{2-}; nitrate, NO_3^-; sulfate, SO_4^{2-}; phosphate, PO_4^{3-}; and others. For example, iron commonly forms $Fe(OH)_2$, $Fe(OH)_3$, $FeCO_3$, $Fe(NO_3)_2$, $Fe(NO_3)_3$, $FeSO_4$, $Fe_2(SO_4)_3$, $Fe_3(PO_4)_2$, and $FePO_4$. Some of these oxygen-containing compounds can be converted to the metal by a combination of thermal decomposition followed by reduction. For example, iron(II) carbonate, $FeCO_3$, is thermally decomposed upon strong heating to iron(II) oxide, FeO,

$$FeCO_3(s) \xrightarrow{\text{heat}} FeO(s) + CO_2(g)$$

which is reduced by heating with natural gas to the metal.

$$3\ FeO(s)\ +\ CH_4(g)\ \xrightarrow{\text{heat}}\ 3\ Fe(s)\ +\ 2\ H_2O(g)\ +\ CO(g)$$

Experimentally these two reactions can be performed simultaneously by passing natural gas through a heated tube containing $FeCO_3$ just as you would to reduce the simple oxide. A possible overall reaction can be obtained by adding the two equations above, using appropriate coefficients.

$$3\ FeCO_3(s)\ \longrightarrow\ 3\ FeO(s)\ +\ 3\ CO_2(g)$$
$$3\ FeO(s)\ +\ CH_4(g)\ \longrightarrow\ 3\ Fe(s)\ +\ 2\ H_2O(g)\ +\ CO(g)$$

$$\overline{3\ FeCO_3(s)\ +\ CH_4(g)\ \longrightarrow\ 3\ Fe(s)\ +\ 2\ H_2O(g)\ +\ 3\ CO_2(g)\ +\ CO(g)}$$

In this experiment a weighed amount of an unknown oxygen-containing compound or mixture is heated in a stream of natural gas, and the water vapor and oxides of carbon are allowed to escape into the air. The metal that remains is weighed, and the loss in mass is that of oxygen and any other nonmetals originally present in the oxide. The experimental data for the mass of metal in the compound or mixture can be used to calculate the percent composition by mass of the metal. You will perform duplicate determinations to assess the reproducibility of your results.

REFERENCES

(1) Kotz, J. C., and Purcell, K. F., <u>Chemistry and Chemical Reactivity</u>, Saunders College Publishing, Philadelphia, 1987, sections 2.5, 2.12, 4.1, and 4.2.

(2) Masterton, W. L., Slowinski, E. J., and Stanitski, C. L., <u>Chemical Principles</u>, 6th ed., Saunders College Publishing, Philadelphia, 1985, sections 2.5, 3.1-3.3, 3.6, and 3.7.

EXPERIMENTAL PROCEDURE

(Study this section and the PRE-LABORATORY QUESTIONS before coming to the laboratory. **Wear safety goggles when performing this experiment**.)

Obtain a 20 x 150-mm test tube and an appropriately fitting (#2) 2-hole rubber stopper Ⓐ from the equipment bench or the stockroom. The glass tubing in FIGURE 4.1 may have already been made in Experiment 1.

① 18-cm straight tubing

② 12-cm tubing with a 90° bend 6 cm from one end

CAUTION: **If you must insert or remove glass tubing into or from the rubber stopper or rubber connectors, follow the instructions in Laboratory Methods F very carefully.** Otherwise, the glass tubing will already be fitted in the 2-hole rubber stopper.

Make certain that the 20 x 150-mm test tube is clean and dry. Weigh the empty test tube on an analytical balance (Laboratory Methods **C**), and record the mass in TABLE 4.1. Obtain a sample of an unknown oxygen-containing compound or mixture from your instructor. Put about half of the sample into the test tube. Tap the test tube gently so that all the solid goes to the bottom. Then weigh the test tube and sample, and record the mass in TABLE 4.1.

Assemble the apparatus shown in FIGURE 4.1. This arrangement allows you to pass the natural gas over the sample and then burn the excess gas in a Bunsen burner. Connect the longer straight tubing to your gas jet with a piece of rubber tubing. Connect the shorter bent tubing to your Bunsen burner with a second piece of rubber tubing. Make certain that all joints are tight, and then <u>have your instructor inspect your apparatus</u>.

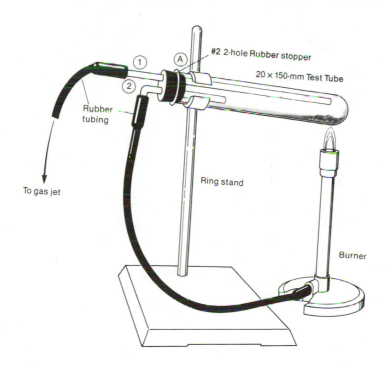

FIGURE 4.1. Apparatus for decomposition and reduction.

Adjust the collar on your Bunsen burner to close the air supply (Laboratory Methods D). Then light the burner, and let the flame burn until a cool luminous yellow flame is obtained. This procedure assures that all the air has been swept from the tube and that it is safe to begin the reduction reaction.

Adjust the collar on your Bunsen burner again so that a hot nonluminous blue flame is obtained. Heat the sample, at first gently and then strongly. **CAUTION: <u>Do not heat the tube so hot that it melts or sags</u>.** The reduction reaction will require at least 10. and up to 30. minutes of heating. As the reaction proceeds, both the color of the solid and the form of the crystals may change. Continue to heat for about 5 minutes after it appears that the reaction is complete.

Stop heating the tube by putting the Bunsen burner to the side **<u>without shutting it off</u>**. Keeping the burner lit ensures that any toxic carbon monoxide remaining in the test tube will be flushed through the burner where it is converted to carbon dioxide according to the following equation.

$$2\ CO(g)\ +\ O_2(g)\ \text{--->}\ 2\ CO_2(g)$$

Moreover, letting the tube and metal cool to room temperature in the presence of natural gas ensures that there is no chance of the hot metal being reoxidized upon exposure to air. When the tube is cool, turn off the gas and the Bunsen burner, and take the apparatus apart. Weigh the test tube and metal on the analytical balance, and record the mass in TABLE 4.1.

If you have time, check to see that the reduction reaction was complete by heating the sample again for about 5 minutes in a stream of natural gas. As you carefully reassemble the apparatus, tap the test tube gently on your laboratory bench to break up the caked metal and any remaining oxygen-containing compound as much as possible. Reweigh the test tube and metal on an analytical balance after cooling, and record the mass in TABLE 4.1. The first and second weighings should differ by no more than 0.001 g if the reduction reaction was complete the first time.

Clean and dry your test tube, and repeat the same procedure with the second half of your sample.

Before leaving the laboratory, find out from your instructor what was the metal in your unknown sample, and record it in TABLE 4.1.

Perform the calculations in TABLE 4.2 including sample calculations for one run and one additional question (if designated by your instructor) on the back of TABLE 4.2.

TABLE 4.1. Mass data for decomposition and reduction.

Unknown number __	Run 1	Run 2
Mass of empty test tube (g)		
Mass of test tube and unknown sample (g)		
Mass of test tube and metal after first heating (g)		
Mass of test tube and metal after second heating (g)		
Metal in unknown sample		

Instructor's initials _____

C A L C U L A T I O N S

TABLE 4.2. Masses and percents by mass of metal and other elements.

	Run 1	Run 2
Mass of unknown sample reacted (g)		
Mass of metal produced from unknown sample (g)		
Mass of other element(s) in unknown sample (g)		
Percent by mass of metal in unknown sample (%)		
Average percent by mass of metal in unknown sample (%)		
Percent by mass of other elements in unknown sample (%)		
Average percent by mass of other elements in unknown sample (%)		

Experiment 4

Sample calculations for Run __

    Mass of unknown sample reacted

    Mass of metal produced from unknown sample

    Mass of other elements in unknown sample

    Percent by mass of metal in unknown sample

    Percent by mass of other elements in unknown sample

ANSWER ONE OF THE FOLLOWING AS DESIGNATED BY YOUR INSTRUCTOR.
Based on your experimental masses, can your unknown sample be a known
(1) oxide, (2) hydroxide, (3) carbonate, or (4) nitrate, or (5) a mixture of
a known carbonate and the metal. Why, or why not? Indicate clearly your
calculations and reasoning.

1. Write a balanced complete equation for the reduction of Pb_3O_4 by each of the following reducing agents.

    a. Hydrogen gas.

    b. Coke, assuming that carbon monoxide is the only carbon-containing product formed.

    c. Aluminum.

    d. Natural gas, assuming that natural gas is pure ethane, C_2H_6, and that carbon dioxide is the only carbon-containing product formed.

2. After the apparatus in FIGURE 4.1 is assembled and the Bunsen burner with its air supply closed is first lit, the Bunsen burner has a relatively nonluminous blue flame. Why should this be so when the air supply is closed? Why is a luminous yellow flame obtained a short time later without making any changes in the collar of the Bunsen burner?

3. Why would it not be wise to start the reduction reaction without first having observed the phenomena described in PRE-LABORATORY QUESTION 2?

4. At the end of the reduction reaction, why is the reduced metal cooled in a stream of natural gas rather than in a stream of air?

5. Write a balanced complete equation for the thermal decomposition of each of the following to form lead(II) oxide, PbO.

    a. Lead(II) hydroxide.

    b. Lead(II) carbonate.

    c. Lead(II) nitrate.

POST-LABORATORY QUESTIONS

1. Indicate clearly whether your calculated percent by mass of metal in the unknown sample would be higher than, lower than, or unchanged from the true value, and why, for each of the following.

    a. The unknown sample contained some inert impurity.

    b. The unknown sample contained some metal as well as its oxygen-containing compound.

    c. Some of the unknown sample was spilled as you were loading it into the test tube.

    d. The reduction by natural gas was incomplete.

e. The metal product was cooled in a stream of air rather than in a stream of natural gas.

f. Some of the metal product that was formed was spilled before the final weighing.

g. The test tube was damp when it was weighed initially.

h. The thermal decomposition of a carbonate salt as the unknown oxygen-containing sample was incomplete.

Analysis of an Unknown Mixture
by Fractional Crystallization

PURPOSE OF EXPERIMENT: Separate three solids (sand, $K_2Cr_2O_7$, and NaCl) from each other by fractional crystallization, and determine the percent composition by mass of an unknown mixture.

A mixture is composed of two or more substances that do not react with each other. The composition of a mixture may be variable, sometimes over a wide range. Mixtures are encountered frequently both in nature and in the laboratory. Salt beds, salt brines, and sea water - major sources of many important salts - are mixtures. The desired product in most chemical syntheses is part of a mixture composed also of undesired side products as well as left over reactants. A heterogeneous mixture like the unknown in this experiment has an irregularly variable composition, structure, and physical and chemical behavior such that abrupt discontinuities or boundaries may be observed. Cement and topsoil are examples of heterogeneous mixtures. Each component retains its chemical integrity, and the components can be separated by taking advantage of differences of physical properties of the components. In this experiment the separation is based on the pure components having different solubilities in a given solvent at different temperatures.

The process of fractional crystallization is a technique by which chemists separate and purify many substances. This process takes advantage of differing variations with temperature of the solubilities of several components in a given solvent. If one of the components is insoluble at all temperatures whereas other components are soluble at some temperature, the insoluble component can be separated easily by filtration. If the solubility of one of the soluble components in the mixture increases rapidly with temperature whereas the solubility of the other soluble component increases only slightly with temperature, an appropriate temperature can be chosen such that one of the components will be only slightly soluble whereas the other will be almost entirely in solution. With one component present primarily in the solid phase and the other in solution, a separation can again be achieved easily by filtration. The solvent used may dissociate components that are salts, but must not react further chemically with the components. The solvent must also allow the components to separate as well formed crystals and must be easily removed from the components, for example, by evaporation. Water serves as an appropriate solvent in this experiment. In many industrial operations mixed solvent systems must be used.

In this experiment you will start with a solid mixture of sand (SiO_2), potassium dichromate $(K_2Cr_2O_7)$, and sodium chloride (NaCl) and will attempt to separate and obtain each of these components in a pure form by fractional crystallization from water.

REFERENCES

(1) Kotz, J. C., and Purcell, K. F., *Chemistry and Chemical Reactivity*, Saunders College Publishing, Philadelphia, 1987, sections 1.3, 2.12, and 12.2.

(2) Masterton, W. L., Slowinski, E. J., and Stanitski, C. L., *Chemical Principles*, 6th ed., Saunders College Publishing, Philadelphia, 1985, sections 1.6, 3.4-3.6, and 12.3.

EXPERIMENTAL PROCEDURE

(Study this section and the PRE-LABORATORY QUESTIONS before coming to the laboratory. **Wear safety goggles when performing this experiment.**)

Sand, $K_2Cr_2O_7$, and NaCl are easily distinguished by color because sand is light tan, $K_2Cr_2O_7$ crystals are orange, and NaCl crystals are white. Sand is insoluble in water whereas $K_2Cr_2O_7$ and NaCl are both soluble. However, as shown in TABLE 5.1, $K_2Cr_2O_7$ shows a 21-fold increase in solubility between 0.°C and 100.°C whereas NaCl exhibits little change in solubility with temperature.

TABLE 5.1. Solubilities (grams of anhydrous solute per 100. mL of water).

Temperature (°C)	$K_2Cr_2O_7$	NaCl
0.0	4.6	35.6
10.0	6.6	35.7
20.0	12.2	35.8
30.0	18.0	36.0
40.0	26.0	36.3
50.0	37.2	36.7
60.0	46.5	37.1
70.0	58	37.5
80.0	70	38.0
90.0	82	38.5
100.0	97	39.4

Given a water solution containing comparable masses of $K_2Cr_2O_7$ and NaCl it should be clear that $K_2Cr_2O_7$ may crystallize in nearly pure form from the solution at temperatures near 0°C and that NaCl may crystallize in nearly pure form from the solution at temperatures near 100°C. You will be able to follow the fractional crystallization visually because of the difference in color of $K_2Cr_2O_7$ and NaCl. Using your mass data, and given solubility data, you will determine the percent by mass of each component in your unknown mixture. Your instructor will judge the quality of your work by considering both the amount and purity of the crystals that you obtain; therefore, work carefully, avoiding loss of solution or solid wherever possible. Be sure to use the amounts specified.

Obtain from your instructor a vial containing an unknown mixture of sand, $K_2Cr_2O_7$, and NaCl. <u>You will be given only one sample.</u>

Using a beam balance (Laboratory Methods **C**), weigh your vial and unknown mixture, and record the mass in TABLE 5.2. With the help of a spatula, add your unknown mixture to a clean 100-mL or larger beaker. Reweigh the empty vial, and record the mass in TABLE 5.2.

A. Dissolving the Soluble Salts

Add to the beaker 50. mL of distilled water, which should be enough to dissolve the <u>soluble solids</u> (Laboratory Methods **K**). Warm the mixture of your unknown and water gently with a Bunsen burner (Laboratory Methods **D**). Stir the solution with a glass rod to make sure that the soluble solids are completely dissolved.

B. Separating Sand

Weigh a piece of filter paper for either a long-stem funnel or a Buchner funnel, as designated by your instructor, and record the mass in TABLE 5.2. When the beaker containing your mixture is cool enough to handle, remove the sand by filtering the solution through the preweighed filter paper by either gravity filtration or vacuum filtration (Laboratory

Methods I). Use a rubber policeman to transfer as much as possible of the sand to the paper. <u>Is the filtrate colored</u>? <u>What must be in the filtrate</u>? <u>Why</u>?

Use four successive 5-mL portions of distilled water to wash any sand remaining in your beaker onto the filter paper and to wash the sand thoroughly (Laboratory Methods I). When no more water is being removed, lift the filter paper and sand from the funnel with the help of a spatula, and put the filter paper on a watch glass. Dry the filter paper and sand by heating either on a steam bath, with a heat lamp, or in an oven (Laboratory Methods M), as designated by your instructor.

C. Separating $K_2Cr_2O_7$

Heat the orange filtrate to its boiling point in a 250-mL or larger beaker, and boil it <u>gently</u> until the solution looks cloudy from the appearance of white NaCl crystals. CAUTION: <u>The solution will have a tendency to bump, so do not heat it very strongly. Hot dichromate solution can give you a bad burn. If any spills or spatters occur onto your skin or clothing, rinse the affected area thoroughly with water</u>. When NaCl crystals are clearly apparent, stop heating, and add 5 mL of distilled water to the solution. This is sufficient water to dissolve the NaCl and to prevent it from crystallizing later along with $K_2Cr_2O_7$. Draw some solution from the beaker into a medicine dropper, and wash any crystallized solid from the walls of the beaker. Stir the solution with a glass rod to dissolve the solids. If necessary, you may heat the solution, but <u>do not boil it</u>.

Cool the solution to near 0°C first in air, then in a water bath, and finally in an ice bath. <u>What are the crystals that form</u>? <u>How do you know</u>?

Stir the cold slurry for several minutes. Place a <u>preweighed</u> piece of filter paper in either a long-stem funnel or a Buchner funnel. Chill the funnel by adding 100. mL of ice-cold distilled water and then pouring the water off or drawing the water through the funnel and discarding it after about one minute. Filter the mixture in the beaker through the cold funnel by either gravity filtration or vacuum filtration. Use your rubber policeman to help transfer the last of the crystals. Press the crystals dry with a clean piece of filter paper, and scrape any crystals sticking to the second piece of filter paper back onto the first piece. When no more water is being removed, pour the filtrate, which contains most of the NaCl from the mixture, into your larger graduated cylinder. Record its volume in TABLE 5.2, and then pour the filtrate into a clean 100-mL or larger beaker. Heat the filtrate to boiling, and boil gently.

While you are waiting for the filtrate to boil, wash the $K_2Cr_2O_7$ crystals in the funnel with 3 mL of ice-cold distilled water. If vacuum filtration is used, let the cold water remain in contact with the crystals for only about 10. seconds before applying suction. The water removed will contain most of the NaCl impurity. When no more water is being removed, lift the filter paper and crystals from the funnel with the help of a spatula, and put the filter paper on a watch glass. Dry the filter paper and $K_2Cr_2O_7$ by heating either on a steam bath, with a heat lamp, or in an oven. Heat the sample for about 10. minutes after the solid starts to crack.

Experiment 5

D. Separating NaCl

Continue to boil the solution containing the NaCl until the volume of NaCl crystals (which appear as the boiling proceeds) is about equal to half of the volume of liquid above the crystals. Again, draw some solution from the beaker into a medicine dropper, and wash any crystallized solid from the walls of the beaker. Stir the solution with a glass rod to dissolve any solid $K_2Cr_2O_7$ that may be present. Reheat the solution to the boiling point.

Use your cloth or a "hot hand" (Laboratory Methods **L**) to hold the hot beaker, and filter the hot solution through a <u>preweighed</u> piece of filter paper using either gravity filtration or vacuum filtration. Transfer as much of the solid NaCl crystals as possible to the filter, first by swirling the slurry as you pour it and then by using your rubber policeman. Press the crystals dry with a piece of filter paper, and scrape any crystals sticking to the second piece of filter paper back onto the first one. At this point the NaCl will appear yellow because of the presence of residual dichromate solution.

Pour the filtrate into your larger graduated cylinder. Record its volume in TABLE 5.2. Wash the NaCl crystals with two 1-mL portions of 6 **M** hydrochloric acid, HCl, solution. **CAUTION: <u>HCl of this concentration causes burns on your skin and holes in your clothing. If any spills or spatters occur onto your skin or clothing, rinse the affected area thoroughly with water</u>.** If this operation is done properly, your purified NaCl crystals will be nearly white, and the filtrate should contain most of the yellow contaminant. When no more water is being removed, lift the filter paper and crystals from the funnel with the help of a spatula, and put the filter paper on a watch glass. Dry the filter paper and NaCl by heating either on a steam bath, with a heat lamp, or in an oven. Heat the sample for about 10. minutes after the sample starts to crack.

Weigh your dry sand and filter paper, your dry $K_2Cr_2O_7$ and filter paper, and your dry NaCl and filter paper, and record the masses in TABLE 5.2. Empty the dry sand into an appropriately marked bottle on the reagent bench. Clean and dry the vial in which you received your original mixture; empty the $K_2Cr_2O_7$ into it; label the vial with your name, date, section, instructor, and mass of $K_2Cr_2O_7$; and give it to your laboratory instructor for evaluation. Show your dry NaCl sample to your instructor for evaluation, and then empty it into an appropriately marked bottle on the reagent bench.

Perform the calculations in TABLE 5.3 including sample calculations on the back of TABLE 5.3.

Analysis of the Reaction Products
after Reacting Magnesium with Air

PURPOSE OF EXPERIMENT: Determine the percent by mass of the products, Mg_3N_2 and MgO, after magnesium is reacted with air.

The two major components of air are nitrogen (~78%) and oxygen (~21%). Magnesium can react with both of these diatomic molecules to form magnesium nitride, Mg_3N_2, and magnesium oxide, MgO, which both form colorless crystals or white powders at room temperature.

$$2\ Mg(s)\quad +\quad O_2(g)\quad \overset{heat}{\text{--->}}\quad 2\ MgO(s) \tag{1}$$

$$3\ Mg(s)\quad +\quad N_2(g)\quad \overset{heat}{\text{--->}}\quad Mg_3N_2(s) \tag{2}$$

Note that one mole of MgO is formed per mole of Mg reacting in equation (1) whereas one mole of Mg_3N_2 is formed per three moles of Mg reacting in equation (2). For either reaction the mass of product is greater than the mass of reacting Mg; however, a given mass of Mg will yield a greater mass of MgO than Mg_3N_2. In addition, Mg_3N_2 can be converted to MgO by reaction with water and subsequent heating to drive off ammonia, NH_3, and excess water.

$$Mg_3N_2(s)\quad +\quad 3\ H_2O(l)\quad \text{--->}\quad 3\ MgO(s)\quad +\quad 2\ NH_3(g) \tag{3}$$

There is an increase in mass when Mg_3N_2 is converted to MgO because we already found that a given mass of Mg produces a greater mass of MgO than Mg_3N_2.

Your task in this experiment is to determine the percent composition by mass of the mixture of MgO and Mg_3N_2 formed by reacting Mg with air. In any mixture problem the same number of mathematical equations are required as there are components in the mixture - two in this case. One of these equations always involves the total mass of the mixture being the sum of its parts.

$$\text{total mass of mixture}\ =\ \text{mass of MgO}\ +\ \text{mass of } Mg_3N_2 \tag{4}$$

The second equation is usually based on conservation of mass of a common element or compound across the chemical reactions that are occurring. There are two possible mass balance equations in this experiment, for example, the pair based on the mass of Mg. Thus, the mass of Mg in the components of the mixture, MgO and Mg_3N_2, equals either the initial mass of Mg or the final mass of Mg in dried total MgO.

$$\text{mass of Mg in MgO}\ +\ \text{mass of Mg in } Mg_3N_2\ =\ \text{mass of Mg} \tag{5}$$

$$\text{mass of Mg in MgO}\ +\ \text{mass of Mg in } Mg_3N_2\ =$$
$$\text{mass of Mg in dried total MgO} \tag{6}$$

Note that the mass of Mg in MgO is determined by multiplying the mass of MgO times the proportion of that mass which is Mg.

$$\text{mass of Mg in MgO}\ =\ (m_{MgO})\left(\frac{M_{Mg}}{M_{MgO}}\right)$$

95

Experiment 6

A similar expression can be written for the mass of Mg in Mg_3N_2, keeping in mind that there are three moles of Mg per mole of Mg_3N_2.

$$\text{mass of Mg in } Mg_3N_2 = (m_{Mg_3N_2})\left(\frac{3\ M_{Mg}}{M_{Mg_3N_2}}\right)$$

If equation (4) is then substituted into equation (5) or equation (6), the masses of MgO and Mg_3N_2 can be determined. The percent by mass of either component in the mixture is then the mass of either component divided by the total mass of the mixture and multiplied by 100 to convert to percent.

In this experiment a sample of magnesium will be reacted with air to form MgO and Mg_3N_2. The Mg_3N_2 will then be converted into MgO by reaction with water followed by drying. The original mass of magnesium, the mass of the product mixture, and the mass of dried total MgO are then used to perform two independent calculations of the percent composition by mass of the product mixture.

REFERENCES

(1) Dingledy, D. P., "Synthesis and Determination of the Formula of a Compound," STQ1-168 Modular Laboratory Program in Chemistry, Willard Grant Press, Boston, 1976; Sherman, A., Sherman, S., and Russikoff, L., "Laboratory Experiments for Basic Chemistry," 3rd ed., Houghton-Mifflin Company, Boston, 1984; Leary, J. J., and Gallaher, T. N., J. Chem. Educ., 1983, 60, 673; Gallaher, T. N., Moody, F. P., Burkholder, T. R., and Leary, J. J., J. Chem. Educ., 1985, 62, 626.

(2) Kotz, J. C., and Purcell, K. F., Chemistry and Chemical Reactivity, Saunders College Publishing, Philadelphia, 1987, sections 2.5, 2.12, 3.3a, and 4.1.

(3) Masterton, W. L., Slowinski, E. J., and Stanitski, C. L., Chemical Principles, 6th ed., Saunders College Publishing, Philadelphia, 1985, sections 2.5 and 3.1-3.7.

EXPERIMENTAL PROCEDURE

(Study this section and the PRE-LABORATORY QUESTIONS before coming to the laboratory. **Wear safety goggles when performing this experiment.**)

Support a clean porcelain crucible with the cover slightly tilted on a wire triangle, and heat the crucible as hot as possible for 5 minutes (Laboratory Methods D). The bottom of the crucible should glow a dull red (red heat) for the full 5 minutes. Use only tongs to handle the crucible and cover after it has been heated. From this point on in the experiment the crucible and cover must not be touched with your hands. Place the hot crucible and cover in a dessicator, if available, and allow it to cool to room temperature. Then weigh the crucible and cover on an analytical balance (Laboratory Methods C), and record the mass in TABLE 6.1.

Reheat the crucible and cover a second time, and permit them to cool in the dessicator, if available. Reweigh them, and record the mass in TABLE 6.1. The mass of the crucible and cover should not have changed by more than 0.001 g. If you observe a mass change of greater than 0.001 g, the crucible must be reheated, cooled, and reweighed until no mass change greater than 0.001 g is observed.

96

Clean a 50.-cm length of magnesium ribbon by rubbing it _gently_ with steel wool. Weigh the magnesium ribbon on an analytical balance, and record the mass in TABLE 6.1. _Loosely_ pack the magnesium into the bottom of the preweighed crucible using a stirring rod and forceps instead of your fingers. _Remember - never touch the crucible or cover with your hands_. Reweigh the crucible and cover, and record the mass in TABLE 6.1.

Place the crucible on the wire triangle in a slightly tilted position. Place the cover partially over the crucible, but tilted slightly in the opposite direction from the crucible to permit access of air. Heat slowly at first, then more strongly, _making certain that the opening between the crucible and its top is not enveloped in the flame_. You don't want the magnesium to "burn" or burst into flame. If the magnesium starts to burn and smoke is given off, cover the crucible completely, and remove the flame for a short time. Start reheating after the magnesium stops burning. Continue to heat the crucible until the contents no longer glow brightly and no smoke is given off. Then heat strongly for an additional 5 minutes with the cover tilted open and the crucible bottom glowing dull red. Allow the crucible and cover to cool in the dessicator, if available, and then weigh them, and record the mass in TABLE 6.1.

Push the fluffy product down into the bottom of the crucible with a stirring rod. Try to minimize the amount of material which adheres to the stirring rod. Add 10. drops of distilled water to the crucible. Check carefully for the odor of ammonia, $NH_3(g)$. It may be necessary to heat the crucible gently in order to detect the odor of ammonia. Heat the crucible with the cover tilted as before, gently at first and then at red heat for 5 minutes. Allow the crucible and cover to cool in the dessicator, if available. Reweigh them, and record the mass in TABLE 6.1.

Perform the calculations in TABLE 6.2 including sample calculations on the back of TABLE 6.2.

Student ID number _____

Section _____ Date _____

Instructor _____   D A T A

TABLE 6.1. Mass data for reacting magnesium with air.

Mass of magnesium (g)	
Mass of crucible and cover after 1st heating and cooling (g)	
Mass of crucible and cover after 2nd heating and cooling (g)	
Mass of crucible and cover after 3rd heating and cooling, if necessary (g)	
Mass of crucible, cover, and MgO-Mg$_3$N$_2$ product mixture (g)	
Mass of crucible, cover, and final total MgO (g)	

Instructor's initials _____

CALCULATIONS

TABLE 6.2. Calculation of percent composition by mass of product.

	Based on Mg	Based on final MgO
Mass of MgO-Mg$_3$N$_2$ product mixture (g)		
Mass of final total MgO (g)		
Mass of Mg$_3$N$_2$ in product (g)		
Mass of MgO in product (g)		
Percent Mg$_3$N$_2$ by mass in product (%)		
Percent MgO by mass in product (%)		

Experiment 6

Sample calculations

Write one pair of mathematical equations which will be used to calculate the masses of MgO and Mg_3N_2 formed by reaction of Mg with air. Indicate all units.

Substitute experimental quantities and other known factors into the equations, and solve for the mass of Mg_3N_2 formed by reaction of Mg with air.

Mass of MgO formed by reaction of Mg with air

Percent Mg_3N_2 by mass in the product mixture

Percent MgO by mass in the product mixture

PRE-LABORATORY QUESTIONS

1. Write a balanced chemical equation for the reaction of magnesium with oxygen (O_2).

2. Write a balanced chemical equation for the reaction of magnesium with nitrogen (N_2).

3. Using a mathematical equation, describe the stoichiometric relationship between Mg_3N_2 and MgO upon reaction of Mg_3N_2 with water.

4. Show clearly by calculations that 1.0 g of magnesium will produce a greater mass of MgO (upon reaction with O_2) than Mg_3N_2 (upon reaction with N_2).

5. Explain clearly why the product mixture containing MgO and Mg_3N_2 increases in mass after reaction with water and subsequent drying.

6. Why should you never touch the crucible or cover with your hands?

7. Why might it be a problem during heating if the magnesium bursts into flame and considerable smoke is evolved?

Name _____

Student ID number _____

Section _____ Date _____

Instructor _____

POST—LABORATORY QUESTIONS

1. Compare values of percents Mg_3N_2 and MgO by mass calculated from the starting mass of Mg versus values calculated from the final total mass of MgO. Suggest reasons for differences in your values.

2. Compare your percent composition by mass of products to the percent composition by mass of N_2 and O_2 in air. Why are they different?

3. If, after adding water to the reaction products, you failed to remove all of the excess water, how would your percent composition by mass of MgO change in comparison to that calculated using good technique? Justify your answer.

4. If you had failed to clean the magnesium ribbon before reaction, how would your percent MgO by mass change in comparison to that calculated using good technique? Justify your answer.

Gravimetric Analysis of a

Two-Component Mixture

PURPOSE OF EXPERIMENT: Determine the percent composition by mass of a two-component mixture of only $NaHCO_3$ and Na_2CO_3.

Mixtures are composed of two or more components which do not react with each other. (The Introduction to Experiment 5 contains a discussion of additional characteristics of mixtures.)

Stoichiometry deals with the mass relationships between reactants and products in chemical reactions. The primary bases of stoichiometry are the balanced chemical equation and the mole concept. In this experiment the concepts of stoichiometry will be used to calculate the percent composition of a mixture composed of sodium hydrogen carbonate (sodium bicarbonate), $NaHCO_3$, and sodium carbonate, Na_2CO_3. The number of moles of reactants and products will be calculated using only experimental mass measurements. When an analytical procedure that is used to determine the stoichiometry of a reaction involves only mass measurements, the analysis is called a gravimetric analysis.

The percent composition of a mixture of $NaHCO_3$ and Na_2CO_3 can be calculated from gravimetric analysis data by knowing the following five factors.

(1)   The identity of the components in the mixture.
(2)   The total mass of the mixture used for analysis.
(3)   The identity of common products to which both components can be converted.
(4)   The two balanced chemical equations which describe the reactions used to convert each component to the common products.
(5)   The mass of one of the products from the above reactions.

A mixture of $NaHCO_3$ and Na_2CO_3 reacts with hydrochloric acid, HCl, solution to form three common products, sodium chloride, NaCl(s), carbon dioxide, CO_2(g), and water, H_2O(l) according to the following balanced equations.

$$NaHCO_3(s) \; + \; HCl(aq) \; ---> \; NaCl(s) \; + \; CO_2(g) \; + \; H_2O(l) \qquad (1)$$

$$Na_2CO_3(s) \; + \; 2\,HCl(aq) \; ---> \; 2\,NaCl(s) \; + \; CO_2(g) \; + \; H_2O(l) \qquad (2)$$

One of the products after evaporation is solid NaCl which can easily be dried and weighed. Its mass can then be used to calculate the percent composition of the mixture. Since the mixture contains two components, two equations must be solved simultaneously to determine the percent composition.

The two simultaneous equations which will be used to calculate the masses of each component in the mixture must have identical unknown quantities including units. The first important mass relationship equates the mass of the mixture to the masses of the two components.

$$\text{Mass of mixture (g)} \; = \; \text{Mass of } NaHCO_3 \text{ (g)} \; + \; \text{Mass of } Na_2CO_3 \text{ (g)} \qquad (3)$$

The second important mass relationship is derived from the two balanced equations which describe the reactions of each component to the common set of products. The important product related to your mass measurements in this experiment is NaCl. Equations (1) and (2) show that one mole of $NaHCO_3$ produces only one mole of NaCl whereas one mole of Na_2CO_3 gives two moles of NaCl. Therefore, a mass relationship between moles of NaCl formed and moles of $NaHCO_3$ and moles of Na_2CO_3 in the component mixture exists.

Moles of NaCl =
2(moles of Na_2CO_3 in mixture) + 1(moles of $NaHCO_3$ in mixture)   (4)

However, because equation (3) and equation (4) have different units, grams versus moles, one of these equations must be changed by the use of unit conversion factors. Since the goal of this experiment is to determine the percent composition by mass, an appropriate common unit for both equations is grams. The percent composition of the two components is then calculated from the masses of the two components.

In this experiment a sample of the mixture in a dry tared crucible is treated with an excess of HCl solution. The CO_2 produced is immediately displaced into the atmosphere. The H_2O produced in the reaction and the excess acid are then evaporated by gently heating the mixture. The dry residue which is pure NaCl is finally weighed. The mass of the original mixture and the mass of the dry residue after reaction are used to calculate the percent composition of the mixture.

REFERENCES

(1)   Kotz, J. C., and Purcell, K. F., _Chemistry and Chemical Reactivity_, Saunders College Publishing, Philadelphia, 1987, sections 1.3, 2.5, 2.11a, 2.12, and 4.1.

(2)   Masterton, W. L., Slowinski, E. J., and Stanitski, C. L., _Chemical Principles_, 6th ed., Saunders College Publishing, Philadelphia, 1985, sections 1.6, 2.5, 3.1-3.3, and 3.7.

EXPERIMENTAL PROCEDURE

(Study this section and the PRE-LABORATORY QUESTIONS before coming to the laboratory. **Wear safety goggles when performing this experiment**.)

Support a clean crucible and cover on a wire triangle, and heat them for 5 minutes until they are thoroughly dry (Laboratory Methods **M**). Allow the crucible and cover to cool to room temperature. Then weigh them on an analytical balance (Laboratory Methods **C**), and record the mass in TABLE 7.1. Use tongs to handle the crucible.

Transfer to this crucible approximately 0.50 g of an unknown mixture of $NaHCO_3$ and Na_2CO_3. Determine the mass of the crucible, cover, and mixture on the analytical balance, and record the mass in TABLE 7.1.

Place the crucible in the wire triangle again, and place the entire apparatus under the hood. Carefully measure 3 mL of concentrated (12 $\underline{M}$) hydrochloric acid, HCl, solution into a 10-mL graduated cylinder. **CAUTION: Concentrated HCl is quite corrosive. Avoid getting it on your skin or clothes. If you do, wash it off with water immediately**. Working under the hood, add the HCl solution **one drop at a time** to the $NaHCO_3$-Na_2CO_3 mixture in the crucible. Wait for the reaction to subside before adding the next drop of acid to avoid spattering and subsequent loss of material. After all the acid has been added, **heat gently** to drive off the excess acid and water. It is best to wave the flame under the crucible and to keep the cover partially open. The rate of heating must be gentle enough to prevent the acid from boiling and thus spattering the mixture. When the solid that is formed appears dry, heat the crucible in the full flame of the Bunsen burner for about 10. minutes. Allow the crucible and its contents to cool for 15 minutes. Then weigh the crucible, cover, and residue on the analytical balance, and record the mass in TABLE 7.1. Again use tongs to handle the crucible.

To ensure that all the volatile substances have evaporated, reheat the crucible, cover, and residue a second time for 5 minutes; allow them to cool for 15 minutes; and reweigh. If there is a change in mass of more than 0.001 g, reheat, cool, and reweigh. The process of heating, cooling, and weighing must be continued until a constant mass ($\pm$0.001 g) is obtained. You will use this constant mass to calculate the percent composition of the mixture.

If your instructor indicates that a duplicate run is to be accomplished, repeat the experimental procedure with a second crucible and cover, beginning at the time when the first crucible, cover, and contents are cooling after the reaction.

Perform the calculations in TABLE 7.2 including sample calculations for one run on the back of TABLE 7.2.

TABLE 7.1. Mass data for reaction of $NaHCO_3$-Na_2CO_3 mixture with HCl.

Unknown number __	Run 1	Run 2
Mass of crucible and cover (g)		
Mass of crucible, cover, and mixture (g)		
Mass of crucible, cover, and residue after reaction with HCl (g) 1st weighing		
2nd weighing		
3rd weighing, if necessary		
4th weighing, if necessary		

Instructor's initials _____

CALCULATIONS

TABLE 7.2. Calculation of percent $NaHCO_3$ and Na_2CO_3 in an unknown mixture.

	Run 1	Run 2
Mass of unknown mixture used (g)		
Mass of NaCl formed (g)		
Mass of $NaHCO_3$ in unknown mixture (g)		
Mass of Na_2CO_3 in unknown mixture (g)		
Percent $NaHCO_3$ by mass in unknown mixture (%)		
Percent Na_2CO_3 by mass in unknown mixture (%)		

Experiment 7

Sample calculations for Run ___

Write the two mathematical equations using the experimental quantities of your experiment. You must remember that both equations must have identical units for both sides of the two equations.

Solve the two equations and calculate the mass of $NaHCO_3$ and the mass of Na_2CO_3.

Calculate the percent $NaHCO_3$ in the mixture.

Student ID number _____

Section _____ Date _____

Instructor _____   ## PRE-LABORATORY QUESTIONS

1. What is a gravimetric analysis?

2. Define a <u>mole</u> of sodium carbonate.

3. Write the two simultaneous mathematical equations [equations (3) and
    (4)] using common <u>mass</u> units. Show all units.

4. Write the two simultaneous mathematical equations [equations (3) and (4)] using common <u>mole</u> units. Show all units.

5. Why do you use tongs at all times to handle the crucible?

6. Why do you reheat the crucible a second time at the end of the experiment?

POST-LABORATORY QUESTIONS

1. Explain how the calculated percent $NaHCO_3$ by mass would be affected if some of the mixture is lost while either (a) adding HCl solution or (b) heating to drive off excess HCl and water. Indicate clearly your reasoning.

2. Explain how the calculated percent Na_2CO_3 by mass would be affected if the water and excess HCl had not been completely driven off by the time the residue is weighed. Indicate clearly your reasoning.

3. Explain how the calculated percent $NaHCO_3$ and Na_2CO_3 by mass would be affected if the initial mixture were wet with water. Indicate clearly your reasoning.

4. Explain how the calculated percent $NaHCO_3$ and Na_2CO_3 by mass would be affected if the initial mixture contained a third component which did not react with HCl. Indicate clearly your reasoning.

Analysis of an Unknown
Mixture by Dehydration

PURPOSE OF EXPERIMENT: Study some properties of hydrates, and determine the percent composition by mass of an unknown mixture of a hydrated salt and an anhydrous salt.

Crystalline solids may contain water as adsorbed water or as hydrate water. Adsorbed water simply adheres to the surface of the crystals and arises during precipitation of crystals from aqueous solutions or by exposing crystals to atmospheric moisture. Adsorbed water is usually small in amount and is not present in stoichiometric amounts. It can be removed easily by a process called drying using any reasonable drying procedure (Laboratory Methods M) that causes evaporation of water from the crystalline surface. Hydrate water is bound through ion-dipole interactions to the cation or anion of an ionic salt or is hydrogen bonded either to the anion or to other water molecules that are already bound to the cation. Hydrate water arises when solids are crystallized from aqueous solutions. Hydrate water is usually larger in amount than adsorbed water and may be present in stoichiometric amounts. Hydrate water can be removed in a process called dehydration with varying difficulty depending on how strongly it is bound within the crystals. Some hydrates lose part or all of their hydrate water to the atmosphere just on standing at room temperature because the water vapor pressure of the hydrate is higher than the partial pressure of water vapor in the atmosphere. This process is called efflurescence. Thus, sodium sulfate decahydrate, $Na_2SO_4 \cdot 10H_2O$, is stable if the partial pressure of water vapor in the air is greater than 14 mmHg but loses water if the partial pressure of water vapor is less than 14 mmHg. Other hydrates are stable when the humidity is high but lose water when it is low. For example, red-violet cobalt(II) chloride hexahydrate, $CoCl_2 \cdot 6H_2O$, is stable in moist air; violet cobalt(II) chloride dihydrate, $CoCl_2 \cdot 2H_2O$, is stable at intermediate humidities; and blue anhydrous cobalt(II) chloride, $CoCl_2$, is the dominant form in dry air. These color changes have been used in simple hydrometers to indicate relative humidity. Some hydrates such as zinc nitrate hexahydrate, $Zn(NO_3)_2 \cdot 6H_2O$, decompose thermally on heating to other compounds either before or after all the water is removed.

Dehydration of hydrates is a reversible process, that is, adding water to an anhydrous salt reforms a hydrated salt. Thus, blue copper(II) sulfate pentahydrate, $CuSO_4 \cdot 5H_2O$, can be dehydrated on heating to white, anhydrous copper(II) sulfate, $CuSO_4$, which can be rehydrated on addition of water to blue $CuSO_4 \cdot 5H_2O$. Some anhydrous compounds such as calcium chloride, $CaCl_2$, form hydrates just on exposure to air. These are termed hygroscopic and can be used as drying agents or desiccants. Some water soluble hygroscopic solids such as sodium hydroxide, NaOH, even remove sufficient water from the air to dissolve completely and form water solutions. These are called deliquescent compounds.

Some compounds that do not contain either adsorbed or hydrate water still give off water on heating as a result of decomposition of the compounds. This decomposition frequently involves the splitting out of water between two molecules of the compound in what are called condensation reactions. There is at least one example, shown below, that illustrates a loss of hydrate water followed by three condensation reactions.

115

$$2\ H_3PO_4 \cdot H_2O(s) \xrightarrow{\text{heat}} H_2O(g)\ +\ 2\ H_3PO_4(s) \qquad\qquad \text{dehydration}$$

orthophosphoric acid

$$2\ H_3PO_4(s) \xrightarrow{\text{heat}} H_2O(g)\ +\ H_4P_2O_7(s) \qquad\qquad \text{condensation}$$

pyrophosphoric acid

$$H_4P_2O_7(s) \xrightarrow{\text{heat}} H_2O(g)\ +\ 2\ HPO_3(s) \qquad\qquad \text{condensation}$$

metaphosphoric acid

$$4\ HPO_3(s) \xrightarrow{\text{heat}} 2\ H_2O(g)\ +\ P_4O_{10}(s) \qquad\qquad \text{condensation}$$

phosphorus(V) oxide

These reactions are reversible, that is, P_4O_{10} can be converted to HPO_3, $H_4P_2O_7$, and H_3PO_4 by the required amounts of water. However, carbohydrates undergo loss of water in decomposition reactions that are not reversible. For example, glucose loses water and appears to char on heating, but forms an amber solution that does not contain glucose when water is added.

For those hydrates that don't decompose thermally, the extent of hydration can be determined by weighing the solid sample before and after heating. The mass after heating is for the anhydrous salt, and the difference in the two masses gives the mass of water originally present. Moles of both the anhydrous salt and water can then be calculated, and the mole ratio of water to anhydrous salt can be determined. These ideas are illustrated with a hydrate of barium chloride, $BaCl_2 \cdot xH_2O$.

before heating		after heating		

$$BaCl_2 \cdot xH_2O(s) \xrightarrow{\text{heat}} BaCl_2(s)\ +\ x\ H_2O(g)$$

0.244 g	−	0.208 g	=	0.036 g
		208.27 g/mol		18.015 g/mol
		0.00100 mol		0.0020 mol

$$\frac{\text{mol } H_2O}{\text{mol } BaCl_2} = \frac{0.0020 \text{ mol}}{0.00100 \text{ mol}} = 2.0, \quad \text{and the original formula is } BaCl_2 \cdot 2H_2O.$$

Similar stoichiometric calculations can be used to determine the percent composition by mass of a mixture of barium chloride dihydrate, $BaCl_2 \cdot 2H_2O$, and anhydrous barium chloride, $BaCl_2$. Only masses before and after heating are required. The difference in the two masses again gives the mass of water initially present from which moles of H_2O can be calculated. Half as many moles of $BaCl_2 \cdot 2H_2O$ must have been present initially since all the water comes from $BaCl_2 \cdot 2H_2O$, and 1 mole of $BaCl_2 \cdot 2H_2O$ produces 2 moles of H_2O. The moles of $BaCl_2 \cdot 2H_2O$ can be converted to mass, and the percent $BaCl_2 \cdot 2H_2O$ by mass is simply this mass divided by the total mass of the original sample and multiplied by 100. The remainder of the original sample must have been anhydrous $BaCl_2$.

In this experiment you will study some of the properties of hydrates including effluorescence, deliquescence, and the reversibility of dehydration and hydration. In addition, you will determine the percent composition by mass of an unknown mixture of $BaCl_2 \cdot 2H_2O$ and $BaCl_2$.

REFERENCES

(1)   Kotz, J. C., and Purcell, K. F., <u>Chemistry and Chemical Reactivity</u>, Saunders College Publishing, Philadelphia, 1987, sections 2.5, 2.11c, 2.12, and 4.1.

(2)   Masterton, W. L., Slowinski, E. J., and Stanitski, C. L., <u>Chemical Principles</u>, 6th ed., Saunders College Publishing, Philadelphia, 1985, sections 2.5, 3.1-3.3, and 3.7.

EXPERIMENTAL PROCEDURE

(Study this section and the PRE-LABORATORY QUESTIONS before coming to the laboratory. **Wear safety goggles when performing this experiment.**)

A.   Properties of Hydrates

1.   Dehydration and hydration. Place about 0.1 g of each of the compounds listed below in separate 10 x 75-mm or larger test tubes. These samples need <u>not</u> actually be weighed. This is just enough sample to fill the curved portion of a 10 x 75-mm test tube or is about the volume of a pea if you are using a larger test tube. Hold each test tube at about a 30° angle to the benchtop, and heat the base of the test tube gently over a Bunsen flame (Laboratory Methods **D**). Note carefully any condensation of water droplets on the cool upper portion of the test tube and any changes of color or crystalline form of each sample. Then let each test tube cool, and try to dissolve each residue in several mL of distilled water. Stir the mixtures, and heat gently, if necessary (Laboratory Methods **K**). Record your observations in TABLE 8.1.

Copper(II) nitrate trihydrate, $Cu(NO_3)_2 \cdot 3H_2O$

Nickel(II) chloride hexahydrate, $NiCl_2 \cdot 6H_2O$

Orthoboric acid, H_3BO_3

Sodium chloride, NaCl

Sucrose, $C_{12}H_{22}O_{11}$

2.   Effluorescence and deliquescence. Place a few crystals of each of the compounds listed below in separate spots in the concave dip of a watch glass. Allow them to stand for the remainder of the laboratory period. At about half-hour intervals note carefully any changes in crystalline form, color, or dampness for each sample. Record your observations in TABLE 8.2.

Iron(III) chloride hexahydrate, $FeCl_3 \cdot 6H_2O$

Phosphorus(V) oxide, P_4O_{10} (**CAUTION:** **<u>Avoid contact with your skin</u>**.)

Sodium acetate trihydrate, $NaC_2H_3O_2 \cdot 3H_2O$

Sodium carbonate monohydrate, $Na_2CO_3 \cdot H_2O$

Sodium hydroxide, NaOH   (**CAUTION:** **<u>Avoid contact with your skin</u>**.)

Sodium thiosulfate pentahydrate, $Na_2S_2O_3 \cdot 5H_2O$

B. **Dehydration of an Unknown Sample**

Thoroughly wash a porcelain crucible and its cover. Then hold each with clean crucible tongs; dip each for a few seconds into 6 $\underline{M}$ nitric acid, HNO_3, solution provided in a beaker in the hood; and rinse each thoroughly with distilled water. CAUTION: <u>HNO_3 of this concentration is corrosive and causes yellow discoloration and burns on your skin or holes in your clothing. If any spills occur onto your skin or clothing, rinse the affected area thoroughly with water.</u> Any stains on the crucible that are not removed by this treatment will not cause problems in this experiment. **Handle the crucible and cover throughout each run only with clean crucible tongs.**

Place the crucible on a wire triangle supported by an iron ring mounted on a ring stand (Laboratory Methods **D**). Place the cover on the crucible slightly ajar as shown in FIGURE 8.1. Heat the crucible and cover with a Bunsen burner flame, gently at first, and then strongly so that the bottom of the crucible glows cherry red for about 2 minutes. Allow the crucible and cover to cool. Weigh the crucible and cover on an analytical balance (Laboratory Methods **C**), and record the mass in TABLE 8.3.

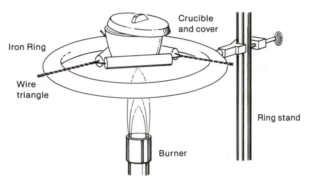

FIGURE 8.1. Crucible with cover slightly ajar.

Obtain an unknown sample from your instructor. Record your unknown number in TABLE 8.3 if it has one. Use a spatula to place about 2 g of unknown sample into your crucible. Weigh your crucible, cover, and sample on the analytical balance, and record the mass in TABLE 8.3.

Place the crucible with its cover slightly ajar back on a wire triangle supported by an iron ring mounted on a ring stand. The cover should be displaced to one side and open just enough to let gas escape as shown in FIGURE 8.1. Heat the crucible gently in a cool, luminous Bunsen burner flame for about 5 minutes. Gentle heating prevents spattering of the sample as water vapor is initially driven off. Then gradually heat more intensely, causing the bottom of the crucible to glow cherry red for about 10. minutes. This strong heating ensures complete removal of water of hydration.

Remove the flame, center the cover on the crucible, and allow them to cool to room temperature. Begin another run during this period of cooling. Finally, weigh the cooled crucible, cover, and dehydrated sample on the analytical balance, and record the mass in TABLE 8.3. Use a spatula to scrape the dehydrated sample from the crucible into a bottle on the reagent bench marked DEHYDRATED $BaCl_2$.

Perform two or three runs (as designated by your instructor) to check the reproducibility of your results, beginning each later run while dehydrated sample from the previous run is cooling.

Perform the calculations in TABLE 8.4 including sample calculations for one run on the back of TABLE 8.4.

Name ______________________________

Student ID number ________________

Section __________ Date __________

Instructor ________________________

Experiment 8

DATA

TABLE 8.1. Behavior toward heating followed by addition of water.

Copper(II) nitrate trihydrate, $Cu(NO_3)_2 \cdot 3H_2O$

Nickel(II) chloride hexahydrate, $NiCl_2 \cdot 6H_2O$

Orthoboric acid, H_3BO_3

Sodium chloride, NaCl

Sucrose, $C_{12}H_{22}O_{11}$

TABLE 8.2. Behavior toward water vapor in the air.

Iron(III) chloride hexahydrate, $FeCl_3 \cdot 6H_2O$

Phosphorus(V) oxide, P_4O_{10}

Sodium acetate trihydrate, $NaC_2H_3O_2 \cdot 3H_2O$

Sodium carbonate monohydrate, $Na_2CO_3 \cdot H_2O$

Sodium hydroxide, NaOH

Sodium thiosulfate pentahydrate, $Na_2S_2O_3 \cdot 5H_2O$

Name _____   **Experiment 8**

Student ID number _____

Section _____ Date _____

Instructor _____   **D A T A**

TABLE 8.3. Mass data for dehydration.

Unknown number ___	Run 1	Run 2	Run 3
Mass of crucible and cover (g)			
Mass of crucible, cover, and unknown sample before heating (g)			
Mass of crucible, cover, and dehydrated sample after heating (g)			

Instructor's initials _____

C A L C U L A T I O N S

TABLE 8.4. Masses, moles, and percent compositions.

	Run 1	Run 2	Run 3
Mass of unknown sample before heating (g)			
Mass of dehydrated sample after heating (g)			
Mass of water lost (g)			
Moles of water lost (mol)			
Moles of $BaCl_2 \cdot 2H_2O$ (mol)			
Mass of $BaCl_2 \cdot 2H_2O$ (g)			
Percent $BaCl_2 \cdot 2H_2O$ by mass (%)			
Average percent $BaCl_2 \cdot 2H_2O$ by mass (%)			
Percent anhydrous $BaCl_2$ by mass in the original sample (%)			
Average percent anhydrous $BaCl_2$ by mass (%)			

Experiment 8

Sample calculations for Run ___

Mass of water lost

Moles of water lost

Moles of $BaCl_2 \cdot 2H_2O$

Mass of $BaCl_2 \cdot 2H_2O$

Percent $BaCl_2 \cdot 2H_2O$ by mass

Percent $BaCl_2$ by mass

Section _____ Date _____

Instructor _____   # PRE-LABORATORY QUESTIONS

1. Define clearly in your own words the difference between

    a. Adsorbed water and hydrate water.

    b. Hydration and dehydration.

    c. Effluorescence and deliquescence.

    d. Dehydration and condensation.

2. Why do you handle the crucible and cover with crucible tongs at all times during this experiment?

Experiment 8

3. Describe clearly how you adjust a Bunsen burner to obtain a

   a. Cool, luminous flame.

   b. Hot, nonluminous flame.

4. A sample of zinc sulfate heptahydrate, $ZnSO_4 \cdot 7H_2O$, was dehydrated using procedures in part **B** of this experiment except that the sample was heated strongly only for 5 minutes. If the following data were obtained, was the sample completely dehydrated? Indicate clearly your calculations and reasoning.

   Mass of crucible and cover (g)   21.132

   Mass of crucible, cover, and hydrated sample before heating (g)   23.241

   Mass of crucible, cover, and dehydrated sample after heating (g)   22.384

POST-LABORATORY QUESTIONS

1. Suppose that you had an unknown hydrated salt, $CA \cdot xH_2O$ (where C and A represent cation and anion, respectively), on a watch glass on an analytical balance. How could you determine easily whether the sample is effluorescent, hygroscopic, or deliquescent? Describe clearly what behavior you would expect for each.

2. Indicate clearly whether your calculated percent $BaCl_2 \cdot 2H_2O$ by mass would be higher than, lower than, or unchanged from the true value, and why, for each of the following.

   a. The unknown sample contained a second inert component.

   b. The unknown sample contained a second hydrated component.

c. Some of the unknown sample spattered from the crucible because the sample was heated too hot initially.

d. The crucible was damp when it was weighed initially.

e. The unknown sample was heated so strongly that some thermal decomposition of the dehydrated sample to volatile products occurred after initial dehydration.

3. Answer part of POST-LABORATORY QUESTION 2 (designated by your instructor) for either calculated moles of water lost upon dehydration or calculated percent $BaCl_2$ by mass in place of calculated percent $BaCl_2 \cdot 2H_2O$ by mass.

Analysis of an Unknown Mixture

Using the Ideal Gas Law

PURPOSE OF EXPERIMENT: Determine the percent composition by mass of an unknown mixture of sodium nitrite, $NaNO_2$, and sodium chloride, $NaCl$, after collecting a gas evolved by reaction of $NaNO_2$.

A mixture is composed of two or more substances that do not react with each other. The possible compositions of some mixtures may vary over a wide range. Mixtures are encountered frequently both in nature and in the laboratory. Salt beds, salt brines, and sea water - major sources of many important salts - are mixtures. The desired product in most chemical syntheses is part of a mixture composed of other reaction products as well as undesired side products and left over reactants. A heterogeneous mixture like the unknown in this experiment can have an irregularly variable composition, structure, and physical and chemical behavior such that abrupt discontinuities or boundaries may be observed. Cement and topsoil are examples of heterogeneous mixtures.

One of the methods of determining the percent composition by mass of a mixture is to measure quantitatively the amount of product formed by reaction of one of the components under conditions where the other component is inert and remains unreacted. If one of the products is a gas, that gas can be collected, its volume can be determined under measured conditions of pressure and temperature, and the number of moles of gaseous product can be calculated using a rearranged form of the Ideal Gas Law. The moles of product can be related back to the moles and mass of reactant from which the product was formed. By knowing the mass of one component which reacted, the mass of the other component can then be determined by difference from the total mass of unknown mixture used. Finally, percent composition by mass can be calculated.

In this experiment you will react a mixture of sodium nitrite, $NaNO_2$, and sodium chloride, $NaCl$, of unknown composition with an excess of sulfamic acid, HSO_3NH_2. The $NaCl$ is unreactive under these conditions whereas the $NaNO_2$ reacts to form nitrogen gas by the following equation.

$$NO_2^-(aq) + HSO_3NH_2(aq) \longrightarrow HSO_4^-(aq) + H_2O(l) + N_2(g)$$

You will collect the N_2 gas at atmospheric pressure and room temperature and calculate from your data the moles of N_2, moles of $NaNO_2$, mass of $NaNO_2$, mass of $NaCl$, and percent composition by mass of $NaNO_2$ and $NaCl$.

REFERENCES

(1) Kotz, J. C., and Purcell, K. F., <u>Chemistry and Chemical Reactivity</u>, Saunders College Publishing, Philadelphia, 1987, sections 1.3, 2.5, 2.12, 3.4b, 4.1, 4.2, and 6.1-6.5.

(2) Masterton, W. L., Slowinski, E. J., and Stanitski, C. L., <u>Chemical Principles</u>, 6th ed., Saunders College Publishing, Philadelphia, 1985, sections 1.6, 2.5, 3.6-3.8, 6.1-6.4, and 23.2.

EXPERIMENTAL PROCEDURE

(Study this section and the PRE-LABORATORY QUESTIONS before coming to the laboratory. **Wear safety goggles when performing this experiment.**)

Following rigorously the procedures given in Laboratory Methods F for inserting glass tubing into rubber stoppers, carefully insert a medicine dropper into a #1 1-hole rubber stopper Ⓐ. Then insert a second medicine dropper into a #0 1-hole rubber stopper Ⓑ if you are using a measuring tube in FIGURE 9.1 or into a #00 1-hole rubber stopper if you are using a buret in FIGURE 9.1.

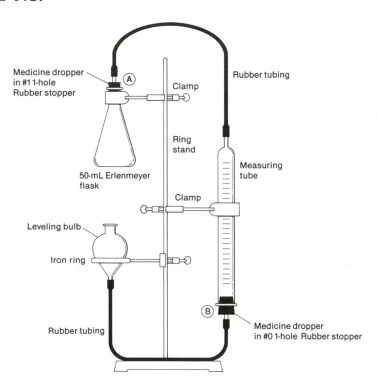

FIGURE 9.1. Apparatus for reaction of a mixture.

Assemble the apparatus shown in FIGURE 9.1. Note that a buret may be used as the measuring tube if designated by your instructor. The stopcock must be open if a buret with a stopcock is used. Add sufficient water to the leveling bulb, raising it as appropriate, so that the measuring tube is filled to within 1 mL of the top calibration when the water level in the bulb is the same as in the measuring tube. Check the apparatus for leaks by lowering and raising the leveling bulb. If all the joints are tight, the level of the water in the measuring tube will return to its original level when the leveling bulb is raised to the same original level.

Weigh 0.18 g of sulfamic acid, HSO_3NH_2, on a beam balance (Laboratory Methods C), remove stopper Ⓐ, and add the sulfamic acid to the 50-mL Erlenmeyer flask. Also add 10. mL of distilled water. **CAUTION: Sulfamic acid and sodium nitrite must never be mixed together as solids. In the presence of traces of water the solids react to evolve nitrogen and heat so rapidly as to be dangerous.**

Using an analytical balance, weigh into a <u>dry</u> 1-dram vial just over 0.1 g of your unknown sample. Record your masses in TABLE 9.1. Add 1 mL of distilled water to the vial, and lower the vial carefully, sliding it down the side of the tilted Erlenmeyer flask. It should rest against the bottom and side of the Erlenmeyer flask without allowing any solution to either enter or leave the vial. Then insert stopper Ⓐ securely into the Erlenmeyer flask. **CAUTION: <u>Be sure that the rubber tubing connecting the Erlenmeyer flask and the measuring tube does not have a kink in it</u>.** Finally, have your instructor check your apparatus.

Before beginning the reaction, make certain that the buret stopcock is open if a buret is being used, and adjust the leveling bulb once again so that the water levels in the leveling bulb and the measuring tube are at exactly the same level. Then read the initial volume indicated by the meniscus in the measuring tube (Laboratory Methods **B**), and record the initial volume in TABLE 9.1.

Tip the 50-mL Erlenmeyer flask so that <u>some</u> mixing of the sulfamic acid solution with the solution inside the vial occurs. As gas evolution decreases, tip the Erlenmeyer flask more to achieve additional mixing. Continue to mix the solutions by gentle shaking and swirling of the flask until there is no further evolution of nitrogen gas. Then wait about 5 minutes to be certain that the contents of the flask have returned to room temperature.

Adjust the leveling bulb so that water levels in the leveling bulb and the measuring tube are at exactly the same level. At this point, the pressure of the gas is equal to atmospheric pressure. Read the final volume of gas in the measuring tube, and record it in TABLE 9.1. Also record in TABLE 9.1 both the barometric pressure and room temperature.

Clean the Erlenmeyer flask and vial carefully, and repeat the experiment once or twice more as designated by your instructor.

Perform the calculations in TABLE 9.2 including sample calculations for one run on the back of TABLE 9.2.

Molecular Weight of an
Unknown Volatile Liquid

PURPOSE OF EXPERIMENT: Determine the molecular weight of an unknown volatile liquid.

A substance, whether solid, liquid, or gas, has a number of characteristic properties that either alone or in combination can be determined experimentally in order to identify the substance. One characteristic property is the molecular weight which is numerically equal to the mass in grams of one mole or of the Avogadro number of molecules of the substance.

For a volatile liquid the molecular weight determination can be made on the vapor. The pressure, P, volume, V, number of moles, n, and temperature, T, of an ideal gas are related by

$$P V = n R T \qquad (1)$$

in which R is a proportionality constant called the universal gas constant. Substituting in the ideal gas equation

$$n = \frac{m}{M} \qquad (2)$$

in which m is the mass and M is the molecular weight of the gas, and rearranging, one obtains

$$M = \frac{m R T}{P V} \qquad (3)$$

Thus, you can calculate the molecular weight of a liquid's vapor if the vapor behaves ideally and if the mass, temperature, pressure, and volume of the vapor are measured experimentally.

In this experiment an amount of liquid more than sufficient to fill the flask when vaporized is placed in a flask of measured volume and mass. The flask is then heated in a boiling water bath so that the liquid vaporizes completely. The vapor drives air out of the flask and fills the flask at barometric pressure and the temperature of the water bath, both of which can be measured. The pressure is low enough and the temperature is high enough for the gas to behave ideally. The flask is weighed after cooling to condense the vapor. You will then calculate the molecular weight using equation (3). You will repeat the determination a couple times to assess the reproducibility of your results.

REFERENCES

(1) Kotz, J. C., and Purcell, K. F., Chemistry and Chemical Reactivity, Saunders College Publishing, Philadelphia, 1987, sections 2.10, 6.1, 6.3, and 11.3.

(2) Masterton, W. L., Slowinski, E. J., and Stanitski, C. L., Chemical Principles, 6th ed., Saunders College Publishing, Philadelphia, 1985, sections 2.5 and 6.2.

EXPERIMENTAL PROCEDURE

(Study this section and the PRE-LABORATORY QUESTIONS before coming to the laboratory. **Wear safety goggles when performing this experiment.**)

Fill a 400-mL or larger beaker about two-thirds full with water. (A 600-mL or larger beaker is required if a 250-mL Erlenmeyer flask is used later to contain the sample.) Start heating the water to boiling (Laboratory Methods **D**).

In the meantime, obtain a square of aluminum foil, and weigh it together with a <u>clean</u>, <u>dry</u> 125-mL or larger Erlenmeyer flask on an analytical balance (Laboratory Methods **C**), recording the mass in TABLE 10.1. Check this mass because it is the basis of all of your runs. If your instructor wants you to calculate maximum error (APPENDIX B) as part of your report, also enter estimated errors in TABLE 10.1 for each of your measurements.

Pour into the flask 3 mL of an unknown liquid obtained from your instructor. **CAUTION:** <u>Do not inhale the vapor from the flask</u>. Crimp the square of aluminum foil down over the mouth of the flask to form a cap. Fold the edges of the foil so that no more than one-quarter inch of the neck of the flask is covered, making sure that the edges of the foil are pressed tightly against the glass. Use a pin or other device provided by your instructor to punch a small pinhole in the foil.

Refer to FIGURE 10.1 as you set up the apparatus. The iron ring should be about 4 inches above the top of your Bunsen burner. Clamp the flask firmly by the mouth rather than by the neck and tilted at a slight angle. Tilting the flask allows you to perceive more easily when all the liquid has evaporated. Place the flask in the boiling water in the beaker, holding the flask down with the clamp so that the flask has 0.5-cm minimum clearance with the bottom of the beaker and as much as possible of the flask is immersed in the water bath. Add more hot water to the beaker, if necessary, to get the proper immersion. The unknown liquid will begin to vaporize immediately. **CAUTION:** <u>The vapor is flammable. If vapor exiting through the pinhole catches fire, you can easily blow out the flame as you would a match.</u>

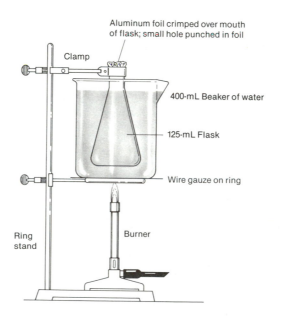

Aluminum foil crimped over mouth of flask; small hole punched in foil

Clamp

400-mL Beaker of water

125-mL Flask

Wire gauze on ring

Ring stand

Burner

FIGURE 10.1. Apparatus for molecular weight determination.

Continue heating the water bath for at least 1 minute after you perceive that all of the liquid has evaporated. The vapor will force air from the flask, and the flask will finally contain only vapor. During this period, take the temperature of the water, placing the bulb of the thermometer on a level with the middle of the flask. This temperature can be recorded in TABLE 10.1 as the temperature of the vapor in the flask.

Remove the flask from the water, and while it is still warm, dry it and the outside surface of the foil with your towel. Allow the flask to cool for 10. minutes with the foil still in place. Weigh the flask, foil, and contents on the analytical balance, and record the mass in TABLE 10.1.

Record the barometric pressure (either from a barometer or as provided by your instructor) in TABLE 10.1. This equals the total pressure of gases in the flask throughout your experiment. Also measure room temperature, and record it in TABLE 10.1.

Remove the foil cap, discard your condensed liquid in a waste bottle designated for your unknown, and add a second 3-mL portion of your unknown liquid to the flask. Replace the foil cap, and repeat EXPERIMENTAL PROCEDURE except for the first three paragraphs until you have three error-free runs. New squares of aluminum foil are available if you need them, but remember that you must reweigh the dry flask and foil if a new piece is used.

Finally, remove the foil cap, discard your condensed liquid in the waste bottle designated for your unknown, and wash your flask. Fill your flask completely with water, tapping it to make sure there are no bubbles of air trapped in it. Dry the outside of the flask. Measure the volume of water in the flask by pouring the water into a large graduated cylinder on the equipment bench, if available, or into the largest graduated cylinder in your desk set in several portions. Repeat this volume measurement twice more, recording values in TABLE 10.1. The average of these volumes equals the volume that the vapor occupied in the boiling water bath and also the volume of air in the flask at the first weighing.

Before leaving the laboratory, obtain from your instructor a value for the vapor pressure of your unknown liquid at room temperature. Record that value in TABLE 10.1.

CORRECTING FOR LOSS OF AIR

The mass of vapor filling the flask at the elevated temperature is equal to the mass of vapor condensed upon cooling. It appears from the headings in TABLE 10.1 that one can simply subtract (mass of flask and foil) from (mass of flask, foil, and condensed vapor) and obtain (mass of condensed vapor). However, the simple headings overlook the fact that there is air in the flask during each weighing, but not the same mass of air. The following headings and resulting mathematics illustrate these ideas.

final weighing	m_{flask} + $m_{air\ (final)}$ + $m_{condensed\ vapor}$
initial weighing	$-[m_{flask}$ + $m_{air\ (initial)}]$

| difference | $[m_{air\ (final)}$ - $m_{air\ (initial)}]$ + $m_{condensed\ vapor}$ |

It is explained below that air is lost between the initial and final weighings and that $m_{air\ (initial)}$ is greater than $m_{air\ (final)}$. Therefore, $[m_{air\ (final)}$ - $m_{air\ (initial)}]$ is a negative quantity, and the difference between initial and final weighings gives only an uncorrected mass of condensed vapor that is too small by the absolute value of

$[m_{air \ (final)} - m_{air \ (initial)}]$. Therefore, in the determination of molecular weight by this method, there is a significant correction which should be made for the mass of air lost between initial and final weighings.

In the initial weighing of the dry flask and foil, the flask was filled with air at barometric pressure and room temperature. This air contributed to the total mass of your flask. This could be verified experimentally if the flask were first evacuated and then weighed. This mass of air during the initial weighing can be calculated according to the following procedure.

Rearranging equation (3) and solving for density, d, gives

$$d = \frac{m}{V} = \frac{P \, M}{R \, T} \qquad (4)$$

The density of an ideal gas is directly proportional to pressure and inversely proportional to absolute temperature. Thus, the density of dry air under standard conditions (1.2929 g/L at STP) can be corrected for both pressure and temperature to conditions of your experiment. Since the volume of the flask was measured, volume and density can be used to calculate the mass of air during the initial weighing.

Heating and vaporizing the unknown liquid expelled the air and filled the flask with vapor of the unknown at the temperature of the water bath and at barometric pressure. Upon cooling, much of this vapor condensed to a liquid of negligible volume, and air entered back through the pinhole to take its place. However, with a rather volatile unknown, not all of its vapor condensed; some remained in the flask to contribute the equilibrium vapor pressure of the liquid at room temperature. This prevented the same partial pressure of air from entering the flask. Since the total pressure in the flask is the sum of the partial pressures of air and vapor, you can calculate the final partial pressure of air if you have previously measured the vapor pressure of your unknown liquid or if the value is provided by your instructor.

The mass of air in the flask is directly proportional to the partial pressure of air, which is clearly smaller for the final weighing than for the initial weighing. Thus, the mass of air in the flask at the final weighing can be calculated from the mass of air at the initial weighing corrected by a ratio of the partial pressures.

$$m_{air \ (final)} = \left[m_{air \ (initial)}\right] \left[\frac{P_{air \ (final)}}{P_{air \ (initial)}}\right] \qquad (5)$$

The overall effect was a loss of air equal to the difference between $m_{air \ (final)}$ and $m_{air \ (initial)}$.

The mass lost must be added to the <u>uncorrected</u> mass of condensed vapor to obtain a <u>corrected</u> mass of condensed vapor that can be used in equation (3) to calculate the molecular weight of the unknown liquid.

Perform the calculations in TABLE 10.2 including sample calculations for one run on the back of TABLE 10.2. In addition, if your instructor wants you to calculate maximum error as part of your molecular weight calculations in TABLE 10.2, use the data from <u>one</u> of your runs and the procedures explained in APPENDIX B.

Vapor Pressure, Boiling Point, and Heat of Vaporization of an Unknown Liquid

PURPOSE OF EXPERIMENT: Measure the equilibrium vapor pressure of an unknown liquid at several temperatures, determine the boiling point of that liquid, and calculate the heat of vaporization of that liquid.

If water is placed in an open beaker around room temperature, the water evaporates slowly until the beaker is empty. This behavior occurs because liquid molecules are in constant motion, and at any given temperature, the distribution of kinetic energies is such that a fraction of the molecules has sufficient energy to escape into the vapor phase. Moreover, as the more highly energetic molecules escape, heat is absorbed from the surroundings to maintain constant temperature and average kinetic energy. Therefore, evaporation continues. Meanwhile, little or no condensation occurs because the molecules of vapor diffuse away from the beaker into the room.

If, on the other hand, water is placed in a closed container, the amount of water decreases at first, but then remains constant. Evaporation occurs for the same reasons given above, and no condensation of vapor to liquid occurs initially. However, as an increasing number of water molecules accumulate in the vapor phase with no chance of escaping from the container, collision of molecules of vapor with the liquid becomes more probable, and condensation occurs at an increasing rate. A point is finally reached at which the rates of evaporation and condensation are equal, and a condition of dynamic equilibrium is achieved. Neither process stops, but rather both occur at the same rate with the result that the number of molecules in each phase remains constant. The gaseous water molecules, in the course of their rapid, random diffusion, bombard the walls of the container and exert a pressure called the <u>equilibrium vapor pressure</u> (frequently called <u>vapor pressure</u> with equilibrium assumed).

The magnitude of the vapor pressure depends both on the nature of the attractive forces in the liquid and on the temperature. Liquids with relatively large forces of attraction between molecules have small tendencies to escape into the vapor phase and have low vapor pressures. Thus, methyl alcohol, CH_3OH, with fairly strong dipole-dipole and hydrogen bonding intermolecular forces, has a relatively low vapor pressure at room temperature. Liquids with relatively small forces of attraction have large tendencies to escape into the vapor phase and have high vapor pressures. Thus, diethyl ether, $CH_3CH_2OCH_2CH_3$, with only weak dipole-dipole and London intermolecular forces, has a high vapor pressure at room temperature. Liquids or solids having high vapor pressures at room temperature are said to be <u>volatile</u>.

When the temperature of a liquid is increased, the average kinetic energy of its molecules and the number of highly energetic molecules increase. Thus, more molecules have sufficient kinetic energy to escape to the vapor phase, and the vapor pressure increases. Conversely, vapor pressure decreases with decreasing temperature. If the temperature is increased sufficiently, the <u>boiling point</u>, at which the vapor pressure equals the prevailing atmospheric pressure, is reached. At this point the atmosphere is pushed aside, and liquid is converted to vapor through the formation of bubbles within the interior of the liquid. Note that a boiling point measurement is thus also a vapor pressure measurement (barometric pressure) at a specific temperature (the boiling point). Like vapor pressure, boiling point also depends on the nature of the attractive forces in the liquid except that it is a direct dependence. Hence, liquids with relatively small forces of attraction between molecules have low boiling points, whereas liquids with relatively large forces of attraction between molecules have high boiling points.

When heat is added to a liquid at its boiling point, the temperature of the liquid does not increase as long as any liquid is present. Instead, the added heat energy is used to overcome the attractive forces between the molecules of the liquid so that the molecules can separate into independent entities characteristic of the vapor state. The quantity of heat energy required to convert one mole of liquid to the vapor state is called the molar heat of vaporization and is designated $\overline{\Delta H}_{vap}$.

Your experimental data will show that vapor pressures do not vary linearly either with Celsius temperatures or with Kelvin temperatures. Many experiments with liquids have shown, instead, that the following logarithmic relationship between vapor pressure, P, and Kelvin temperature, T, is approximately valid.

$$\ln P = \left(-\frac{\overline{\Delta H}_{vap}}{R}\right)\left(\frac{1}{T}\right) + \text{constant} \tag{1}$$

where $\overline{\Delta H}_{vap}$ is the molar heat of vaporization and R is the gas constant (but not in units of L atm/mol K used for ideal gas law calculations). $\overline{\Delta H}_{vap}$ has units of kJ/mol when a value of 8.314×10^{-3} kJ/mol K is used for R.

Equation (1) has the form of a straight line equation, $y = mx + b$, where $y = \ln P$, $m = -\overline{\Delta H}_{vap}/R$, $x = 1/T$, and $b = $ constant. Therefore, a plot of ln P versus 1/T should give a straight line with slope = $-\overline{\Delta H}_{vap}/R$. $\overline{\Delta H}_{vap}$ can then be calculated from the slope since R is known.

In this experiment you will measure the vapor pressure of your unknown liquid at two temperatures, and you will measure its boiling point. Each measurement will be repeated several times to check the reproducibility of your results. You will then plot your data and determine from the slope of the resulting line the equilibrium vapor pressure at 25°C and the molar heat of vaporization, $\overline{\Delta H}_{vap}$, for your unknown liquid.

REFERENCES

(1) Kotz, J. C., and Purcell, K. F., Chemistry and Chemical Reactivity, Saunders College Publishing, Philadelphia, 1987, sections 5.1, 5.4, 6.1-6.3, 6.5, 11.2, and 11.3.

(2) Masterton, W. L., Slowinski, E. J., and Stanitski, C. L., Chemical Principles, 6th ed., Saunders College Publishing, Philadelphia, 1985, sections 5.1, 6.1, 6.2, 6.4, 11.1, and 11.4.

EXPERIMENTAL PROCEDURE

(Study this section and the PRE-LABORATORY QUESTIONS before coming to the laboratory. **Wear safety goggles when performing this experiment.**)

You will work in pairs in this experiment unless your instructor states otherwise. Your instructor may also decide that you will perform only part **A** or part **B**, and either your instructor or another pair of students will provide data for the other part. If your instructor wants you to calculate maximum error (APPENDIX B) for any part of your report, enter estimated errors in TABLES 11.1 and 11.2 for each of your measurements.

A. Measuring Vapor Pressure

Record the barometric pressure and its estimated error provided by your instructor in TABLE 11.1. Measure room temperature, estimate its error and record them in TABLE 11.1. Half fill two 1-L beakers with tap water. Add hot or cold tap water to one of the beakers to adjust the temperature of the water to within 1°C of room temperature. Add about a dozen ice cubes to the second beaker to prepare an ice water bath.

Obtain the apparatus shown in FIGURE 11.1 from your instructor or from the stockroom. It will already be constructed for you, and your instructor will demonstrate its use. Make certain that the apparatus is clean and dry.

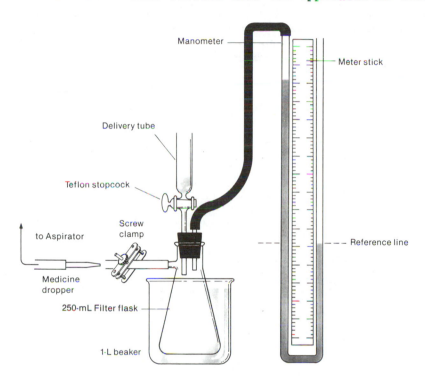

FIGURE 11.1. Vapor pressure apparatus.

Assemble the apparatus carefully, making certain that all connections are tight and that you do not tip the manometer which is secured to a ring stand and mounting board. Insert the 2-hole rubber stopper into the filter flask with a twisting motion. Then close the Teflon stopcock. Close the screw clamp onto the thin-walled rubber tubing. Insert the medicine dropper from the aspirator tubing into the thin-walled rubber tubing, and turn the aspirator ON fully. Loosen the screw clamp <u>slowly</u> to evacuate the flask partially until the difference in mercury heights between the two arms of the manometer is about 500 mmHg or slightly more. Then close the screw clamp tightly. Disconnect the medicine dropper from the thin-walled tubing before turning the aspirator OFF to avoid water backing up into the filter flask.

Note the heights of the mercury columns on both sides of the manometer. If there is no leak, the mercury levels will not change over a period of several minutes, and you are ready to proceed. If there is a leak, it must be stopped. Normally this can be accomplished by checking and tightening the seal first at the stopcock, then at the rubber stopper, and finally at the screw clamp.

When the mercury heights are stable and you are certain that there are no leaks, immerse the filter flask in your room temperature water bath. Clamp the flask so that it is under water as far as possible. Allow a few minutes for air remaining in the flask to reach thermal equilibrium with the water bath. Then measure the temperature and the mercury heights, and record this data with their estimated errors in TABLE 11.1.

Obtain your unknown liquid from your instructor. Pour 2 mL of your unknown liquid into the delivery tube above the stopcock. **CAUTION: <u>Do not inhale the vapor</u>**. Open the Teflon stopcock <u>slowly</u>, and allow <u>some</u> of the liquid to enter the flask. <u>The liquid enters the flask very rapidly; make certain that you close the stopcock while there is still about 0.5 mL of liquid remaining in the delivery tube above the stopcock</u>. If any air also enters the flask at this point, you must start the run again!

Check the mercury heights at 3-minute intervals until consecutive readings agree within 3 mmHg. Then record the mercury heights and their estimated errors in TABLE 11.1.

Remove the flask from the water bath, and immerse the flask in the ice water bath. Add more ice cubes at this point if little ice is present. Allow a few minutes for the flask and contents to reach thermal equilibrium with the ice water bath. Then check the mercury heights at 3-minute intervals until consecutive readings agree within 3 mmHg. Finally, record the mercury heights and their estimated errors in TABLE 11.1.

Loosen the screw clamp <u>slowly</u>, and allow the contents of the flask to return to atmospheric pressure. Open the Teflon stopcock, and drain remaining unknown liquid into the flask. Remove the rubber stopper from the flask, and pour any unknown liquid into a waste bottle designated for your unknown liquid. No water should be used to clean the flask, but the flask should be dried between runs. This is easily accomplished by clamping the flask upside down to the ringstand and evaporating any remaining liquid from the flask by playing the aspirator tubing and medicine dropper around the inside of the flask while the aspirator is ON fully.

Repeat the procedure in part **A** until you have three error-free runs.

B. Determining Boiling Point

Capillary tubes, 3 cm in length and sealed at one end, may have already been made in Experiment 1 (MANIPULATING GLASS TUBING). If not, you will either make them now, or they will be available at the equipment bench. If you need to make them, obtain a 20-cm length of soft glass tubing from the equipment bench. Use the procedure in Experiment 1 for drawing out glass tubing to make a long capillary tube. Then cut six 3-cm long capillary tubes, and seal each carefully at one end. Use more than one 20-cm piece, if necessary, to make six capillary tubes. Your instructor will demonstrate this procedure if you have not already done Experiment 1.

Obtain from the equipment bench or the stockroom the extra equipment that is not part of your desk set but is required to construct the apparatus shown in FIGURE 11.2. Note that the Bunsen burner is intentionally offset to one side of the 400-mL beaker, and the 16 x 150-mm test tube is offset to the other side but is not touching the wall of the beaker. This arrangement sets up convection currents in the beaker and makes it unnecessary for you to stir the water to achieve thermal equilibrium. Insert two capillary tubes, <u>open end down</u>, into the test tube. During heating, air bubbles on which bubbles of vapor can form, escape from the capillary tubes, allowing safe, even boiling rather than "bumping" or sudden spurts of boiling.

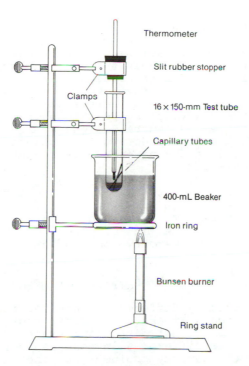

Thermometer

Slit rubber stopper

16 × 150-mm Test tube

Clamps

Capillary tubes

400-mL Beaker

Iron ring

Bunsen burner

Ring stand

FIGURE 11.2.   Apparatus for determining boiling point.

Obtain your unknown liquid from your instructor if you have not already done so for part **A**. **CAUTION**: <u>Your unknown liquids are flammable. Keep open containers of unknown liquids, both yours and your neighbor's, away from Bunsen burner flames</u>.

Place approximately 1 mL of your unknown liquid in the test tube. Adjust the test tube so that the liquid is about 1 cm below the surface of the water. Adjust the thermometer so that the bulb is about 0.5 cm above the surface of the liquid, the thermometer does not touch the wall of the test tube, and the thermometer scale can be seen through the slit in the rubber stopper. Heat the water bath gently until the liquid boils as indicated by condensation and dripping from the bulb of the thermometer, by a ring of condensing vapor appearing at least 2 cm below the top of the test tube, and by a steadying of the temperature on the thermometer. It is easy to miss these observations if you either watch the liquid rather than the upper portion of the test tube and the thermometer or heat too fast. Air will be released from the capillary tubes before boiling begins and will aid in keeping the liquid boiling smoothly. The steady temperature attained under these conditions is the boiling point of the liquid at the prevailing atmospheric pressure. Stop heating, and record that temperature and its estimated error as the boiling point in TABLE 11.2. Also record the barometric pressure and its estimated error provided by your instructor if you have not already done so while completing part **A**.

Add water to the water bath to cool it below the boiling temperature of your liquid. Pour any unknown liquid into a waste bottle designated for your unknown liquid, and discard the capillary tubes in solid waste containers designated for GLASS. Clean the test tube, and repeat the procedure in part **B** until you have three error-free runs.

Check the calibration of your thermometer by measuring the temperatures of a well-stirred bath of ice and distilled water (0.°C) and a boiling distilled water bath (100.°C). Record your measured temperatures in TABLE 11.2, and use these measurements to adjust the estimated errors in your previously measured boiling points, if necessary.

Experiment 11

CALCULATING VAPOR PRESSURES

To understand the calculations you must have a clear picture of the pressures exerted on the right and left mercury columns at the dashed reference line in FIGURE 11.1. The right side of the mercury manometer is open to the atmosphere. This side is the lower side in all of your measurements and serves as a basis for the dashed reference line. The only pressure exerted on the right mercury column is atmospheric pressure which is equal to the barometric pressure and can be designated P_{atm}. It is then only a matter of equating P_{atm} to the pressures exerted on the left mercury column at the dashed reference line during each measurement to obtain the equations required for your calculations.

At your first measurement near room temperature (rt), no unknown liquid is present, but a certain pressure of air, $P_{air(rt)}$, remains because the flask is only partially evacuated. Both $P_{air(rt)}$ and a column of mercury equal to the difference in measured mercury heights, P_{Hg}, are pushing down on the left mercury column at the dashed reference line. Therefore,

$$P_{atm} = P_{air(rt)} + P_{Hg} \qquad (2)$$

$P_{air(rt)}$ can be calculated by difference since the other two quantities are measured. At your second measurement after your unknown liquid has been added to the flask, an additional partial pressure of your unknown liquid, $P_{vp(rt)}$, as well as a new difference in measured mercury heights, P_{Hg}', are pushing down on the left mercury column at the dashed reference line. Therefore,

$$P_{atm} = P_{air(rt)} + P_{Hg}' + P_{vp(rt)} \qquad (3)$$

$P_{vp(rt)}$ can be calculated by difference since $P_{air(rt)}$ was calculated from equation (2), and the other two quantities are measured. At your third measurement after the flask has been cooled to a lower temperature (lt) in the ice bath, an equation like equation (3) is still valid after two changes. The partial pressure of air, $P_{air(lt)}$, is reduced in proportion to the decrease of Kelvin temperature (a Charles' law calculation since the number of moles, n, is constant, and the volume, V, is approximately constant).

$$P_{air(lt)} = \left[P_{air(rt)}\right]\left(\frac{T_{lt}}{T_{rt}}\right) \qquad (4)$$

In addition, a new difference in measured mercury heights, P_{Hg}'', is appropriate. Therefore,

$$P_{atm} = P_{air(lt)} + P_{Hg}'' + P_{vp(lt)} \qquad (5)$$

$P_{vp(lt)}$ can be calculated by difference since $P_{air(lt)}$ was calculated in equation (4), and the other two quantities are measured.

Perform the calculations in TABLES 11.3 and 11.4 including sample calculations for one run on the backs of those tables. TABLE 11.4 requires a careful plot of ln P versus 1/T on good quality graph paper (at least 10 divisions per inch) followed by calculation of the slope of the resulting line. In addition, if your instructor wants you to calculate maximum error as part of your vapor pressure calculations in TABLE 11.3, use the data from one of your runs in TABLE 11.1 and the procedures explained in APPENDIX B. For maximum error calculations in TABLES 11.4 and 11.5, use estimated errors from TABLE 11.2 and appropriate maximum errors for one run in TABLE 11.3.

Student ID number _____

Section _____ Date _____

Instructor _____   D A T A

TABLE 11.1.   Temperatures and mercury heights.

Unknown number __	Run 1		Run 2		Run 3		Estimated error
	Left	Right	Left	Right	Left	Right	
Water bath temperature (^{O}C)							±
Hg heights with no unknown liquid (mmHg)							±
Hg heights with unknown liquid present (mmHg)							±
Ice water bath temperature (^{O}C)							±
Hg heights with unknown liquid present (mmHg)							±
Atmospheric pressure, P_{atm} (mmHg)							±

TABLE 11.2.   Boiling points and calibration points.

Unknown number __	Run 1	Run 2	Run 3	Estimated error
Boiling point (^{O}C)				±
Atmospheric pressure (mmHg)				±
Temperature of ice bath (^{O}C)				
Temperature of boiling water bath (^{O}C)				

Instructor's initials _____

Name _____

Student ID number _____

Section _____ Date _____

Instructor _____

CALCULATIONS

TABLE 11.3. Pressures exerted by mercury and air, and vapor pressure of unknown liquid.

	Run 1	Run 2	Run 3	Maximum error
Water bath temperature, $T_{(rt)}$ (K)				$\pm$
Mercury column pressure with no unknown liquid present, P_{Hg} (mmHg)				$\pm$
Air pressure with no unknown liquid present, $P_{air(rt)}$ (mmHg)				$\pm$
Mercury column pressure with unknown liquid present, P_{Hg}' (mmHg)				$\pm$
Vapor pressure of unknown liquid, $P_{vp(rt)}$ (mmHg)				$\pm$
Average vapor pressure of unknown liquid (mmHg)				
Average deviation (mmHg)				
Ice water bath temperature, $T_{(lt)}$ (K)				$\pm$
Air pressure with no unknown liquid present, $P_{air(lt)}$ (mmHg)				$\pm$
Mercury column pressure with unknown liquid present, P_{Hg}'' (mmHg)				$\pm$
Vapor pressure of unknown liquid, $P_{vp(lt)}$ (mmHg)				$\pm$
Average vapor pressure of unknown liquid (mmHg)				
Average deviation (mmHg)				

Experiment 11

Sample calculations for Run ___

Water bath temperature

Mercury column pressure with no unknown liquid present

Air pressure with no unknown liquid present

Mercury column pressure with unknown liquid present

Vapor pressure of unknown liquid

Ice water bath temperature

Air pressure with no unknown liquid present

Mercury column pressure with unknown liquid present

Vapor pressure of unknown liquid

TABLE 11.4.   Coordinates for plotting graph and for calculating slope.

	ln P	Maximum error	1/T (K^{-1})	Maximum error
Low temperature		±		±
Room temperature		±		±
Boiling point		±		±
Coordinates chosen to calculate slope		±		±
		±		±

TABLE 11.5.   Vapor pressure and heat of vaporization.

	Results	Maximum error
ln P_{25^oC} (from graph)		±
Vapor pressure of unknown liquid at 25°C (mmHg)		±
Slope (K)		±
$\overline{\Delta H}_{vap}$ (kJ/mol)		±

Experiment 11

Sample calculations

   Maximum error for ln P at boiling point

   Maximum error for 1/T at boiling point

   Slope

   Maximum error for slope

   $\overline{\Delta H}_{vap}$

   Maximum error for $\overline{\Delta H}_{vap}$

1. If no unknown liquid has been added and only air is present in the
    filter flask in FIGURE 11.1, write an equation expressing the pressure
    on each mercury column at the dashed reference line.

2. Why do you check mercury heights at 3-minute time intervals at several
    points in part **A** of this experiment?

3. Why are the test tube, beaker, and Bunsen burner in FIGURE 11.2 not
    directly over each other?

4. Why shouldn't the thermometer bulb simply be immersed in the unknown
    liquid in the test tube in FIGURE 11.2?

5. Given the following data for an unknown volatile liquid obtained by procedures in part **A**, calculate each of the following.

24°C	left	right
Mercury heights with no unknown liquid present (mmHg)	720	211
Mercury heights with unknown liquid present (mmHg)	670	263

4°C	left	right
Mercury heights with unknown liquid present (mmHg)	709	223

Barometric pressure (mmHg)　　752

a. Partial pressure of air at 24°C.

b. Vapor pressure of the liquid at 24°C.

c. Partial pressure of air at 4°C.

d. Vapor pressure of the liquid at 4°C.

e. $\overline{\Delta H}_{vap}$.

Name ______________________________ **Experiment 11**

Student ID number ______________

Section __________ Date __________

Instructor ______________________

POST-LABORATORY QUESTIONS

1. You are given the following vapor pressure data for compounds listed in alphabetical order and having similar molecular weights.

Ammonia, NH_3	100. mmHg at -68.4°C
Lithium fluoride, LiF	100. mmHg at 4373°C
Methane, CH_4	100. mmHg at -181.4°C
Water, H_2O	100. mmHg at 51.6°C

a. Arrange the compounds above in order of increasing volatility.

b. Arrange the compounds above in order of expected increase in boiling point.

c. Arrange the compounds above in order of expected increase in H_{vap}.

d. Explain clearly why the trend in c. would be expected based on the nature of the intermolecular forces in each compound.

2. Would each of the following procedural errors cause an increase, a decrease, or no change in your calculated vapor pressure at room temperature? Indicate clearly your reasoning in each case.

   a. Some air entered the filter flask while unknown liquid was being added through the stopcock.

   b. About 5 mL of unknown liquid was added rather than less than 2 mL.

   c. Air leaked around the rubber stopper and into the filter flask as you were immersing the flask into the ice water bath.

3. A student attempted to determine the boiling point of an unknown liquid that should boil at about 110.°C using the procedure in part B. He did not observe a ring of condensing vapor on the test tube, but he observed the temperature steadying at 99.6°C and reported that as his boiling point.

   a. Why was this reported boiling point incorrect?

   b. How could he modify the procedure in part B to obtain a correct boiling point for his unknown liquid?

Analysis of an Alloy

PURPOSE OF EXPERIMENT: Determine the percent composition of an aluminum-zinc alloy by measuring the H_2 formed upon reaction with acid.

An alloy is a solution of two or more metals. Amalgams are alloys which involve mercury as one component. A silver amalgam as used in dental work is a solution of silver and mercury. The pewter used to make silverware in Revolutionary War times is an alloy with the composition 85.5% tin, 6.8% copper, 6% bismuth, and 1.7% antimony. Most alloys usually have physical properties which are different from those of the individual components. In contrast, the chemical properties of the alloy are related to those of the individual components.

Aluminum-zinc, Al-Zn, alloys react with aqueous solutions of strong acids such as hydrochloric acid, HCl, to produce $Al^{3+}(aq)$, $Zn^{2+}(aq)$, and H_2 gas.

$$Zn(s) + 2 H^+(aq) ---> Zn^{2+}(aq) + H_2(g)$$

$$2 Al(s) + 6 H^+(aq) ---> 2 Al^{3+}(aq) + 3 H_2(g)$$

These balanced net ionic equations show that 1 mole of Zn produces 1 mole of H_2 whereas 1 mole of Al yields 1.5 mole of H_2. Therefore, if an alloy composed of only Al and Zn reacts completely with HCl, the volume of H_2 produced can be used to determine the percent composition of the alloy. The following expression which relates the number of moles of H_2 to the number of moles of Al and Zn is used to derive an equation to calculate the percent composition of the mixture.

$$n_{H_2} = \left(n_{Zn} \times \frac{1 \text{ mol } H_2}{1 \text{ mol } Zn} \right) + \left(n_{Al} \times \frac{3 \text{ mol } H_2}{2 \text{ mol } Al} \right)$$

In this experiment you will react a weighed sample of an aluminum-zinc alloy of unknown composition with hydrochloric acid and collect the hydrogen gas that is formed. The number of moles of hydrogen formed can be calculated from the volume, partial pressure of H_2, and temperature of the gas sample. Using this data, the percent composition of the alloy can be calculated.

REFERENCES

(1) Masterton, W. L., J. Chem. Educ., **1961**, _38_, 558.

(2) Kotz, J. C., and Purcell, K. F., Chemistry and Chemical Reactivity, Saunders College Publishing, Philadelphia, 1987, sections 1.3, 2.5, 2.12, 3.2, 3.4, 4.1, 6.1-6.5, 20.1, and 21.1.

(3) Masterton, W. L., Slowinski, E. J., and Stanitski, C. L., Chemical Principles, 6th ed., Saunders College Publishing, Philadelphia, 1985, sections 2.5, 3.6, 3.7, 4.3, 6.1-6.4, 8.4, and 13.1.

EXPERIMENTAL PROCEDURE

(Study this section and the PRE-LABORATORY QUESTIONS before coming to the laboratory. **Wear safety goggles when performing this experiment.**)

The glass tubing in FIGURE 12.1 may have already been made in Experiment 1.

① two 12-cm glass tubes with a 90°-bend 6 cm from one end

If so, obtain the required rubber stoppers (one #6 2-hole and one #0 or #00 1-hole) from the equipment bench or the stockroom, and skip to the next paragraph. If not, the glass tubing, thistle tube, and medicine dropper (or short glass tube) will be already fitted in the appropriate stoppers, and you can skip the next paragraph. Note that your instructor has the option of designating that a 250-mL Erlenmeyer flask be used as the bottle and that a buret be used as the measuring tube in FIGURE 12.1.

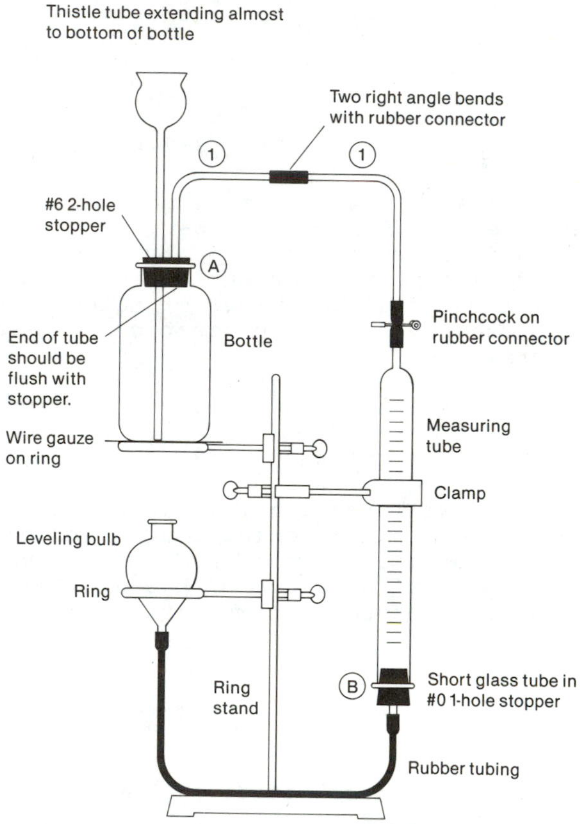

Thistle tube extending almost to bottom of bottle

Two right angle bends with rubber connector

① ①

#6 2-hole stopper

Ⓐ

End of tube should be flush with stopper.

Bottle

Pinchcock on rubber connector

Measuring tube

Wire gauze on ring

Clamp

Leveling bulb

Ring

Ring stand

Ⓑ Short glass tube in #0 1-hole stopper

Rubber tubing

FIGURE 12.1. Apparatus for analysis of an alloy.

Following rigorously the procedures for inserting glass tubing into rubber stoppers that are given in Laboratory Methods F, carefully insert one end of one 12-cm glass tube with a 90°-bend ① into the #6 2-hole rubber stopper Ⓐ until it is flush with the inside of the rubber stopper. Then insert the thistle tube into the same rubber stopper Ⓐ until it projects about 11 cm below the rubber stopper and nearly reaches the bottom of the bottle when the stopper is inserted snugly into the mouth of the bottle.

Insert a medicine dropper (or short glass tube) into a #0 1-hole rubber stopper Ⓑ if you are using a measuring tube in FIGURE 12.1 or into a #00 1-hole rubber stopper if you are using a buret in FIGURE 12.1.

Construct the apparatus shown in FIGURE 12.1. Note that no pinchcock is required if a buret with a stopcock is used as the measuring tube. The stopcock of the buret serves the same function as the pinchcock.

Using an analytical balance (Laboratory Methods C), weigh onto weighing paper a slightly greater than 0.100-g sample of the unknown alloy. Record your masses in TABLE 12.1.

Obtain a clean piece of fine copper gauze from the reagent bench, and wrap your unknown alloy in it. Disconnect the bottle, fill it half-full of water, drop in the unknown wrapped in gauze, and shake the bottle to remove any air bubbles adhering to the gauze. Fill the bottle to the top with water, and place it on the wire gauze on the ring. Fill the leveling bulb with water, and raise it until the water runs out of the end of the glass bend at the #6 rubber stopper Ⓐ. Close the pinchcock, and place the rubber stopper Ⓐ in the bottle, pushing it in firmly and being careful to trap no air bubbles in the bottle. The entire apparatus will now be full of water, from the thistle tube to the leveling bulb. Confirm that the rubber stoppers and tubing are tight. No water should leak from any part of the apparatus. Ask your instructor to check your apparatus.

Lower the leveling bulb until the surface of the water in it is at the level of the top of the stem of the thistle tube, and open the pinchcock.

Pour 10. mL of concentrated (12 M) hydrochloric acid, HCl, solution into the thistle tube. CAUTION: <u>Concentrated</u> (12 M) <u>HCl causes burns on your skin or holes in your clothing. If any spills or spatters occur onto your skin or clothing, rinse the affected area thoroughly with water</u>. A vigorous evolution of hydrogen should result. CAUTION: <u>The level of the liquid in the thistle tube should always be kept above the top of the stem</u>. This condition may be maintained by adjusting the position of the leveling bulb, which should be gradually lowered as the hydrogen is evolved.

When the unknown sample has completely disappeared, drive all of the hydrogen in the bottle over into the measuring tube by pouring water into the thistle tube. The bottle should be tapped to free any bubbles of hydrogen that are held in the copper gauze. When the two right-angle bends are filled with water, close the pinchcock (or stopcock), and raise the leveling bulb until the water stands at the same level in the bulb and in the measuring tube. Record in TABLE 12.1 the volume of hydrogen collected in the measuring tube, the room temperature, and the barometric pressure.

Empty the bottle, and rinse it, the thistle tube, and the copper gauze with water. Make a second determination, using another weighed unknown sample. When you have finished, return the copper gauze to the reagent bench.

If you used a buret to collect H_2 in FIGURE 12.1, determine the uncalibrated volume between the 50-mL mark and the stopcock or pinchcock of the buret by filling that volume with water, draining it into a 10-mL graduated cylinder, measuring the volume, and recording it in TABLE 12.1. Remember that this must be added to your volume of H_2 collected to get total volume of H_2 collected in TABLE 12.2.

Perform the calculations in TABLE 12.2 including sample calculations for one run on the back of TABLE 12.2.

Name _____

Student ID number _____

Section _____ Date _____

Instructor _____

TABLE 12.1. Data for reacting alloy with acid.

Unknown number __	Run 1	Run 2
Mass of weighing paper (g)		
Mass of unknown and weighing paper (g)		
Volume of H_2 gas collected (mL)		
Uncalibrated volume of buret, if used (mL)		
Room temperature ($^{\circ}$C)		
Barometric pressure (mmHg)		

Instructor's initials _____

TABLE 12.2. Masses and percent compositions of aluminum and zinc.

	Run 1	Run 2
Mass of unknown used (g)		
Total volume of H_2 collected (mL)		
Partial pressure of water at room temperature (from APPENDIX C) (mmHg)		
Partial pressure of H_2 collected (mmHg)		
Moles of H_2 collected (mol)		
Mass of aluminum in the alloy (g)		
Mass of zinc in the alloy (g)		
Percent by mass of aluminum (%)		
Average percent by mass of aluminum (%)		
Percent by mass of zinc (%)		
Average percent by mass of zinc (%)		

Experiment 12

Sample calculations for Run ___

Partial pressure of H_2 collected

Moles of H_2 collected

Mass of aluminum in the alloy

Mass of zinc in the alloy

Percent by mass of aluminum in the alloy

Percent by mass of zinc in the alloy

Name _____

Student ID number _____

Section _____ Date _____

Instructor _____

PRE-LABORATORY QUESTIONS

1. Define or state Dalton's Law of Partial Pressures. Explain in your own words the meaning of the Law.

2. Will Dalton's Law of Partial Pressures be used in any step of the calculations for determining the percent composition of the alloy? If so, specify where, how, and why Dalton's Law is used.

3. Derive a mathematical equation which relates the moles of H_2 to the mass of aluminum and the mass of alloy. The mass of zinc should not appear in this equation.

4. Could the procedure in this experiment be used to determine the composition of a sodium-potassium alloy? Justify your answer clearly, stating how if it could and why not if it could not.

5. Could the procedure in this experiment be used to determine the composition of an amalgam? Justify your answer clearly, stating how if it could and why not if it could not.

1. Calculate the total mass of anhydrous $AlCl_3$ plus $ZnCl_2$ which could be recovered from the reaction of your sample of alloy with HCl in Run 1.

2. Will each of the following experimental errors increase, decrease, or cause no change in your calculated percent aluminum in the alloy? Indicate clearly your reasoning.

   a. The temperature of the water was significantly lower than room temperature.

   b. The mass of alloy used in the experiment was actually less than the mass used in the calculations.

    c. The alloy was contaminated by a component which did not react with HCl.

    d. A large bubble was trapped under rubber stopper Ⓐ when it was inserted into the bottle just before beginning the run.

    e. You neglected to correct for the partial pressure of water in the measuring tube and assumed that the total pressure was due to H_2.

Preparation of Alum

PURPOSE OF EXPERIMENT: Prepare $KAl(SO_4)_2 \cdot 12H_2O$, an aluminum alum, from aluminum metal. (Note: Experiment 14 describes the gravimetric analysis of the sulfate content in alum.)

Alums have the general formula $M^+M^{3+}(SO_4)_2 \cdot 12H_2O$, where M^+ is commonly Na^+, K^+, Tl^+, NH_4^+, or Ag^+ and M^{3+} is Al^{3+}, Fe^{3+}, Cr^{3+}, Ti^{3+}, or Co^{3+}. Samples of these compounds have a variety of uses in our everyday lives. Sodium aluminum alum, $NaAl(SO_4)_2 \cdot 12H_2O$, is used in baking powder. When water is added to baking powder, a mixture of Na_2CO_3 and aluminum alum, the aluminum ion reacts with H_2O to form H^+, which in turn combines with HCO_3^- to form H_2CO_3. Eventually, CO_2 is released which makes the cake "rise". Potassium aluminum alum, $KAl(SO_4)_2 \cdot 12H_2O$, is used in water purification, sewage treatment, and fire extinguishers. Ammonium aluminum sulfate, $NH_4Al(SO_4)_2 \cdot 12H_2O$, is used in pickling cucumbers, whereas chrome alum, $KCr(SO_4)_2 \cdot 12H_2O$, is used in tanning leather, that is, converting hides into leather for shoes, wallets, and other leather goods.

In this experiment you will prepare potassium aluminum sulfate dodecahydrate, $KAl(SO_4)_2 \cdot 12H_2O$, a double salt, starting with aluminum metal. Aluminum metal reacts rapidly with a hot aqueous solution of potassium hydroxide, KOH, to produce a soluble potassium aluminate salt.

$$2\ Al(s) + 2\ K^+(aq) + 2\ OH^-(aq) + 10\ H_2O(l) \longrightarrow$$

$$2\ K^+(aq) + 2\ [Al(H_2O)_2(OH)_4]^-(aq) + 3\ H_2(g)$$

When this salt is reacted with sulfuric acid, H_2SO_4, aluminum hydroxide, $Al(H_2O)_3(OH)_3$, precipitates

$$2\ K^+(aq) + 2\ [Al(H_2O)_2(OH)_4]^-(aq) + 2\ H^+(aq) + SO_4^{2-}(aq) \longrightarrow$$

$$2\ K^+(aq) + SO_4^{2-}(aq) + 2\ Al(H_2O)_3(OH)_3(s)$$

and then dissolves as additional acid is added.

$$2\ Al(H_2O)_3(OH)_3(s) + 6\ H^+(aq) + 3\ SO_4^{2-}(aq) \longrightarrow$$

$$2\ Al^{3+}(aq) + 3\ SO_4^{2-}(aq) + 12\ H_2O(l)$$

Since the resulting solution contains K^+, Al^{3+}, and SO_4^{2-}, potassium aluminum sulfate precipitates as octahedrally shaped crystals when a nearly saturated solution is cooled to $0^\circ C$.

$$K^+(aq) + Al^{3+}(aq) + 2\ SO_4^{2-}(aq) \longrightarrow KAl(SO_4)_2 \cdot 12H_2O(s)$$

REFERENCES

(1) Kotz, J. C., and Purcell, K. F., Chemistry and Chemical Reactivity, Saunders College Publishing, Philadelphia, 1987, sections 4.1-4.3, 21.1a, and 22.2c.

(2) Masterton, W. L., Slowinski, E. J., and Stanitski, C. L., Chemical Principles, 6th ed., Saunders College Publishing, Philadelphia, 1985, sections 3.7, 3.8, 4.3, 21.1, 21.2, and 26.5.

Experiment 13

EXPERIMENTAL PROCEDURE

(Study this section and the PRE-LABORATORY QUESTIONS before coming to the laboratory. **Wear safety goggles when performing this experiment.**)

Using a beam balance (Laboratory Methods C), weigh a piece of weighing paper, and then weigh about 1 g of aluminum metal onto it. Record the masses in TABLE 13.1. Add the metal to a 250-mL or larger beaker. Place the beaker on a wire gauze and iron ring connected to a ring stand so that it can be heated **under the hood**. Carefully add 25 mL of 3.0 $\underline{M}$ potassium hydroxide, KOH, solution to the beaker containing the aluminum. **CAUTION: KOH is exceedingly caustic. Do not spatter the resulting mixture. If any spills or spatters occur onto your skin or clothing, rinse the affected area thoroughly with water.** Begin heating the beaker **gently** with a small flame (Laboratory Methods D). Since hydrogen gas is being evolved, the beaker is not covered and is placed **under the hood**. Continue heating with frequent stirring until all of the aluminum has reacted. If necessary, carefully add distilled water to the beaker to maintain nearly 25 mL of solution. After all of the aluminum has reacted, add 25 mL of distilled water. Filter the warm solution into a 150-mL or larger beaker by gravity filtration (Laboratory Methods I), using a thin layer of Pyrex glass wool placed in the long-stem funnel or Buchner funnel.

Permit the resulting clear solution containing K^+, $[Al(H_2O)_2(OH)_4]^-$, and excess OH^- to cool to room temperature. Slowly acidify the solution by adding small amounts of 6.0 $\underline{M}$ sulfuric acid, H_2SO_4, solution until a total of 30.0 mL has been added. As you add the H_2SO_4, a precipitate first forms, and then redissolves as more acid is added. If all of the precipitate does not redissolve after the addition of 30.0 mL of acid is complete, heat the mixture gently. You must have a clear, colorless solution. Continue to stir, heat the solution to boiling, and reduce the volume of solution to 50. mL (Laboratory Methods L).

Cool the solution in an ice bath for 25-30 minutes without any agitation. If the solution is maintained as motionless as possible, well-formed octahedral crystals of alum should grow in the beaker. (If no crystals form, reheat the solution to evaporate an additional 10. mL of water, and recool the solution in the ice bath.) After all of the solid alum has formed, filter the mixture using either gravity filtration or suction filtration, and wash the resulting crystals with 15 mL of a 50:50 (by volume) water-alcohol solution (Laboratory Methods I). Weigh a clean, dry watch glass on a beam balance, and record the mass in TABLE 13.1. Transfer the crystals to the watch glass, and permit them to air dry in your laboratory desk drawer until the following laboratory period.

Weigh the <u>dry</u> crystals and the watch glass on a beam balance, and record the mass in TABLE 13.1. The crystals of alum can be used for the analysis of alum (Experiment 14).

Calculate the theoretical yield (grams) and the experimental percent yield of aluminum alum based on the mass of aluminum, and record your values in TABLE 13.2. Begin this calculation by writing under sample calculations on the back of TABLE 13.2 one chemical equation for the conversion of Al to $KAl(SO_4)_2 \cdot 12H_2O$. Show clearly the steps in your calculations.

Analysis of Alum

PURPOSE OF EXPERIMENT: Determine the sulfate content in alum by forming and weighing dry $BaSO_4$.

Alums have the general formula $M^+M^{3+}(SO_4)_2 \cdot 12H_2O$. Since aluminum alum, $KAl(SO_4)_2 \cdot 12H_2O$, is soluble in water, the sulfate content can be determined by precipitating, drying, and then weighing an insoluble metal sulfate by means of a gravimetric analysis. An appropriate insoluble metal sulfate for gravimetric analysis is barium sulfate, $BaSO_4$. In this experiment a weighed sample of alum is dissolved in water, and then a soluble barium compound, barium chloride, $BaCl_2$, is added to precipitate the very insoluble $BaSO_4$.

$$Ba^{2+}(aq) + SO_4^{2-}(aq) \longrightarrow BaSO_4(s)$$

This insoluble salt can be readily isolated by filtration, dried without decomposition by heating in a Bunsen flame, and weighed. Even though other negative ions such as the carbonate ion, CO_3^{2-}, also form insoluble barium salts, appropriate procedures can be used to remove such interfering species before the precipitation of $BaSO_4$ occurs. For example, the addition of acid leads to the decomposition of the carbonate ion.

$$CO_3^{2-}(aq) + 2 H^+(aq) \longrightarrow CO_2(g) + H_2O(l)$$

To reduce errors arising from incomplete precipitation, an excess of $BaCl_2$ is added, and the solution is maintained near the boiling point for sufficient time to permit the precipitation to reach equilibrium. Then, cooling of the mixture leads to the quantitative formation of $BaSO_4$. The complete equation for the reaction of an aluminum alum solution with a barium chloride solution is

$$KAl(SO_4)_2 \cdot 12H_2O(aq) + 2 BaCl_2(aq) \longrightarrow$$

$$2 BaSO_4(s) + KCl(aq) + AlCl_3(aq) + 12 H_2O(l)$$

The net ionic equation is simply

$$Ba^{2+}(aq) + SO_4^{2-}(aq) \longrightarrow BaSO_4(s)$$

REFERENCES

(1) Kotz, J. C., and Purcell, K. F., <u>Chemistry and Chemical Reactivity</u>, Saunders College Publishing, Philadelphia, 1987, sections 2.11, 2.12, 4.1, and 4.2.

(2) Masterton, W. L., Slowinski, E. J., and Stanitski, C. L., <u>Chemical Principles</u>, 6th ed., Saunders College Publishing, Philadelphia, 1985, sections 3.2, 3.7, and 3.8.

Experiment 14

EXPERIMENTAL PROCEDURE

(Study this section and the PRE-LABORATORY QUESTIONS before coming to the laboratory. **Wear safety goggles when performing this experiment.**)

Weigh 0.5 to 1.0 g of aluminum alum, $KAl(SO_4)_2 \cdot 12H_2O$, crystals on an analytical balance (Laboratory Methods **C**), and record the mass in TABLE 14.1. Place this sample in a clean 400-mL or larger beaker, and add 200. mL of distilled water to completely dissolve the alum. After the alum has dissolved, add 5 mL of 6 **M** hydrochloric acid, HCl, solution. **CAUTION: <u>HCl of this concentration is corrosive and causes burns on your skin or holes in your clothing. If any spills or spatters occur onto your skin or clothing, rinse the affected area thoroughly with water</u>**. Then add 50. mL of 0.1 **M** barium chloride, $BaCl_2$, solution (**CAUTION:** <u>Poisonous</u>.) with constant stirring. After the addition of $BaCl_2$ is complete, carefully heat the beaker until the contents boil (Laboratory Methods **D**). Boil the mixture <u>gently</u> for 10. minutes, and then allow it to cool with stirring or agitation. (Boiling increases the size of the $BaSO_4$ particles so that the precipitate can be more easily isolated by filtration.)

After the mixture has cooled to room temperature and the precipitate has settled, carefully decant (Laboratory Methods **H**) the clear solution, and discard this clear solution. Next, use <u>ashless</u> filter paper, and filter the remaining mixture of $BaSO_4$ and remaining clear solution using either gravity filtration or vacuum filtration (Laboratory Methods **I**). Use distilled water from your wash bottle to assist transfer of all $BaSO_4$ from the beaker onto the ashless filter paper.

During the filtration, heat a clean crucible and cover, held by a wire triangle (Laboratory Methods **D**), as hot as possible for approximately 5 minutes. Permit the crucible to cool to room temperature and then weigh it on an analytical balance. Record the mass in TABLE 14.1.

Transfer the wet filter paper with the $BaSO_4$ to the crucible, and fold the paper so that it is contained entirely within the crucible. Place the crucible and cover on the wire triangle <u>under the hood</u>. Leave the cover slightly ajar so that some air can enter the crucible to enable the paper to burn. Heat the crucible gently at first to char the filter paper. The filter paper should not flame. (In the event that the paper does flame, cover the crucible with the cover.) Finally, heat the crucible as strongly as possible. When heating is complete, the filter paper should be burned off completely, and only white $BaSO_4$ should remain in the crucible. Allow the crucible to cool completely to room temperature. Weigh the crucible and $BaSO_4$ and record the mass in TABLE 14.1.

Reheat the crucible and $BaSO_4$ with the full flame for a second time, allow the crucible to cool to room temperature, and reweigh. Record the mass in TABLE 14.1. The mass of the crucible and cover should <u>not</u> change by more than 0.001 g. If you have a mass change of greater than 0.001 g, reheat the crucible a third time, cool it, and weigh it.

Determine the mass of $BaSO_4$ precipitate, and calculate the percent sulfate by mass in your sample of alum. Compare the percent sulfate in your alum with the theoretical percent for pure alum. Complete TABLE 14.2 including the sample calculations on the back of TABLE 14.2.

Name ______________________________ Experiment 14

Student ID number ________________

Section __________ Date __________

Instructor ________________________

DATA

TABLE 14.1. Mass data for analysis of alum.

Mass of weighing paper for alum (g)	
Mass of weighing paper plus alum sample (g)	
Mass of empty heated and cooled crucible (g)	
Mass of crucible plus $BaSO_4$ after first heating (g)	
Mass of crucible plus $BaSO_4$ after second heating (g)	
Mass of crucible plus $BaSO_4$ after third heating (g)	

Instructor's initials ____________

CALCULATIONS

TABLE 14.2. Percent SO_4^{2-} in alum.

Mass of alum (g)	
Mass of $BaSO_4$ (g)	
Percent SO_4^{2-} in your alum sample (%)	
Theoretical percent SO_4^{2-} in pure alum (%)	
Analytical purity of your alum sample (%)	

Experiment 14

Sample calculations

Moles of $BaSO_4$ formed from your alum sample

Mass of SO_4^{2-} in your alum sample

Percent SO_4^{2-} by mass in your alum sample

Theoretical percent SO_4^{2-} by mass in potassium aluminum alum

Name ______________________________ **Experiment 14**

Student ID number ________________

Section __________ Date __________

Instructor _______________________

PRE-LABORATORY QUESTIONS

1. Write two balanced net ionic equations to demonstrate how the addition of H^+ leads to the stepwise decomposition of CO_3^{2-} to liberate $CO_2(g)$.

2. a. What mass of your alum sample would be required to react stoichiometrically with the amount of $BaCl_2$ used in this experiment? Show clearly your calculations and reasoning.

   b. What mass of $BaSO_4$ could be formed? Show clearly your calculations and reasoning.

3. Why is the reaction mixture boiled after $BaSO_4$ has been precipitated?

4. What ions are present in the solution which is removed by decantation?

5. Why do you reheat the crucible and $BaSO_4$ a second time?

6. What is meant by the term, gravimetric analysis?

Name ______________________________ Experiment 14

Student ID number ________________

Section __________ Date __________

Instructor ________________________

POST-LABORATORY QUESTIONS

1. How would the following affect (increase, decrease, or not change) the percent SO_4^{2-} by mass in your alum? Justify your answer.

   a. Your alum was wet.

   b. You forgot to add HCl to your precipitation mixture, and the precipitate contained both $BaCO_3$ and $BaSO_4$. Assume the CO_3^{2-} was not present in your weighed sample.

   c. You forgot to add HCl to your precipitation mixture, and the precipitate contained both $BaCO_3$ and $BaSO_4$. Assume the CO_3^{2-} was present as an impurity in your weighed sample of alum.

d. You did not transfer all of the $BaSO_4$ from the beaker to the filter paper.

e. The filter paper was not totally removed during the ignition process.

f. Your crucible containing $BaSO_4$ picked up moisture while cooling and before the final weighing.

Preparation and Analysis

of an Iron Complex

PURPOSE OF EXPERIMENT: Prepare potassium tris(oxalato)ferrate(III) trihydrate, $K_3[Fe(C_2O_4)_3] \cdot 3H_2O$, and analyze it for oxalate ion and iron.

The synthesis and characterization of a compound is a primary task of a research chemist. Complex ions or coordination complexes are interesting to prepare in the laboratory because their formation is usually accompanied by one or more color changes. Coordination complexes contain a metal ion and some specific number of ligands, usually four or six. Ligands, the groups bound to the metal ion, can be negatively charged or neutral, and can have one or sometimes more sites to bond to the metal ion. Some common monodentate ligands, ligands which have only one bonding site, include water, H_2O; ammonia, NH_3; halide ions, X^-; cyanide, CN^-; and nitrite, NO_2^-. Bidentate ligands bond to the metal at two positions. Examples include ethylenediamine, $H_2NCH_2CH_2NH_2$, and oxalate ion, $C_2O_4^{2-}$. Another important feature of coordination complexes is the charge of the complex, the algebraic sum of the charge of the metal ion plus the charge of the ligands.

In this experiment you are going to prepare potassium tris(oxalato)-ferrate(III) trihydrate, $K_3[Fe(C_2O_4)_3] \cdot 3H_2O$, and then quantitatively analyze your product for $C_2O_4^{2-}$ and iron to prove its composition and purity. The complex ion in $K_3[Fe(C_2O_4)_3] \cdot 3H_2O$ is $[Fe(C_2O_4)_3]^{3-}$ with the iron having octahedral geometry. Your synthetic sequence will start with metallic iron, which you will oxidize to Fe^{2+}. The addition of oxalic acid, $H_2C_2O_4$, leads to the formation of iron(II) oxalate, $FeC_2O_4 \cdot 2H_2O$, which you will then oxidize with hydrogen peroxide, H_2O_2, in the presence of potassium oxalate, $K_2C_2O_4$, to form the desired compound, $K_3[Fe(C_2O_4)_3] \cdot 3H_2O$.

The process of titration may be used for the analysis of solutions of oxidizing and reducing agents, provided a suitable method for observing the endpoint of the titration is available. When potassium permanganate, $KMnO_4$, solution is used as the oxidizing agent in acidic solution, the endpoint is easily apparent because the reduction of the brilliantly purple colored MnO_4^- anion gives the pale pink manganese(II) ion, Mn^{2+}, which in dilute solution is practically colorless. Early in the titration a pink color can be seen where the drop hits because of the localized excess of MnO_4^- and the kinetically slow reduction of MnO_4^-. The pink color disappears slowly with swirling of the flask. Therefore, you will titrate until a drop of MnO_4^- produces the first permanent pink color throughout the solution, indicating the consumption of all the reducing agent and the presence of excess MnO_4^-. In this experiment permanganate titrations will be used for the analysis of both oxalate ion and iron.

REFERENCES

(1) Kotz, J. C., and Purcell, K. F., Chemistry and Chemical Reactivity, Saunders College Publishing, Philadelphia, 1987, sections 2.5, 2.11a, 3.2, 3.4, 4.1, 4.4, 25.1-25.4, and 25.6..

(2) Masterton, W. L., Slowinski, E. J., and Stanitski, C. L., Chemical Principles, 6th ed., Saunders College Publishing, Philadelphia, 1985, sections 2.5, 2.6, 3.2-3.7, 4.3, 8.6, 21.1-21.3, 23.1, 23.2, 25.1, and 25.2.

EXPERIMENTAL PROCEDURE

(Study this section and the PRE-LABORATORY QUESTIONS before coming to the laboratory. **Wear safety goggles when performing this experiment.**)

A. Preparation of $K_3[Fe(C_2O_4)_3] \cdot 3H_2O$

Using a beam balance (Laboratory Methods **C**), weigh a 0.70-g sample of iron powder into a 200-mL or larger beaker. Record your masses in TABLE 15.1A. Add 15 mL of 3 **M** sulfuric acid, H_2SO_4, solution, and place a watch glass over the beaker. **CAUTION:** $\underline{H_2SO_4}$ $\underline{\text{of this concentration is corrosive}}$ $\underline{\text{and causes burns on your skin or holes in your clothing. If any spills or}}$ $\underline{\text{spatters occur onto your skin or clothing, rinse the affected area}}$ $\underline{\text{thoroughly with water}}$. Heat the mixture very gently (Laboratory Methods **D**) until all of the iron has reacted. You should play the flame of the Bunsen burner under the beaker until the iron begins to react, and then withdraw the flame. When the reaction slows, gently reheat the reaction mixture as before. Continue until all evidence of reaction has disappeared.

Filter the solution by gravity filtration into a 200-mL or larger beaker, or by suction filtration followed by transfer to a 200-mL or larger beaker (Laboratory Methods **I**). Slowly add 1 **M** sodium hydroxide, NaOH, solution to your filtrate until your solution has a pH of 4. Use wide range pH paper to determine the pH (Laboratory Methods **P**). If you add too much NaOH solution, and a gelatinous precipitate of iron(II) hydroxide, $Fe(OH)_2$, forms, add 3 **M** H_2SO_4 dropwise until the pH is 4. After adjusting the pH of your solution, add 25 mL of 1.0 **M** oxalic acid, $H_2C_2O_4$, solution. **CAUTION:** $\underline{\text{Wash your hands after using } H_2C_2O_4 \text{ or oxalate solutions}}$. Heat the mixture to boiling, and stir it constantly to prevent bumping. Cool the mixture, and allow the precipitate of $FeC_2O_4 \cdot 2H_2O$ to settle. Decant the supernatant liquid (Laboratory Methods **H**), and then wash the precipitate with 20. mL of distilled water (Laboratory Methods **I**). Warm the mixture to aid washing. Allow the precipitate to settle, and decant again. **Save the precipitate.**

Add 10. mL of a saturated solution of potassium oxalate, $K_2C_2O_4$, to your beaker, and heat the mixture to 40°C. Slowly add 20. mL of 3% hydrogen peroxide, H_2O_2, solution using your buret (Laboratory Methods **B**). Continuously stir the solution, and keep the temperature as close to 40°C as you can. After all the H_2O_2 has been added, heat the mixture to boiling. Add 5 mL of 1.0 **M** oxalic acid, $H_2C_2O_4$, solution all at once; then add 3 mL more dropwise. Keep the temperature of the mixture near boiling. You should have an emerald green solution.

Filter the hot solution by gravity filtration into a 100-mL or larger beaker, or by suction filtration followed by transfer to a 100-mL or larger beaker. Add 10. mL of ethyl alcohol, C_2H_5OH, and warm the mixture to redissolve any crystals that may have formed. Tie a short piece of thread to a wooden splint, and suspend the thread in the solution. The thread must be in the solution, not just lying on the top of the solution. Store your beaker in your desk until the next laboratory period. Then remove the green crystals, dry them between pieces of filter paper, and weigh them on a beam balance. Record your masses in TABLE 15.1A. Calculate the percent yield of product on the basis of the iron used. If white oxalic acid, $H_2C_2O_4$, crystals form, discard them.

B. Analysis for Oxalate Ion

Using an analytical balance (Laboratory Methods **C**), weigh precisely about a 0.3-g sample of your product, potassium tris(oxalato)ferrate(III) trihydrate, $K_3[Fe(C_2O_4)_3] \cdot 3H_2O$, on a piece of weighing paper. Record your masses in TABLE 15.1B. Transfer the sample quantitatively to a 250-mL or larger beaker, and dissolve it in about 20. mL of 3 **M** sulfuric acid, H_2SO_4, solution. **CAUTION:** $\underline{H_2SO_4}$ $\underline{\text{of this concentration is corrosive and}}$

causes burns on your skin or holes in your clothing. If any spills or spatters occur onto your skin or clothing, rinse the affected area thoroughly with water. Then add about 20. mL of distilled water.

Set up a ring stand and buret clamp, clean a buret carefully with soap solution, and rinse it thoroughly with distilled water. Practice reading the buret and manipulating the stopcock or pinch clamp (Laboratory Methods B) before continuing this analysis. Finally, drain the buret.

Fill your buret with an approximately 0.02 $\underline{M}$ potassium permanganate, $KMnO_4$, solution. Record the exact molarity of the $KMnO_4$ solution in TABLE 15.1B. Heat the solution of $K_3[Fe(C_2O_4)_3] \cdot 3H_2O$ to 55-60°C (Laboratory Methods E). **Temperature control is very important for the oxalate analysis.** (At temperatures below 55°C the reaction between MnO_4^- and $H_2C_2O_4$ is too slow to give a good endpoint, but at approximately 55°C the oxalic acid reacts at a suitable rate. Above 60.°C, oxalic acid decomposes. The reaction is also autocatalytic; that is, a product of the reaction, Mn^{2+}, is a catalyst.) Titrate the hot solution using the following procedure. Place a white sheet of paper under the beaker so that the color of the solution may be seen clearly. Add the MnO_4^- solution dropwise from the buret, swirling the sample constantly, until the last drop leaves a permanent pink color. Rinse the inside wall of the beaker with distilled water from your wash bottle to make sure all the $H_2C_2O_4$ reacts (Laboratory Methods B). The pink color of the MnO_4^- should persist about 30. seconds at the endpoint. Record the initial and final volumes of MnO_4^- in TABLE 15.1B. Save the solution from the titration because you will use it for the analysis for iron.

Repeat the analysis, using a second sample of $K_3[Fe(C_2O_4)_3] \cdot 3H_2O$.

C. Analysis for Iron(III)

The solutions from the analyses for the oxalate ion will be used in turn to make duplicate runs for the Fe^{3+} analysis. If the color of the MnO_4^- from the oxalate analysis persists in your solution, add a drop of 3% hydrogen peroxide, H_2O_2, solution, and stir the solution. Allow the solution to stand until it becomes colorless. Evaporate your solution by boiling (Laboratory Methods L) until you have a volume of about 25 mL. To this hot solution, add dropwise a 0.5 $\underline{M}$ solution of tin(II) chloride, $SnCl_2$, to reduce iron(III) to iron(II). The first drop of $SnCl_2$ will make your solution yellow, owing to the formation of $FeCl^{2+}$. Continue to add $SnCl_2$ until the yellow color disappears, then add **one drop** in excess. Cool your solution to room temperature. Add 10. mL of a saturated solution of mercury(II) chloride, $HgCl_2$, all at once and with vigorous stirring. A slight white precipitate of mercury(I) chloride, Hg_2Cl_2, should form after a few minutes. The $HgCl_2$ has oxidized all of the excess tin(II) to tin(IV). Add 10. mL of a prepared solution of phosphoric acid, H_3PO_4; sulfuric acid, H_2SO_4; and manganese(II) sulfate, $MnSO_4$; which prevents the formation of $FeCl^{2+}$ during your MnO_4^- titration (the yellow color of $FeCl^{2+}$ would obscure the endpoint).

Titrate the resulting solution with the approximately 0.02 $\underline{M}$ potassium permanganate, $KMnO_4$, solution at room temperature. Record the initial and final volumes of MnO_4^- in TABLE 15.1C.

Repeat the procedure for run 2.

Perform the calculations in TABLE 15.2 including the sample calculations following TABLE 15.2.

TABLE 15.1. Data for preparation and analysis of $K_3[Fe(C_2O_4)_3] \cdot 3H_2O$.

		Run 1	Run 2
A	Mass of beaker (g)		
	Mass of beaker and iron powder (g)		
	Mass of weighing paper (g)		
	Mass of weighing paper and product (g)		
B	Mass of weighing paper (g)		
	Mass of weighing paper and $K_3[Fe(C_2O_4)_3] \cdot 3H_2O$ (g)		
	Molarity of MnO_4^- solution ($\underline{M}$)		
	Initial buret reading (mL)		
	Final buret reading (mL)		
C	Initial buret reading (mL)		
	Final buret reading (mL)		

Instructor's initials _____

Name _____

Student ID number _____

Section _____ Date _____

Instructor _____

CALCULATIONS

TABLE 15.2. Percent yield and percents by mass of $C_2O_4^{2-}$ and Fe^{3+}.

A			
	Mass of iron powder used (g)		
	Theoretical yield of $K_3[Fe(C_2O_4)_3] \cdot 3H_2O$ (g)		
	Actual yield of $K_3[Fe(C_2O_4)_3] \cdot 3H_2O$ (g)		
	Percent yield of $K_3[Fe(C_2O_4)_3] \cdot 3H_2O$ (%)		
		Run 1	Run 2
B	Mass of sample (g)		
	Volume of MnO_4^- solution (mL)		
	Moles of MnO_4^- reacted (mol)		
	Moles of $H_2C_2O_4$ reacted (mol)		
	Mass of $C_2O_4^{2-}$ reacted (g)		
	Percent by mass of $C_2O_4^{2-}$ in sample (%)		
	Theoretical percent by mass of $C_2O_4^{2-}$ in sample (%)		
	Percent purity of your product based on $C_2O_4^{2-}$ analysis (%)		
	Average percent purity of your product (%)		

199

TABLE 15.2 (continued)

		Run 1	Run 2
C	Mass of sample (g)		
	Volume of MnO_4^- (mL)		
	Moles of MnO_4^- reacted (mol)		
	Moles of Fe^{2+} reacted (mol)		
	Mass of iron reacted (g)		
	Percent by mass of iron in sample (%)		
	Theoretical percent by mass of iron in sample (%)		
	Percent purity of your product based on iron analysis (%)		
	Average percent purity of your product based on iron analysis (%)		
	Experimental ratio of $C_2O_4^{2-}/Fe^{3+}$		

Name _____

Student ID number _____

Section _____ Date _____

Instructor _____

Sample calculations for Run ___

A **Theoretical yield of** $K_3[Fe(C_2O_4)_3] \cdot 3H_2O$

Percent yield of $K_3[Fe(C_2O_4)_3] \cdot 3H_2O$

B **Balanced net ionic equation for the reaction of** MnO_4^- **with** $H_2C_2O_4$ **in an acidic solution**

Moles of MnO_4^- reacted

Moles of $H_2C_2O_4$ which reacted with MnO_4^-

Mass of $C_2O_4^{2-}$ which reacted with MnO_4^-

Percent by mass of $C_2O_4^{2-}$ in sample

Theoretical percent by mass of $C_2O_4^{2-}$ in $K_3[Fe(C_2O_4)_3] \cdot 3H_2O$

Percent purity of your product

Experiment 15

Sample calculations for Run ____ (continued)

C Balanced net ionic equation for the reaction of MnO_4^- with Fe^{2+} in an acidic solution

Moles of MnO_4^- reacted

Moles of Fe^{2+} which reacted with MnO_4^-

Mass of iron which reacted with MnO_4^-

Percent by mass of iron in sample

Theoretical percent by mass of iron in $K_3[Fe(C_2O_4)_3] \cdot 3H_2O$

Percent purity of your product

Ratio of moles of $C_2O_4^{2-}$ found in sample to moles of Fe^{3+} found in sample

Name ______________________________ Experiment 15

Student ID number ________________

Section __________ Date __________

Instructor ________________________

PRE-LABORATORY QUESTIONS

1. Why is H_2O_2 added in the preparation of $K_3[Fe(C_2O_4)_3]\cdot 3H_2O$ in part **A**?

2. What is the oxidation state of iron in $K_3[Fe(C_2O_4)_3]\cdot 3H_2O$? Show clearly your reasoning.

3. Write a balanced net ionic equation for the reaction of MnO_4^- with $H_2C_2O_4$ in acidic solution. This reaction is used for the analysis of $C_2O_4^{2-}$ in part **B**.

4. What is the oxidation state of iron in your solution after you have completed the analysis for $C_2O_4^{2-}$ in part B?

5. Write a balanced net ionic equation for the reaction of MnO_4^- with Fe^{2+} in acidic solution. This reaction is used for the analysis of iron in part **C**.

6. Why is H_2O_2 added in the analysis for Fe^{3+} in part C?

7. Why is $SnCl_2$ added in the analysis for Fe^{3+} in part C?

8. Why is H_3PO_4 added in the analysis for iron in part C?

9. Draw clearly and describe in words the structure of $[Fe(C_2O_4)_3]^{3-}$ in the coordination complex, $K_3[Fe(C_2O_4)_3] \cdot 3H_2O$.

Name _____

Student ID number _____

Section _____ Date _____

Instructor _____

1. What is the color of FeC_2O_4 and $K_3[Fe(C_2O_4)_3] \cdot 3H_2O$?

2. Explain the origin of the color of $K_3[Fe(C_2O_4)_3] \cdot 3H_2O$.

3. Give possible reasons for your observed percent purity based on your $C_2O_4{}^{2-}$ analysis being different from 100%.

4. Give possible reasons for your observed percent purity based on your iron analysis being different from 100%.

5. Draw all possible isomers for $[Fe(C_2O_4)_2(H_2O)_2]^{-1}$.

6. a. Draw the Lewis electron dot structures for oxalic acid and the oxalate ion.

    b. Do either oxalic acid or the oxalate ion have important resonance forms? If so, draw them.

Preparation and Analysis
of a Cobalt Complex

PURPOSE OF EXPERIMENT: Prepare an unknown cobalt complex with ammine and/or chloro ligands, and analyze the complex for anionic chloride and for charge to determine its identity.

Many research chemists are concerned with preparing and characterizing new compounds. A considerable number of these new compounds are comprised of complex ions, which are particularly interesting because their formation is usually accompanied by one or more color changes. Complex ions are polyatomic cations or anions consisting of a central atom or ion (usually a metal atom or ion) surrounded by two or more (frequently four or six) anions or neutral molecules collectively called ligands. Complex ions generally remain intact as polyatomic entities in solution but may dissociate to some degree and undergo other chemical reactions. Complex ions are normally placed in brackets in straight-line formulas to identify them clearly. For example, $K_3[Fe(CN)_6]$ indicates a compound consisting of K^+ cations and the complex anion, $[Fe(CN)_6]^{3-}$. The net charge on a complex ion is the algebraic sum of the charges on the central atom and the surrounding ligands. Thus, the -3 in $[Fe(CN)_6]^{3-}$ arises from the algebraic sum of $+3$ from Fe(III) and (6×-1) from six CN^- ligands. Ligands are often classified according to the number of sites with which they bind to cations. For example, H_2O, NH_3, Cl^-, and CN^- bind through one site and are called monodentate ligands, whereas ethylenediamine, $H_2NCH_2CH_2NH_2$, binds through two sites and is called a bidentate ligand.

In this experiment you will start with an aqueous solution of cobalt(II) chloride, $CoCl_2$. Ammonium chloride, NH_4Cl, will be added to provide a source of additional Cl^-, and aqueous ammonia, NH_3, will be added to provide potential NH_3 ligands. Finallly, hydrogen peroxide, H_2O_2, will be added to oxidize Co(II) to Co(III). After the reaction is complete, you will filter crystals of your cobalt-ammine complex. It is not uncommon in chemical syntheses that several products can be formed from the same set of reactants. In this experiment you should consider $[Co(NH_3)_6]Cl_3$, $[Co(NH_3)_5Cl]Cl_2$, and $[Co(NH_3)_4Cl_2]Cl$ as possible products (See PRE-LABORATORY QUESTION 1). Formation of one of the products can often be favored by subtle changes in the mole ratios of the reactants, the particular catalyst used, or conditions such as temperature. After purification of the primary product, its identity can be determined by analyses or by instrumental techniques. In this experiment, you will use two independent analyses for this purpose.

The first analysis is for the chloride ion which serves to balance the charge of the complex ion. Many complex ions are stable enough that ligand chlorides attached directly to the central metal ion (inside the bracket in the formula) are not precipitated as silver chloride, AgCl, upon the addition of silver nitrate, $AgNO_3$, solution, whereas anionic chlorides (outside the bracket in the formula) are precipitated quantitatively by this technique. For example, only one Cl precipitates from the following complex.

$$[Cr(OH_2)_4Cl_2]Cl(aq) + AgNO_3(aq) \longrightarrow [Cr(OH_2)_4Cl_2]NO_3(aq) + AgCl(s)$$

The AgCl can be filtered, dried, and weighed. From the mass of AgCl you can calculate the moles of AgCl and Cl and the mass of Cl. This also represents the mass of Cl present as an anion in the complex since it all came from that source. Thus, using the original mass of your sample, you can calculate the percent anionic Cl by mass in your complex and compare it with theoretical percent anionic chlorides from possible products. If there is not considerable overlap of theoretical values, you may be able to identify your product from this single analysis.

The second analysis is for net charge on the cation of the complex. The analysis makes use of a technique called ion exchange followed by an acid-base titration. Ion exchange involves the reversible interchange of ions between a solution (the mobile phase) and a high molecular weight, insoluble electrolyte called an ion exchange resin (the stationary phase). This electrolyte usually consists of one very large ion and an oppositely charged, small simple ion which can be exchanged for the ions in the solution in contact with the resin. If the small exchangeable ion is an anion, the resin is called an anion exchange resin. If the small exchangeable ion is a cation, it is called a cation exchange resin. A cation exchange resin from which H^+ is liberated is used in the cation charge analysis in this experiment. Using the same Cr(III) example as was mentioned in the anionic chloride analysis, the following cation exchange occurs.

$$[Cr(OH_2)_4Cl_2]^+(aq) \quad + \quad H\text{-resin}(s) \quad \text{--->} \quad [Cr(OH_2)_4Cl_2]\text{-resin}(s) \quad + \quad H^+(aq)$$

Note that the number of moles of H^+ displaced from the resin is equivalent to the number of moles of positive charge on the complex cation being exchanged. Stated another way, the ratio of the moles of H^+ liberated per mole of complex cation exchanged is a direct measure of the charge on the cation.

$$\text{cation charge} \quad = \quad \frac{\text{moles of } H^+}{\text{moles of complex ion}}$$

Moles of complex ion is determined from the original mass of sample used.

$$\text{moles of } [Cr(OH_2)_4Cl_2]Cl \quad = \quad \frac{\text{mass of } [Cr(OH_2)_4Cl_2]Cl}{\text{molecular weight of } [Cr(OH_2)_4Cl_2]Cl}$$

It only remains to determine moles of H^+ by means of an acid-base titration in which the volume of base of a known molarity required to neutralize the H^+ is measured. An appropriate indicator, in this case, phenolphthalein, is used to detect the endpoint. Since the acid and base react in a 1:1 ratio,

$$H^+(aq) \quad + \quad OH^-(aq) \quad \text{--->} \quad H_2O(l)$$

it follows that

$$\text{moles of } H^+ \quad = \quad \text{moles of } OH^- \quad = \quad (\text{molarity of } OH^-)(\text{volume of } OH^-)$$

In this experiment you will synthesize a cobalt complex and determine experimentally the percent anionic chloride and the charge of the complex ion. From this evidence you will decide which of three possible cobalt complexes you synthesized.

REFERENCES

(1) Kotz, J. C., and Purcell, K. F., _Chemistry and Chemical Reactivity_, Saunders College Publishing, Philadelphia, 1987, sections 2.5, 2.9, 3.2-3.4, 4.1-4.4, 12.1, 17.1, 17.3, 17.4, 25.3, 25.4a, and 25.6.

(2) Masterton, W. L., Slowinski, E. J., and Stanitski, C. L., _Chemical Principles_, 6th ed., Saunders College Publishing, Philadelphia, 1985, sections 2.5, 2.6, 3.1-3.8, 8.6, 12.2, 18.2, 19.6, 19.7, 21.1-21.3, and 25.2.

EXPERIMENTAL PROCEDURE

(Study this section and the PRE-LABORATORY QUESTIONS before coming to the laboratory. **Wear safety goggles when performing this experiment.**)

A. Synthesis of a Cobalt Complex

Using a beam balance (Laboratory Methods C), weigh 5.0 g of cobalt(II) chloride hexahydrate, $CoCl_2 \cdot 6H_2O$, into a 100-mL or larger beaker. Record your masses in TABLE 16.1A. Then add 2.0 g of ammonium chloride, NH_4Cl, into the same beaker, and record your mass in TABLE 16.1A. Add 5 mL of distilled water, and stir until most of the salts are dissolved (Laboratory Methods K). Then add 10. mL of concentrated (15 $\underline{M}$) aqueous ammonia, NH_3. CAUTION: <u>Concentrated NH_3 can cause burns on your skin. Wear rubber gloves when handling it. If any spills occur onto your skin, rinse the affected area thoroughly with water</u>. Finally, add 3.0 mL of 30.% (by mass) hydrogen peroxide, H_2O_2, solution <u>in small amounts</u> and with vigorous stirring for the purpose of oxidizing Co(II) to Co(III). CAUTION: <u>H_2O_2 of this concen-tration causes severe burns and bleaching of your skin or clothing. Wear rubber gloves when handling it. If any spills or spatters occur onto your skin or clothing, rinse the affected area thoroughly with water</u>. The solu-tion will effervesce in a controlled fashion as long as small quantities are added and stirring is maintained. When effervescence ceases, heat the solution <u>gently</u> over a Bunsen burner (Laboratory Methods D), and continue stirring constantly. The temperature should be maintained at about 80°C for a period of 10. minutes. CAUTION: <u>If the solution boils, it is too hot</u>. Allow the product to crystallize as the solution cools to room temperature.

Filter the crystals in the beaker using either gravity filtration or suction filtration (Laboratory Methods I), continuing the process until they are as dry as possible. Then wash the crystals carefully with three separ-ate 3-mL portions of distilled water to wash soluble occluded Cl⁻ ions from the crystals. When no more water is being removed, lift the filter paper and product from the funnel with the help of a spatula, and place the filter paper on a watch glass. Dry the filter paper and product by heating on a steam bath, with a heat lamp, or in an oven (Laboratory Methods M), as designated by your instructor, or allow the filter paper and product to air dry in your drawer for a week. If you are drying by heating, heat the sample for about 10. minutes after it starts to crack, probably about 15 minutes overall.

Using a beam balance, weigh a glass vial, and record your mass in TABLE 16.1A. Transfer your dried crystals to the vial, reweigh the vial, and record your mass in TABLE 16.1A. These dried crystals are your cobalt complex and will be used in the analyses in parts B and C.

B. Analysis for Anionic Chloride

Label three clean, dry 250-mL or larger beakers 1, 2, and 3. Using an analytical balance, weigh the vial containing your cobalt complex, and record the mass in TABLE 16.1B. Tap the vial carefully to transfer to beaker 1 slightly more than 0.1 g of cobalt complex. Then reweigh the vial, and record your mass in TABLE 16.1B. Repeat this transfer procedure and reweighing twice more so that three samples are weighed by difference into beakers 1, 2, and 3.

If necessary with your crucibles, place a circle of glass fiber filter paper into each of three clean, <u>dry</u> Gooch crucibles. Make certain that you can differentiate the three crucibles. Using an analytical balance, weigh each crucible, and record your masses in TABLE 16.1B.

Carry out this and the next paragraph of the analysis consecutively on each weighed sample, completing work on one sample before starting with the next. Add about 20. mL of distilled water to the first sample, and stir to dissolve it. Then add 10. drops of 6 $\underline{M}$ nitric acid, HNO_3, solution. **CAUTION: $\underline{HNO_3}$ of this concentration causes burns on your skin or holes in your clothing. If any spills or spatters occur onto your skin or clothing, rinse the affected area thoroughly with water**. Add slowly with constant stirring 3.0 mL of 0.5 $\underline{M}$ silver nitrate, $AgNO_3$, solution. **CAUTION: $\underline{AgNO_3}$ stains your skin and is toxic. Wash your hands thoroughly after use or if there is any spill**. $AgNO_3$ is also very expensive, so do not waste it. After all the $AgNO_3$ solution has been added, continue stirring vigorously for about 1 minute to coagulate the AgCl precipitate. Allow the precipitate to settle, and then decant the supernatant liquid through the first Gooch crucible with the aspirator ON. Make certain that most of the AgCl remains in the beaker at this point.

Prepare a solution for washing the AgCl precipitate by adding 10. drops of 6 $\underline{M}$ HNO_3 to 100. mL of distilled water. Add 5 mL of this wash solution to the beaker containing the AgCl precipitate, and stir the mixture to wash soluble occluded ions free from the precipitate. Allow the precipitate to settle, and decant the wash solution through the Gooch crucible with the aspirator ON. Repeat this washing procedure twice more before finally transferring the AgCl precipitate from the beaker to the Gooch crucible with the help of a rubber policeman, if available. Rinse the beaker with 10. mL of the wash solution to transfer the last traces of AgCl. When the AgCl precipitate is as dry as possible, remove the rubber hose from the suction flask, turn OFF the aspirator, and remove the Gooch crucible to a dry beaker labeled with your name. Pour the filtrate from your filter flask into a bottle marked USED Ag^+. This solution will be recycled to recover the silver for future use.

Repeat the procedure of the last two paragraphs for your second and third samples of cobalt complex.

Dry the crucibles and precipitates for at least 1 hour in an oven at $110^\circ C$, or preferably overnight at that temperature, as designated by your instructor. When the crucibles have cooled to room temperature, weigh each one on the analytical balance, and record your masses in TABLE 16.1B. Transfer the circles of glass fiber filter paper and the AgCl precipitate to a beaker on the reagent bench marked AgCl PRECIPITATE. This will also be recycled to recover the silver for future use. Finally, clean the crucibles, and return them to the equipment bench.

C. Analysis for Charge

Set up a ring stand and buret clamp, clean a buret carefully with soap solution, and rinse it thoroughly with distilled water. Practice reading the buret and manipulating the stopcock or pinch clamp (Laboratory Methods B) before starting this analysis. Finally, drain the buret.

Place a piece of cotton about the size of a pea in the bottom of the buret, pushing the cotton in gently with the help of a wooden dowel. The cotton wad will hold your ion exchange resin in place and prevent the resin from passing through the stopcock or pinch clamp, while still allowing liquid to pass freely. Place a couple scoops of cation exchange resin in a 250-mL or larger beaker. **Be careful not to spill any of the resin. It is expensive as well as slippery and difficult to clean up**. Add enough distilled water to the beaker to form a slurry with the resin upon stirring, and pour the suspension of cation exchange resin into your buret. After allowing the resin to settle, open the stopcock or pinch clamp to check that the wad of cotton retains the resin but allows free passage of water. **CAUTION: Never allow the water level to drop below the upper level of the resin because the resin may solidify so that it can't be removed from the buret later**. Repeat the process of making and adding a slurry of resin until your buret is filled with settled resin to the 45-mL mark, or some other mark designated by your instructor.

Wash any excess acid from the ion exchange resin by passing distilled water through the column until a drop of the effluent from the buret no longer gives an acid reaction with blue litmus paper (Laboratory Methods P and Q). Then drain the buret until about 2 mL of water remain above the upper level of the resin.

Label three clean, dry 250-mL or larger beakers 1 and 2. Using an analytical balance, weigh the vial containing your cobalt complex, and record the mass in TABLE 16.1C. Tap the vial carefully to transfer to beaker 1 slightly more than 0.15 g of cobalt complex. Then reweigh the vial, and record your mass in TABLE 16.1C. Repeat this transfer and reweighing procedure once more so that two samples are weighed by difference into beakers 1 and 2.

Dissolve one of your weighed samples in about 25 mL of distilled water, warming and stirring vigorously, if necessary. Failure to dissolve the sample completely will lead to an erroneous analysis. Pour the solution onto the ion exchange column in your buret, pass the solution through the resin, and collect the effluent in a 250-mL Erlenmeyer flask. **CAUTION: <u>Do not let the water level drop below the upper level of the resin</u>**. After all of the solution has been added, continue to add distilled water to the top of the column until a drop of the solution passing out of the column and into the Erlenmeyer flask no longer gives an acid reaction with blue litmus paper. Save the solution in the Erlenmeyer flask for titration with standard base. Repeat the procedure of this paragraph with your second weighed sample.

Remove the ion exchange resin from your buret using the following procedure. Half-fill the buret with distilled water, place your thumb over the open end of the buret, and repeatedly invert the buret and then place it upright in order to loosen the resin in the buret. When all the resin is loose, drain it into a 250-mL or larger beaker. Repeat this procedure until all of the resin has been removed from the buret to the beaker. Then pour the suspended resin into a bottle on the reagent bench marked USED CATION EXCHANGE RESIN, rinsing with small quantities of distilled water as appropriate. Finally, force the cotton plug out of your buret <u>but not down the drain</u> by running water from a hose back through your buret tip, and wash and rinse your buret thoroughly.

Obtain in a clean, dry beaker about 50. mL of approximately 0.1 $\underline{M}$ sodium hydroxide, NaOH, solution from the reagent bench. Record the exact concentration of the base in TABLE 16.1C. Rinse your buret with about 10. mL of the standard base. Then add a few milliliters of the standard base to the buret, fill the buret tip completely by manipulating the stopcock or pinch clamp, check to be certain that there are no air bubbles in the buret tip, and add the rest of your standard base to the buret.

Titrate all of your first effluent solution with standard base using the following procedure. Record your initial buret reading in TABLE 16.1C. Add 2 drops of phenolphthalein indicator solution (Laboratory Methods Q). Using a piece of white paper underneath your Erlenmeyer flask as background, add base slowly with swirling of the flask until the first drop that turns the solution a permanent pale pink. You must exercise care so as not to overrun the endpoint. If you pass the endpoint, you may have to repeat the entire analysis! Therefore, add standard base dropwise, swirling between drops, as the pink color starts to persist. Record your final buret reading in TABLE 16.1C. Repeat the procedure of this paragraph with your second effluent solution.

Wash and rinse your buret thoroughly, and return it to the equipment bench.

Perform the calculations in TABLE 16.2 including the sample calculations for one run on the back of TABLE 16.2, and decide which of the three possible cobalt complexes you synthesized.

TABLE 16.1. Mass and volume data.

A		
Mass of beaker (g)		
Mass of beaker and $CoCl_2 \cdot 6H_2O$ (g)		
Mass of beaker and NH_4Cl (g)		
Mass of vial (g)		
Mass of vial and cobalt complex (g)		

	Run 1	Run 2	Run 3
B Mass of vial before removing cobalt complex (g)			
Mass of vial after removing cobalt complex (g)			
Mass of crucible and paper (g)			
Mass of crucible, paper, and AgCl (g)			

	Run 1	Run 2
C Mass of vial before removing cobalt complex (g)		
Mass of vial after removing cobalt complex (g)		
Molarity of standard base ($\underline{M}$)		
Initial volume of base (mL)		
Final volume of base (mL)		

Instructor's initials _____

TABLE 16.2. Percent yield, percent anionic chloride, and charge.

A	Mass of $CoCl_2 \cdot 6H_2O$ used (g)			
	Moles of $CoCl_2 \cdot 6H_2O$ used (mol)			
	Mass of NH_4Cl used (g)			
	Moles of NH_4Cl used (mol)			
	Moles of NH_3 used (mol)			
	Moles of aqueous H_2O_2 used (mol)			
	Assume product is	$[Co(NH_3)_6Cl_3]$	$[Co(NH_3)_5Cl]Cl_2$	$[Co(NH_3)_4Cl_2]Cl$
	Limiting reactant			
	Theoretical yield of cobalt complex (g)			
	Actual yield of cobalt complex (g)			
	Percent yield of cobalt complex (%)			
	Cobalt complex synthesized (based on analyses in parts B and C)			

CALCULATIONS

TABLE 16.2 (continued)

	Run 1	Run 2	Run 3
B Mass of cobalt complex (g)			
Mass of AgCl precipitate (g)			
Mass of Cl in precipitate (g)			
Percent anionic chloride by mass in cobalt complex (%)			
Average percent anionic chloride (%)			
Theoretical percent anionic chloride (%)	$[Co(NH_3)_6]Cl_3$	$[Co(NH_3)_5Cl]Cl_2$	$[Co(NH_3)_4Cl_2]Cl$

	Run 1	Run 2
C Mass of cobalt complex (g)		
Moles of cobalt complex assuming $[Co(NH_3)_6]Cl_3$ (mol)		
Moles of cobalt complex assuming $[Co(NH_3)_4Cl_2]Cl$ (mol)		
Volume of standard base (mL)		
Moles of OH^- used and H^+ titrated (mol)		
Charge of cobalt complex assuming $[Co(NH_3)_6]Cl_3$		
Charge of cobalt complex assuming $[Co(NH_3)_5Cl]Cl_2$		
Charge of cobalt complex assuming $[Co(NH_3)_4Cl_2]Cl$		
Average charge for cobalt complex with best fit		

Experiment 16

Sample calculations for Run __

B Mass of Cl in precipitate

Percent anionic chloride by mass in cobalt complex

C Moles of cobalt complex assuming $[Co(NH_3)_6]Cl_3$

Moles of OH^- used and H^+ titrated

Charge of cobalt complex assuming $[Co(NH_3)_6]Cl_3$

Name _____

Student ID number _____

Section _____ Date _____

Instructor _____

PRE-LABORATORY QUESTIONS

1. Balance the following complete equations describing possible syntheses in part **A**.

   $CoCl_2 \cdot 6H_2O$(aq) + NH_4Cl(aq) + NH_3(aq) + H_2O_2(aq) --->

   $[Co(NH_3)_6]Cl_3$(aq) + H_2O(l)

   $CoCl_2 \cdot 6H_2O$(aq) + NH_4Cl(aq) + NH_3(aq) + H_2O_2(aq) --->

   $[Co(NH_3)_5Cl]Cl_2$(aq) + H_2O(l)

   $CoCl_2 \cdot 6H_2O$(aq) + NH_4Cl(aq) + NH_3(aq) + H_2O_2(aq) --->

   $[Co(NH_3)_4Cl_2]Cl$(aq) + H_2O(l)

2. What is the function of the hydrogen peroxide, H_2O_2, solution in the synthesis in part **A**?

3. Write a balanced net ionic equation describing each of the following.

   a. The precipitation reaction that occurs in the analysis for anionic chloride in part **B**.

   b. The neutralization reaction that occurs in the acid-base titration during the analysis for charge in part **C**.

4. Would the theoretical yields of the cobalt complexes in PRE-LABORATORY QUESTION 1 double if you doubled the mass of cobalt(II) chloride hexahydrate, $CoCl_2 \cdot 6H_2O$, used in the synthesis in part **A**? Indicate clearly your calculations and reasoning.

5. A student set out to determine whether a salt that she had prepared was $[Cr(OH_2)_6]Cl_2$ (230.99 g/mol) or $[Cr(OH_2)_6]Cl_3$ (266.45 g/mol) by determining percent anionic chloride and charge of the complex cation just as you will do in the laboratory for your cobalt-containing complex salt. The following data were obtained.

When a 0.111-g sample was dissolved in water and treated with excess aqueous $AgNO_3$ solution, 0.180 g of solid AgCl (143.321 g/mol) was obtained after filtering and drying.

When a 0.151-g sample was dissolved in water and passed through a cation exchange resin, 13.59 mL of 0.1250 $\underline{M}$ NaOH solution was required to neutralize the effluent.

Which sample had the student prepared? Indicate clearly your calculations and reasoning.

Name _____

Student ID number _____

Section _____ Date _____

Instructor _____

POST-LABORATORY QUESTIONS

1. Would the calculated percent yield for your synthesis in part **A** be increased, decreased, or unchanged from the true value for each of the following mistakes? Indicate clearly your calculations and reasoning.

    a. You forgot to dry your cobalt complex before weighing it.

    b. You misread the beam balance and actually used 6.1 g of $CoCl_2 \cdot 6H_2O$ rather than the 5.1 g that you recorded in TABLE 16.1**A**.

2. Suppose that a standardized barium hydroxide, $Ba(OH)_2$, solution of the same molarity as your standardized NaOH solution was used to titrate the acid solution eluted during passage of your dissolved complex through the ion exchange column. Intuitively, would more or less $Ba(OH)_2$ solution be required compared to the volume of NaOH solution that you used in your titration? Indicate clearly the calculations and reasoning required to determine the actual volume of $Ba(OH)_2$ solution needed so as to verify your intuitive answer.

3. Would the calculated percent anionic chloride in part **B** be increased, decreased, or unchanged from the true value for each of the following mistakes? Indicate clearly your calculations and reasoning.

    a. Your cobalt complex was not properly dried before starting the anionic chloride analysis.

    b. You neglected to scrape all the AgCl precipitate from the beaker into the Gooch crucible during filtration.

    c. Your Gooch crucible was still wet from a previous cleaning when you weighed it at the beginning of the analysis, but was dry when the AgCl precipitate was weighed.

Preparation and Analysis
of a Copper Complex

PURPOSE OF EXPERIMENT: Prepare a copper complex, and analyze the complex for ammonia and copper.

Chemists frequently make a new compound and then analyze the new material in order to determine its formula. In this experiment you will prepare a copper-ammine complex, which is a coordination complex, and then you will analyze the complex for copper and ammonia. Your ultimate goal is to determine the copper-ammine ratio and the probable formula of the coordination complex and of the compound.

The analysis of the copper-ammine complex for the ammonia, NH_3, content will involve an acid-base "back titration." In this analysis, you will react the complex with an accurately measured volume of standardized hydrochloric acid, HCl, solution. The volume of HCl solution should be sufficient to react with all the ammonia and leave some excess HCl. The excess HCl will then be back reacted with standardized sodium hydroxide, NaOH, solution. The difference between the number of moles of HCl added and the number of moles of excess HCl will give the number of moles of ammonia in your measured quantity of copper-ammine complex. For your information, you cannot titrate the ammonia directly with HCl because of problems with seeing the color changes of the appropriate indicator in the presence of the colored copper-ammine complex.

The copper content of the complex will be determined by a disodium ethylenediaminetetraacetic acid (EDTA) titration. Disodium EDTA or "EDTA," in short, is a hexadentate chelating ligand, a ligand which can potentially bond one copper ion at a maximum of six coordination sites.

$$\begin{array}{ccc} & CH_2-CH_2 & \\ & / \qquad \backslash & \\ HOOCHCH_2-N & & N-CH_2COOH \\ | & & | \\ {}^-OOCH_2C & & CH_2COO^- \end{array}$$

The "copper-EDTA" complex will only form when the hydrogen ion concentration of the solution is carefully controlled. Hence, it is important to carefully adjust and buffer the hydrogen ion concentration or pH of the solution before doing the EDTA titration.

REFERENCES

(1) Kotz, J. C., and Purcell, K. F., _Chemistry and Chemical Reactivity_, Saunders College Publishing, Philadelphia, 1987, sections 2.5, 2.11a, 3.2, 4.1, 4.4, 12.1, 17.1-17.4, 25.3, and 25.4a.

(2) Masterton, W. L., Slowinski, E. J., and Stanitski, C. L., _Chemical Principles_, 6th ed., Saunders College Publishing, Philadelphia, 1985, sections 2.5, 2.6, 3.3, 3.6, 3.7, 8.6, 12.2, 19.6, 19.7, 21.1, 21.2, and 25.2.

EXPERIMENTAL PROCEDURE

(Study this section and the PRE-LABORATORY QUESTIONS before coming to the laboratory. **Wear safety goggles when performing this experiment.**)

A. Preparation of Copper-Ammine Complex

Using a beam balance (Laboratory Methods C), weigh 10. g of copper(II) sulfate pentahydrate, $CuSO_4 \cdot 5H_2O$, into a 250-mL or larger beaker, and record your masses in TABLE 17.1A. Crush the bigger lumps as small as possible with your spatula. The smaller the particles of $CuSO_4 \cdot 5H_2O$, the easier it will be to dissolve the solid. Add 12 mL of distilled water to the beaker, and stir until the solid dissolves (Laboratory Methods K). Place the beaker under a hood, and slowly add 100. mL of concentrated (15 $\underline{M}$) aqueous ammonia, NH_3, with vigorous stirring. **CAUTION:** <u>Concentrated aqueous ammonia causes burns on your skin. Wear rubber gloves when handling it. If any spills or spatters occur onto your skin, rinse the affected area thoroughly with water</u>. After all ammonia has been added and stirred, all material should be in solution. Cool the solution for 5 minutes in an ice bath, but be careful that no water from the ice bath gets into your beaker. Slowly add with stirring 50. mL of ethanol, C_2H_5OH. As the ethanol is added, a fine blue precipitate should form. If no precipitate is observed, add 10. mL more of ethanol. After the addition of ethanol is complete, maintain the solution in the ice bath for 10. additional minutes. Filter the mixture using vacuum filtration through a Buchner funnel (Laboratory Methods I). After the solution has been filtered, continue the suction on the solid for an additional 5 to 10. minutes in order to dry the solid as much as possible. Disconnnect the vacuum line from the filter flask, turn off the water, and remove the funnel from the flask. Invert the funnel over a sheet of weighing paper, and tap the funnel against the paper to remove the precipitate and paper. Place the weighing paper, compound, and filter paper on a preweighed watch glass, for which the mass has been recorded in TABLE 17.1A, and allow them to dry until the next laboratory period. Weigh the watch glass and contents on a beam balance at the beginning of the next laboratory period, and record your mass in TABLE 17.1A.

B. Preparation of Disodium Ethylenediaminetetraacetic Acid Salt Solution

(You should work in pairs to prepare this solution.)

Carefully clean a 500-mL volumetric flask, and finally rinse the flask with a small amount (about 10. mL) of distilled water. Weigh 1.9-2.0 g of disodium EDTA dihydrate (372.24 g/mol) onto weighing paper using an analytical balance, recording your masses in TABLE 17.1B. Carefully transfer all of the EDTA to your volumetric flask. Add approximately 300 mL of distilled water to the flask, and dissolve the EDTA (Laboratory Methods B and K). When the EDTA is completely dissolved, adjust the pH to 10 by using 1 $\underline{M}$ sodium hydroxide, NaOH, solution and wide range pH paper (Laboratory Methods P and Q). Dilute the resulting solution to the 500-mL mark with distilled water. Mix the solution thoroughly. Separate the solution into two portions, one for you and one for your partner, by pouring the solution into two, <u>clean</u>, <u>dry</u> 250-mL Erlenmeyer flasks. Carefully stopper each flask, label the flask, and save the solution for the next laboratory period. Calculate the molarity of the EDTA solution, and write the molarity of the solution on your label and in TABLE 17.2B.

C. Analysis of Copper-Ammine Complex for Ammonia

You should do duplicate analyses if you have sufficient compound. Weigh 0.80-0.85 g of your complex onto weighing paper by using your analytical balance, and record your masses in TABLE 17.1C. Place the weighed sample immediately into a labeled 250-mL Erlenmeyer flask.

Set up a ring stand and buret clamp, clean a buret carefully with soap solution, and rinse it thoroughly with distilled water. Practice reading the buret and manipulating the stopcock or pinch clamp (Laboratory Methods B) before starting this analysis. Finally, drain the buret.

Using your buret, add 35-40. mL of standardized hydrochloric acid, HCl, solution. Record the molarity of the HCl solution and the initial and final buret volumes of HCl in TABLE 17.1C. Stir the mixture by swirling the flask. Add 10. drops of Methyl Orange indicator. Titrate the excess HCl with standardized sodium hydroxide, NaOH, solution. (You and your neighbor may find it convenient to share each other's buret. One person's buret contains HCl whereas the other has NaOH. This sharing will avoid the problem of having to empty the buret of HCl, clean the buret, and refill with NaOH. Be sure to record initial volume of NaOH solution in TABLE 17.1C.) The color change of the indicator is from red-orange to yellow. When your titration is complete, record the final volume of NaOH. Save the resulting solution from the "back titration" for your metal analysis in part D. Repeat the analysis if you have the available material.

D. Analysis of Copper-Ammine Complex for Copper

Quantitatively transfer the solution from the previous ammonia analysis in part C to a 250-mL or larger beaker. Add 0.5 $\underline{M}$ aqueous ammonia, NH_3, slowly from your buret until the pH of the solution is 9-10. Add 20. mL of the pH 10. buffer. If the resulting solution is cloudy, slowly with stirring add more buffer until the solution clears or until the total volume of added buffer is 30. mL. Quantitatively transfer the solution to a 500-mL volumetric flask, and dilute the solution with distilled water to the "mark." Mix the solution thoroughly. Rinse and fill your buret with this solution. Measure precisely 50.00 mL of the solution into a clean 250-mL Erlenmeyer flask. Add an additional 125 mL of distilled water to the flask, and heat the solution to 50-60°C (Laboratory Methods D). Add about 0.1 g of Murexide Tablet indicator (0.1 g is approximately the amount if you have 0.5 cm on the end of your spatula) to the flask. Be careful; you should not add excess indicator. Clean the buret, and fill it with your standardized EDTA solution from part B. Titrate the warm metal ion solution with EDTA to a blue endpoint which persists for at least 30. seconds. Record the initial and final volumes of EDTA solution in TABLE 17.1D. Repeat the analysis with a second 50.00-mL portion of the solution from sample 1 in part C. Then repeat the copper analysis in duplicate again if a second sample is available from part C.

E. Optical Spectra of $[Cu(H_2O)_6]SO_4$ and Copper-Ammine Complex

Prepare the Bausch and Lomb Spectronic 20 spectrophotometer for operation as described in Laboratory Methods N.

Prepare 100. mL of 0.015 $\underline{M}$ $CuSO_4$ using copper(II) sulfate pentahydrate, $CuSO_4 \cdot 5H_2O$, as solute and 100. mL of a solution of your copper-ammine solution which is made by dissolving 0.40 g of complex in sufficient water. Measure the % T (transmittance) of $CuSO_4$ and of your copper-ammine complex at 600., 720., and 800. nm, using the following procedure.

Place a cuvette containing distilled water into the cell compartment. (Alternatively, your instructor may designate that an indexed 13 x 100-mm test tube be used as the cuvette. The frosted identification mark on the cuvette, or the mark on your indexed test tube, must be directed toward the front of the instrument.) Adjust the wavelength dial to 600. nm, and then rotate the 100% control dial until the meter reads 100% T. Remove the cuvette, and replace it with the cuvette containing the solution of $[Cu(H_2O)_6]SO_4$. Record the % T in TABLE 17.1E. Remove the cuvette, and replace it with the cuvette containing the solution of 0.5 $\underline{M}$ aqueous ammonia, NH_3, as the "blank" to adjust the 100% T. Remove the cuvette, and replace it with the solution of the copper-ammine complex solution. Record the % T in TABLE 17.1E. Repeat the procedure for the wavelengths 720. and 800. nm. (To repeat, you turn the wavelength dial to 720. nm, place the distilled water-filled cuvette into the cell compartment, adjust the 100% control dial until the meter reads 100% T, then replace the cuvette with $[Cu(H_2O)_6]SO_4$ solution, read % T, then use aqueous ammonia to adjust 100% T, finally, use copper-ammine complex, and read the % T). Do additional wavelength settings, if requested by your instructor.

Perform the calculations in TABLE 17.2 including the sample calculations for one run on the pages following TABLE 17.2. In addition, plot % T (on the y-axis) versus wavelength in nm (on the x-axis) for $[Cu(H_2O)_6]SO_4$ and for your copper-ammine complex, and estimate for each the wavelength of maximum absorbance (least transmittance) and the color which corresponds to that wavelength of maximum absorbance.

Name _____

Student ID number _____

Section _____ Date _____

Instructor _____

Experiment 17

D A T A

TABLE 17.1. Masses, volumes, and percent transmittances.

		Run 1	Run 2
A	Mass of beaker (g)		
	Mass of beaker and $CuSO_4 \cdot 5H_2O$ (g)		
	Mass of watch glass (g)		
	Mass of watch glass and copper complex (g)		
B	Mass of weighing paper (g)		
	Mass of weighing paper and EDTA (g)		
C	Mass of weighing paper (g)		
	Mass of weighing paper and copper complex (g)		
	Molarity of HCl solution (M)		
	Initial buret reading (HCl solution) (mL)		
	Final buret reading (HCl solution) (mL)		
	Molarity of NaOH solution (M)		
	Initial buret reading (NaOH solution) (mL)		
	Final buret reading (NaOH solution) (mL)		

TABLE 17.1 (continued)

D	Sample 1	Run 1	Run 2
	Initial buret reading (EDTA solution) (mL)		
	Final buret reading (EDTA solution) (mL)		
	Sample 2	Run 1	Run 2
	Initial buret reading (EDTA solution) (mL)		
	Final buret reading (EDTA solution) (mL)		
E	% Transmittance of $[Cu(H_2O)_6]SO_4$ at 600 nm		
	% Transmittance of $[Cu(H_2O)_6]SO_4$ at 720 nm		
	% Transmittance of $[Cu(H_2O)_6]SO_4$ at 800 nm		
	% Transmittance of $[Cu(H_2O)_6]SO_4$ at ___ nm		
	% Transmittance of $[Cu(H_2O)_6]SO_4$ at ___ nm		
	% Transmittance of $[Cu(H_2O)_6]SO_4$ at ___ nm		
	% Transmittance of copper-ammine complex at 600 nm		
	% Transmittance of copper-ammine complex at 720 nm		
	% Transmittance of copper-ammine complex at 800 nm		
	% Transmittance of copper-ammine complex at ___ nm		
	% Transmittance of copper-ammine complex at ___ nm		
	% Transmittance of copper-ammine complex at ___ nm		

Instructor's initials _____

Name ______________________________ **Experiment 17**

Student ID number ______________

Section __________ Date __________

Instructor ________________________

CALCULATIONS

TABLE 17.2. Percent yield, moles of NH_3 and Cu, formula of complex, and absorption characteristics.

A	Mass of $CuSO_4 \cdot 5H_2O$ (g)		
	Mass of copper-ammine complex (g)		
	Percent yield of complex (%)		
B	Mass of EDTA (g)		
	Molarity of EDTA solution ($\underline{M}$)		
		Sample 1	Sample 2
C	Mass of copper-ammine complex (g)		
	Volume of HCl solution added (mL)		
	Moles of HCl used (mol)		
	Volume of NaOH solution used (mL)		
	Moles of NaOH used (mol)		
	Moles of NH_3 in complex (mol)		
	Mass of NH_3 in complex (g)		
	Percent NH_3 by mass in complex (%)		
	Average percent NH_3 by mass (%)		

TABLE 17.2 (continued)

D	Sample 1	Run 1	Run 2
	Volume of EDTA solution used (mL)		
	Moles of copper ion (mol)		
	Mole ratio of copper to ammonia		
	Sample 2	Run 1	Run 2
	Volume of EDTA solution used (mL)		
	Moles of copper ion (mol)		
	Mole ratio of copper to ammonia		
	Average mole ratio of copper to ammonia		
	Formula of complex		
E	Wavelength of maximum light absorption for $[Cu(H_2O)_6]SO_4$ (nm)		
	Color of maximum light absorption for $[Cu(H_2O)_6]SO_4$		
	Wavelength of maximum light absorption for copper-ammine complex (nm)		
	Color of maximum light absorption for copper-ammine complex		

Name ____________________

Student ID number ____________

Section ________ Date ________

Instructor ______________

Experiment 17

CALCULATIONS

Sample calculations

A Percent yield of copper-ammine complex (to be calculated after you have determined the formula of the complex in part **D**)

B Molarity of EDTA solution

C Moles of HCl used for neutralization of NH_3 and "back titration"

Moles of NaOH used in "back titration"

Moles of NH_3 in copper-ammine complex

Sample calculations (continued)

   Mass of NH_3 in copper-ammine complex

   Percent NH_3 by mass in copper-ammine complex

D Moles of copper in copper-ammine complex (Remember that you did not titrate the entire sample of copper-ammine complex.)

   Mole ratio of copper to ammonia

   Formula of copper-ammine complex (assuming one Cu per formula)

Name ______________________________ **Experiment 17**

Student ID number ______________

Section __________ Date __________

Instructor ________________________

PRE-LABORATORY QUESTIONS

1. Which parts of this experiment are completed during the first week?

2. Indicate the six coordination sites of disodium EDTA.

3. What is a "back titration"?

4. Indicate specifically why you cannot titrate the ammonia directly with HCl solution.

5. What is the purpose of adding ethanol in the preparation of the copper-
    ammine complex?

6. If you do not stopper the EDTA solution with sufficient tightness, some
    water might evaporate before the solution is used.   What effect would
    this problem have on your copper analysis?

7. Calculate the mass of $CuSO_4 \cdot 5H_2O$ which is required to prepare 100. mL of
    a 0.015 $\underline{M}$ solution of $CuSO_4$.

Name ______________________

Student ID number ______________

Section ________ Date ________

Instructor ____________________

Experiment 17

POST-LABORATORY QUESTIONS

1. Draw the shape of the copper-ammine coordination complex using your calculated formula.

2. Write a balanced net ionic equation for the reaction of the copper-ammine complex with HCl in part **C**.

3. Calculate the concentration of copper ions in the 500-mL volumetric flask after complete dilution of Sample 1 in part **D**. Show clearly your calculations and reasoning.

4. What is the oxidation state of copper in your copper-ammine complex? Justify your answer.

5. Explain the origin of the color of $[Cu(H_2O)_6]^{2+}$.

6. Why does the color of the wavelength of the light which is absorbed to the maximum not correspond to the color of the specific complex?

7. After light has been absorbed by a complex, the complex goes to a higher energy state. How does the complex return to the ground state? Your answer must give a specific type of radiation.

Crystal Structures and
Close-packing of Spheres

PURPOSE OF EXPERIMENT: Investigate and compare the various ways that spheres can pack together to form metals and ionic solids.

Atoms and ions can be considered to be spheres which pack together in special and reproducible patterns to form solid state crystalline materials. The patterns of the spheres repeat in all three directions. The simplest, basic repeating unit in a crystalline solid is called the unit cell. It is of importance to realize that all crystalline solids, no matter how complex, are described by unit cells. The unit cell must be consistent with the chemical formula of the solid, must indicate the coordination number and geometry of each type of atom or ion, and must generate the crystal structure by simple translation or displacement of the unit cell in three dimensions.

The basic approach to the packing of spheres is to fill space as completely and efficiently as possible. It is impossible to pack spheres together and fill all of the available space. Even though you might think that there are many, many ways to pack spheres together, only a very limited number lead to efficient packing with little unused space. The spaces not filled by spheres of a given size are called voids or interstices. The idea is to keep the number and size of the interstices as small as possible. The most efficient arrangements of spheres of a single size involve close-packing of the spheres. The two common types of close-packing are called hexagonal close-packing and cubic close-packing. A characteristic feature of any arrangement of spheres is the coordination number and geometry of the spheres which are in direct contact with a given sphere. Coordination number is the number of spheres in contact with the given sphere. Coordination geometry is the geometrical arrangement of the spheres in contact with the given sphere.

When spheres of a given size are close-packed, the spaces between the layers of spheres (the voids or interstices) can be filled with smaller spheres. If the spheres represent cations and anions, the structures of ionic solids can be visualized. There are two types of interstices between layers of close-packed atoms - tetrahedral holes or interstices and octahedral holes or interstices. Tetrahedral holes are formed when one sphere in a layer fits over or under three spheres in a second layer. Octahedral holes are formed when three spheres in one layer fit over or under three spheres in a second layer. The two types of holes have different numbers per close-packed sphere, different sizes, and different coordination numbers and coordination geometries. The coordination number of the anion would be the number of cations in contact with the anion. The coordination geometry of the anion would be the geometrical arrangement of the cations which surround the anion. Related statements can be made regarding the coordination number and coordination geometry of the cation.

In this experiment you will build models of the simple cubic, body-centered cubic, and face-centered cubic unit cells, of hexagonal close-packed structures and cubic close-packed structures, and of simple ionic solids.

Experiment 18

REFERENCES

(1) Kotz, J. C., and Purcell, K. F., <u>Chemistry and Chemical Reactivity</u>, Saunders College Publishing, Philadelphia, 1987, section 11.4.

(2) Masterton, W. L., Slowinski, E. J., and Stanitski, C. L., <u>Chemical Principles</u>, 6th ed., Saunders College Publishing, Philadelphia, 1985, sections 11.3 and 11.5.

EXPERIMENTAL PROCEDURE

Your laboratory kit contains styrofoam and/or cork balls of three different sizes and wires for connecting the balls. You should carefully insert the wires into the centers of the balls. At the conclusion of the experiment, place the balls and wires in the plastic bag, and return it to your instructor or the stockroom, as appropriate.

As you perform the experiment, answer each question in the space provided. After you have completed all parts of the experiment, remove the DATA section (TABLE 18.1) of the experiment from your laboratory notebook, and submit it to your instructor. The experiment has no PRE-LABORATORY QUESTIONS or POST-LABORATORY QUESTIONS.

TABLE 18.1. Unit cells, close-packed structures, and crystal lattices.

A. Simple Cubic Unit Cell

Using your largest spheres, construct a single simple cubic lattice unit cell. Using dots to represent centers of atoms, draw a diagram to represent your model.

The simple cubic unit cell contains the equivalent of how many atoms? Show how you derived your answer.

_____

What is the edge length (distance from sphere center to sphere center) of your unit cell in terms of the sphere radius "r"?

_____

Calculate the body diagonal of your unit cell in terms of the sphere radius "r". Show your calculations.

_____

What is the volume of the simple cubic unit cell in terms of the sphere radius "r"?

_____

Calculate the total volume occupied by spheres within the simple cubic unit cell in terms of the sphere radius "r". Hint: Volume of sphere = $4\pi r^3/3$.

_____

If your model were extended equal distances in the x, y, and z directions until a total of 27 spheres had been used, how many unit cells would the large cube contain?

_____

TABLE 18.1 (continued)

B. Body-Centered Cubic Unit Cell

Construct a single body-centered cubic lattice unit cell. It might be helpful to start with the center sphere and work outward. Using dots to represent centers of atoms, draw a diagram to represent your model.

Are the spheres in contact along the cube edges? _____

Are the spheres in contact along the body diagonal? _____

Record the body diagonal in terms of sphere radius "r". _____

What is the edge length of this unit cell in terms of the sphere radius "r"?

_____

The body-centered cubic unit cell contains the equivalent of how many atoms? Show how you derived your answer.

_____

What is the volume of the body-centered cubic unit cell in terms of the sphere radius "r"?

_____

Calculate the total volume occupied by spheres within the body-centered cubic unit cell in terms of the sphere radius "r".

_____

C. Face-Centered Cubic Unit Cell

Construct a single face-centered cubic lattice unit cell. Using dots to represent spheres draw a diagram to represent your model.

TABLE 18.1 (continued)

Are the spheres in contact along the face diagonal?   _____

Record the face diagonal in terms of sphere radius "r".   _____

Calculate the edge length of the unit cell in terms of the
sphere radius "r".

_____

The single face-centered cubic unit cell contains the
equivalent of how many atoms? Show how you derived your
answer.

_____

Express the volume of the face-centered cubic unit cell in
terms of the sphere radius "r".

_____

Calculate the total volume occupied by spheres within the
face-centered cubic unit cell in terms of the sphere radius
"r".

_____

D. Close-packed Layers

    Construct a close-packed layer using your largest spheres as illus-
trated in Diagram 1a in FIGURE 18.1.

What is the coordination number in a two-dimensional
close-packed layer? Hint: How many spheres are in contact
with the sphere marked with an asterisk?   _____

    Construct a second close-packed layer of six spheres as shown in
Diagram 1b. Position this layer over the one above so the second layer
nestles tightly over the first.

Each sphere in the upper layer is in contact with how many
spheres in the lower layer?   _____

Note the vacancy or "hole" formed by one sphere fitting
over/under three spheres. This is known as what type of
hole?   _____

How many spheres surround the above hole?   _____

In which layer is the hole located?   _____

TABLE 18.1 (continued)

E. Hexagonal Close-packed Structure

Construct a third close-packed layer of seven spheres as shown in Diagram 1c. Reverse (invert) the first and second layer, and place the third layer on top so each sphere in the top layer is directly above a sphere in the bottom layer. You have constructed three planes of a hexagonally closest-packed lattice. Additional layers would pack: a,b,a,b,a,b,a,b,a,b,...

The sphere marked with the asterisk in your close-packed layer representing Diagram 1a should be in the middle layer.

Count the total number of spheres touching this sphere in all three layers and record your answer. _____

This number is the coordination number of all atoms in a hexagonal close-packed metal.

Locate a vacancy or "hole" formed by three spheres arranged above three others arranged . This is known as what type of "hole"? _____

Which "hole" is larger, an octahedral hole or a tetrahedral hole? _____

How many spheres surround the larger "hole"? _____

What is the coordination number of an ion in a tetrahedral hole? _____

What is the coordination number of an ion in an octahedral hole? _____

F. Cubic Close-packed Structure

Rearrange the top layer so that each sphere fits over an octahedral hole formed by the lower two layers. Displace all three layers horizontally from each other. Position another sphere of the same size on top so that it forms a fourth layer above a sphere in the bottom layer. You have now formed four layers of a cubic close-packed lattice. Additional layers would pack: a,b,c,a,b,c,a,b,c,...

Locate a central sphere and record the coordination number (number of spheres touching this sphere). _____

The layers of a cubic close-packed lattice actually form what unit cell? _____

Is the stacking of close-packed planes along the edge or body diagonal of the cube? _____

TABLE 18.1   (continued)

G. The NaCl Lattice

Use your largest spheres to construct the chloride lattice of NaCl by building two layers as indicated in Diagram 2a and one layer as indicated in Diagram 2b. Arrange the "b" layer between the two layers.

What type of unit cell is formed?   _____

Is there a sphere in the body center of the cube?   _____

This array represents the lattice of elements like copper and the chloride lattice of NaCl.

Complete the NaCl structure by inserting small spheres between the large spheres along the edges of the unit cell. You may have to loosen the spacing of the large spheres to accommodate the small ones which represent Na^+ ions.

Should you place a small sphere at the body center of the cube?   _____

If so, do so.

Record the number of Cl^- ions about each Na^+ ion.   _____

Record the number of Na^+ ions about each Cl^- ion. (Remember that ions on the cube face are coordinated to other ions in adjacent unit cells.)   _____

The unit cell of NaCl consists of how many sites for ions? _____

How many are Cl^- ion sites?   _____

How many are Na^+ ion sites?   _____

Reconcile this fact with the formula for NaCl. Show calculations. Be explicit.

TABLE 18.1 (continued)

H. The CsCl Lattice

Construct two simple cubic layers as shown in Diagram 3a, and position one of these layers directly over the other. You have just constructed the Cl^- framework of CsCl.

Is this arrangement close-packed? _____

Using spheres of the same size, construct a layer of four spheres as shown in Diagram 3b. Position this layer between the other two so that spheres of the middle layer are above and below holes in the other layers. Locate a cube formed by eight spheres in the upper and lower layers. Since the sphere at the body center represents Cs^+ and the eight corner spheres represent Cl^-, this is a simple cubic unit cell of CsCl.

How many unit cells does your model represent? _____

What is the coordination number of Cs^+? _____

What is the coordination number of Cl^-? Hint: Visualize part of adjacent unit cells. _____

Before dismantling the above model, consider all atoms to be the same, such as iron in crystalline iron.

Name the type of lattice structure. _____

How many unit cells does it represent? _____

Do all spheres have the same coordination number? _____

What is the coordination number? _____

Is this lattice close-packed? _____

Replace the four spheres in the middle layer with four spheres of the smallest diameter. Do the smallest spheres fit snugly into these cubic holes? _____

Consider the large spheres to represent large negative ions such as Cl^- and the small ones to represent small positive ions such as Li^+.

Does the structure represent a situation of high or low stability (negative ions in contact with each other, but not in contact with positive ions)? _____

Is the CsCl or NaCl structure more likely for LiCl? _____

I. The CaF_2 Lattice

Reconstruct the CsCl model as in part **H**, but use only two diagonally opposed middle-sized spheres in the middle layer. This model is now part of the CaF_2 structure. Assemble a third layer of the largest spheres as in Diagram 3a and complete the CaF_2 lattice so that two middle-sized spheres fit in diagonally opposed cubic holes between each pair of nine-sphere layers. You have constructed a portion of the CaF_2 lattice with the largest spheres representing F^-.

What is the coordination number of Ca^{2+}? _____

Determine the coordination number of F^-. Hint: Visualize part of an adjacent unit cell. _____

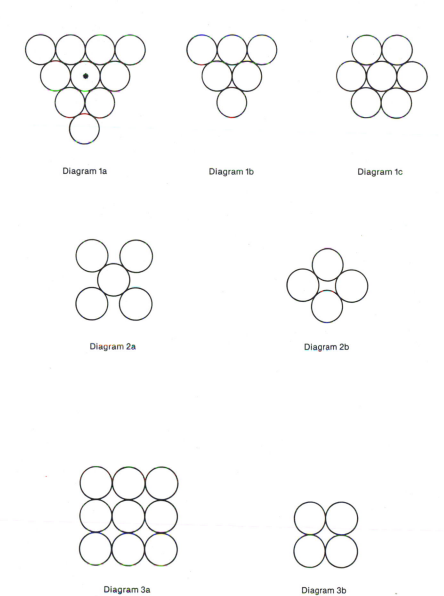

Diagram 1a   Diagram 1b   Diagram 1c

Diagram 2a   Diagram 2b

Diagram 3a   Diagram 3b

Figure 18.1.   Diagrams of layers.

Solubility and Association of Ions

PURPOSE OF EXPERIMENT: Perform a number of chemical reactions involving salts in aqueous solutions to study the effect of formation of a sparingly soluble solid, a sparingly soluble gas, or a weak electrolyte as driving forces for chemical reactions.

When several kinds of positive ions (cations) and several kinds of negative ions (anions) are mixed in aqueous solution, there are a number of different ways in which the cations can combine with the anions to produce different crystalline ionic compounds. The actual association of ions that takes place when a solid separates from a solution at a particular temperature will depend upon the concentration of the ions and the solubilities of the various ionic compounds that can be formed. The first salt for which the concentrations of its ions exceed its solubility limit at the given temperature will normally be the first one to crystallize from solution. These considerations are important when we wish to prepare ionic compounds by crystallization from solutions containing the proper ions. One objective of this experiment is to investigate the factors involved in the association of ions by studying a reaction of two water-soluble salts that place four different ions in solution and produce two other water-soluble salts.

The reaction of two water-soluble salts to produce two other water-soluble salts occurs only because there is a weak driving force for the reaction when one of the product salts is somewhat less soluble than either reactant salt at appropriate temperatures. A small proportion of the product ions are thus removed from solution, and the equilibrium is shifted slightly to the right. A much stronger driving force results for exchange reactions or metathesis reactions in aqueous solution if a large proportion of one or more product ions is removed from solution. The reverse reaction cannot then occur to an appreciable extent, and the equilibrium is shifted strongly toward the right. A second objective of this experiment is to use these principles to devise and carry out simple test tube syntheses of a nickel salt.

A large proportion of one or more product ions is removed from solution in one of three ways, each of which is illustrated below by a complete equation, a complete ionic equation, and a net ionic equation.

Formation of a sparingly soluble solid

$$CuSO_4(aq) + 2\ NaOH(aq) \longrightarrow Cu(OH)_2(s) + Na_2SO_4(aq)$$

$$Cu^{2+}(aq) + SO_4{}^{2-}(aq) + 2\ Na^+(aq) + 2\ OH^-(aq) \longrightarrow$$
$$Cu(OH)_2(s) + 2\ Na^+(aq) + SO_4{}^{2-}(aq)$$

$$Cu^{2+}(aq) + 2\ OH^-(aq) \longrightarrow Cu(OH)_2(s)$$

Formation of a sparingly soluble gas

$$CuSO_3(s) + 2\ HNO_3(aq) \longrightarrow SO_2(g) + Cu(NO_3)_2(aq) + H_2O(l)$$

$$CuSO_3(s) + 2\ H^+(aq) + 2\ NO_3{}^-(aq) \longrightarrow$$
$$SO_2(g) + Cu^{2+}(aq) + 2\ NO_3{}^-(aq) + H_2O(l)$$

$$CuSO_3(s) + 2\ H^+(aq) \longrightarrow SO_2(g) + Cu^{2+}(aq) + H_2O(l)$$

Experiment 19

Formation of a weak electrolyte

$$Cu(OH)_2(s) + 2 HNO_3(aq) ---> 2 H_2O(l) + Cu(NO_3)_2(aq)$$

$$Cu(OH)_2(s) + 2 H^+(aq) + 2 NO_3^-(aq) --->$$
$$2 H_2O(l) + Cu^{2+}(aq) + 2 NO_3^-(aq)$$

$$Cu(OH)_2(s) + 2 H^+(aq) ---> 2 H_2O(l) + Cu^{2+}(aq)$$

Note that a complete ionic equation is obtained by writing all soluble strong electrolytes from a complete equation in their dissociated form. A net ionic equation is obtained by canceling ions that appear on both sides of a complete ionic equation. A complete equation is particularly useful when you want to know what reagents to obtain from the stockroom to carry out a given reaction. A net ionic equation provides in simplest form what species actually take part in a chemical reaction. A complete ionic equation provides the route for getting between a complete equation and a net ionic equation and vice versa. It should be obvious that you must know solubility rules for solids and gases and that you must know common weak acids and bases in order to write the equations described above. In practice you memorized the strong acids and bases and can assume that other acids and bases are weak. A third objective of this experiment is to give you practice observing the formation of various products and writing chemical equations to describe the reactions that occur. To this end your instructor will provide you with a list of reactants that you are to mix.

REFERENCES

(1) C. W. J. Scaife and R. L. Dubs, J. Chem. Educ., **1983**, <u>60</u>, 418.

(2) Kotz, J. C., and Purcell, K. F., <u>Chemistry and Chemical Reactivity</u>, Saunders College Publishing, Philadelphia, 1987, sections 3.1-3.3 and 12.2.

(3) Masterton, W. L., Slowinski, E. J., and Stanitski, C. L., <u>Chemical Principles</u>, 6th ed., Saunders College Publishing, Philadelphia, 1985, sections 3.4-3.6, 12.3, 18.2, and 19.6.

EXPERIMENTAL PROCEDURE

(Study this section and the PRE-LABORATORY QUESTIONS before coming to the laboratory. **Wear safety goggles when performing this experiment.**)

A.   Reaction of Nickel(II) Nitrate and Potassium Chloride

In this experiment you will start with two ionic solids, nickel(II) nitrate hexahydrate, $Ni(NO_3)_2 \cdot 6H_2O$, and potassium chloride, KCl. You will dissolve these solids in water and use either a temperature change or a concentration change by evaporation to crystallize relatively soluble salts from the aqueous solution. The solubilities at different temperatures of the four possible compounds involved are given in TABLE 19.1.

TABLE 19.1.   Solubilities (grams of anhydrous solute per 100. mL of water).

Temperature (°C)	KNO_3	$Ni(NO_3)_2$	KCl	$NiCl_2$
0.0	13.1	78.6	28.3	53.1
20.0	31.3	95.0	34.3	62.2
40.0	63.7	119	40.3	72.6
60.0	109	157	45.5	81.7
80.0	169	190	51.4	86.8
100.0	245	225	55.9	88.1

Obviously you must be able to identify which salt or salts crystallize from solution in order to investigate the effects of solubility in determining what products form. It is easy to distinguish $NiCl_2$ from $Ni(NO_3)_2 \cdot 6H_2O$ because crystals of the former are yellow or yellow-brown whereas crystals of the latter are green. On the other hand, KCl and KNO_3 cannot be distinguished by color because small crystals of both appear white. However, KCl and KNO_3 can be distinguished by the shape of their crystals.

Your instructor will provide crystals of KCl and KNO_3 that were formed under conditions similar to those that you will use in this experiment. Examine carefully the color and shape of each of the crystalline samples, and record your observations in TABLE 19.2A.

Using a beam balance (Laboratory Methods C), weigh 10.9 g of nickel(II) nitrate hexahydrate, $Ni(NO_3)_2 \cdot 6H_2O$, and 5.6 g of potassium chloride, KCl, into a 100-mL or larger beaker. Add 15 mL of distilled water. Warm the mixture with stirring (Laboratory Methods D) until both solids are dissolved completely (Laboratory Methods K).

Let the solution cool to room temperature, swirling it occasionally. Observe, and record in TABLE 19.2A, the color and shape of any crystals that form.

Cool the mixture to near 0°C by placing the beaker in an ice water bath for 5 minutes without stirring or agitation to allow large crystal growth. Then swirl the mixture gently for 3 minutes to achieve temperature equilibrium. (Do not use a stirring rod for stirring.) At the same time chill about 15 mL of distilled water to be used later to wash the crystals.

Filter the cold solution rapidly using either gravity filtration or vacuum filtration (Laboratory Methods I). Rinse the beaker, and wash the crystals first with 4 mL, and then again with 3 mL, of previously chilled distilled water (Laboratory Methods I). Note any changes in the color or shape of the crystals during washing. Use a magnifying glass or microscope, if available. Record your observations in TABLE 19.2A. Let your instructor examine your crystals, and then empty them into a bottle on the reagent bench marked FIRST CROP. Save the filtrate for a later step.

If a solution containing K^+, Ni^{2+}, NO_3^-, and Cl^- ions is evaporated at a given temperature, and water is gradually driven off, the solution will eventually become saturated with respect to one of the compounds that can be formed by ion association. If evaporation is continued, some of that compound will crystallize, removing some of its ions from solution. The other kinds of ions will remain in solution and will increase in concentration as a result of the decrease in volume of water by evaporation.

Transfer the filtrate to an evaporating dish. Evaporate the solution slowly by applying a Bunsen burner flame to one edge of the evaporating dish situated fairly high above the flame (Laboratory Methods L). Record in TABLE 19.2A what you observe just above the edge of the liquid during evaporation.

Continue the evaporation until a thin, crystalline crust forms over the entire surface of the solution. In the meantime, half-fill either a 250-mL or larger beaker or your suction flask with hot water, and insert a 20 x 150-mm test tube into the beaker or suction flask. (A smaller test tube may be required if a smaller than 500-mL suction flask is used.) Grasp the evaporating dish with a towel or "hot hand" (Laboratory Methods L), and rapidly filter the hot mixture <u>into the test tube</u> either by gravity filtration or by vacuum filtration. Collecting the filtrate in the test tube immersed in hot water prevents rapid cooling of the filtrate and allows formation of larger crystals. Let your instructor examine the residue on your filter paper, and then empty the residue into a bottle on the reagent bench marked SECOND CROP.

Remove the test tube from the beaker or the suction flask. Cool the filtrate in the test tube first to room temperature in air and then to near 0.$^{\circ}$C by placing it in an ice water bath for 5 minutes without stirring or agitation to allow large crystal growth. Then swirl the mixture gently for 3 minutes to achieve temperature equilibrium. (Do <u>not</u> use a stirring rod for stirring.) Observe, and record in TABLE 19.2A, the color and shape of any crystals that form.

Filter the cold solution rapidly using either gravity filtration or vacuum filtration, and then wash the crystals with 2 mL of chilled distilled water. Describe any changes in the color or shape of the crystals during washing in TABLE 19.2A. Use a magnifying glass or microscope, if available. Let your instructor examine your crystals, and then empty them into a bottle on the reagent bench marked THIRD CROP.

B. Synthesis of a Nickel Salt

In part **A**, even with the considerable effort for evaporating and causing temperature changes, the conversion was not nearly quantitative. In this part you will use a two-reaction synthesis and a single-reaction synthesis in test tubes to carry out the same overall conversion both quicker and more nearly quantitatively. CAUTION: <u>Use a porcelain casserole for any evaporations; do not evaporate directly from a test tube</u>.

Obtain approval for your procedure in PRE-LABORATORY QUESTION 4 from your instructor, and then carry out your synthesis using a few drops of reagents in a 10 x 75-mm test tube, or whatever is the smallest test tube in your desk set. Record what you observe in TABLE 19.2B.

Obtain approval for your procedure in PRE-LABORATORY QUESTION 5 from your instructor, and then carry out your synthesis using a few drops of reagents in a 10 x 75-mm test tube, or whatever is the smallest test tube in your desk set. Record what you observe in TABLE 19.2B.

C. Exchange Reactions or Metathesis Reactions in Aqueous Solution

For this part your instructor will provide you with a list of reagents that you are to mix. Reagents will be available to you in dropper bottles. Three or four drops of each solution mixed in 10 x 75-mL test tubes, or whatever is the smallest test tube in your desk set, will be sufficient. Watch carefully for any precipitate that is formed, any gas that is evolved, or any other evidence for a reaction, either when the drop of the second reagent first hits or after mixing. Remember that the concentration of the second reagent is initially higher in the region where the drop first hits. Centrifuge (Laboratory Methods J), if necessary, to help you decide whether a precipitate has formed.

Record your observations, and for each case where a reaction occurs, write a balanced complete equation, balanced complete ionic equation, and balanced net ionic equation on the sheet provided by your instructor. Turn in the sheet with your laboratory report.

TABLE 19.2. Colors and shapes of crystals collected and related questions.

(Answer all **QUESTIONS** before handing in your report.)

A Observation Color and shape of KNO_3

Observation Color and shape of KCl

Observation Color and shape of crystals formed on cooling, and changes during washing

QUESTION Based on the color and the shape of the crystals on the filter paper, which compound crystallized? Indicate clearly your reasoning.

QUESTION According to your solubility graph, which compound should have crystallized? Indicate clearly your reasoning.

TABLE 19.2 (continued)

QUESTION   Does your conclusion based on color and shape agree with your conclusion based on your solubility graph? Why, or why not?

Observation   Color of crystals formed during evaporation

QUESTION   Based on the color of the crystals just above the edge of the liquid, which compound crystallized? Indicate clearly your reasoning.

QUESTION   According to your solubility graph, which compound should have crystallized? Indicate clearly your reasoning.

QUESTION   Does your conclusion based on color agree with your conclusion based on your solubility graph? Why, or why not?

Student ID number _____
Section _____ Date _____
Instructor _____   D A T A

TABLE 19.2   (continued)

Observation   Color and shape of crystals formed on cooling, and changes during washing

QUESTION   How do these crystals compare with the first and second crops of crystals?

QUESTION   Based on the color and the shape of the crystals on the filter paper, which compound crystallized? Indicate clearly your reasoning.

QUESTION   According to your solubility graph, which compound should have crystallized? Indicate clearly your reasoning.

QUESTION   Does your conclusion based on color and shape agree with your conclusion based on your solubility graph? Why, or why not?

TABLE 19.2   (continued)

B   **Observation**   Observations for a two-reaction synthesis of $NiCl_2$

QUESTION   Did you obtain the desired product?   What is your evidence?

Observation   Observations for a one-reaction synthesis of $NiCl_2$

QUESTION   Did you obtain the desired product?   What is your evidence?

Instructor's initials _____

Name _____

Student ID number _____

Section _____ Date _____

Instructor _____

PRE-LABORATORY QUESTIONS

1. a. Obtain a good quality piece of graph paper (with at least 10. divisions per inch), and plot on the same graph the solubility-temperature data given in TABLE 19.1 for each of the salts. Solubility should be plotted on the vertical axis. Draw a smooth solubility curve for each compound, and label each curve with the name of the compound.

   b. Use your graph in a., and show clearly your calculations and reasoning to answer each of the following questions.

      (1) Which of the four compounds varies most in solubility (in g solute/100. mL water) with an increase in temperature? Which varies least?

      (2) Which of the four compounds is most soluble, and which is least soluble at room temperature (in g solute/100. mL water)?

      (3) At what temperature do KNO_3 and $NiCl_2$ have the same solubility (in g solute/100. mL water)?

2. a. Write two balanced ionic equations showing how nickel(II) nitrate, $Ni(NO_3)_2$, and potassium chloride, KCl, dissociate into ions when they are dissolved in water.

   b. When aqueous solutions of these two compounds are mixed, what four ions will be present? (Neglect the hydrogen and hydroxide ions from water and ions formed from possible reactions with water.)

   c. Write four balanced net ionic equations showing all of the possible ways in which these ions might associate to form ionic compounds, and name the compounds.

3. Based on your solubility graph, what minimum temperature must be achieved to dissolve 10.9 g of $Ni(NO_3)_2 \cdot 6H_2O$ and 5.6 g of KCl in the initial 15 mL of distilled water, assuming that neither salt affects the solubility of the other? Indicate clearly your calculations and reasoning.

4. Devise a two-reaction test tube synthesis (including detailed procedure) by which you could obtain solid $NiCl_2$ starting with an aqueous solution of $Ni(NO_3)_2$ and other necessary aqueous reagents. Write a balanced complete equation and a balanced net ionic equation describing each reaction.

5. Devise a one-reaction test tube synthesis (including detailed procedure) by which you could obtain solid $NiCl_2$ starting with an aqueous solution of $NiSO_4$ and any other necessary reagent(s). Write a balanced complete equation and a balanced net ionic equation describing the reaction.

Name _____

Student ID number _____

Section _____ Date _____

Instructor_____ POST–LABORATORY QUESTIONS

1. a. Write a balanced complete equation describing the overall reaction that you were able to carry out in part A starting with solid $Ni(NO_3)_2 \cdot 6H_2O$ and solid KCl.

    b. What two variables did you alter to force this reaction to the right?

2. If equal number of moles of the three ions (K^+, NO_3^-, and Cl^-) and half as many moles of Ni^{2+} were in solution and water was removed slowly at $40.^oC$ by evaporation, what compound would you expect to crystallize first? Indicate clearly your calculations and reasoning.

3. Answer POST–LABORATORY QUESTION 2 for a temperature of $90.^oC$ rather than $40.^oC$.

4. What is the molar concentration of each of the ions in part **A** after dissolving 10.9 g of $Ni(NO_3)_2 \cdot 6H_2O$ and 5.6 g of KCl in 15 mL of water if the final volume of solution is 18 mL?

5. You could have prepared $NiCl_2$ using a two-reaction sequence similar to that in the beginning of part **B**, but starting with an aqueous solution of $NiSO_4$ as a reactant in place of $Ni(NO_3)_2$. Write a balanced complete equation and a balanced net ionic equation describing each reaction.

6. You could have prepared $NiCl_2$ using a single reaction similar to that in the end of part **B**, but starting with solid $NiCO_3$ in place of an aqueous solution of $NiSO_4$. Write a balanced complete equation and a balanced net ionic equation describing the reaction.

Freezing-Point Depression
and Molecular Weight

PURPOSE OF EXPERIMENT: Determine the molecular weight of an unknown solid by using the colligative property of freezing point depression.

The colligative properties of a solution, freezing point depression, boiling point elevation, vapor pressure lowering, and osmotic pressure, depend only on the number of solute particles which are dissolved in a given amount of solvent. If both the mass and the number of solute particles in a given amount of solvent are known, the mass of each particle can be calculated. The mass of each particle is also known as the molecular weight, if each particle is a molecule. In this experiment the colligative property of freezing point depression will be used to determine the molecular weight of an unknown molecular solute.

Freezing point depression is the difference in freezing point or freezing temperature between the pure solvent and the solution. The solution will always freeze at a lower temperature than the pure solvent. The freezing point of a pure liquid is the temperature at which the liquid and solid phases are at equilibrium. If heat is slowly removed from a pure liquid, the temperature will decrease until the liquid begins to convert into solid. After this stage is achieved, the temperature remains constant as long as liquid is being converted into solid. When only solid remains, the temperature will begin to decrease again.

When a solution is slowly cooled, pure solid from the solvent will begin to form at some temperature. This temperature at which solid first begins to form is the freezing point of the solution. It is important to note the temperature at which solid first forms because, unlike the case of freezing pure solvent, the temperature of the solution continues to drop at a slower constant rate as additional solvent freezes from the solution. This slow constant temperature drop occurs because the molality of the solution (and thus the freezing point depression) increases as more solvent freezes and the same amount of solute remains dissolved in less and less solvent. The freezing point of the solution can be determined graphically by plotting temperature versus time. If heat is removed from the solution at a constant rate, the temperature will decrease at a constant rate. When solid forms from the solution, the rate of cooling will be constant, but it will be less than when no solid forms or when only the solid cools. Thus, the temperature of the intersection of the line representing the rate of cooling of the solution with the line of lower slope representing the rate of cooling as solid forms will be the freezing point. The difference between the freezing point of the solvent and of the solution, ΔT_{FP}, will depend upon the freezing point depression constant of the specific solvent used, K_{FP}, and the number of moles of solute particles dissolved in one kilogram of solvent, the molality of the solution, $\underline{m}$.

$$\Delta T_{FP} = K_{FP}\underline{m}$$

Thus, if a mole is a number of particles, and if molality is the number of moles of particles per kilogram of solvent, then freezing point depression is used to count the number of solute particles per unit amount of solvent.

In this experiment, you will measure the freezing point of pure naphthalene and the freezing point of a solution of an unknown solid nonelectrolyte in naphthalene. The freezing point depression constant of naphthalene is 6.85°C/molal of solute. These data are then used to calculate the molecular weight of the unknown solid.

REFERENCES

(1) Kotz, J. C., and Purcell, K. F., <u>Chemistry and Chemical Reactivity</u>, Saunders College Publishing, Philadelphia, 1987, sections 12.1-12.3.

(2) Masterton, W. L., Slowinski, E. J., and Stanitski, C. L., <u>Chemical Principles</u>, 6th ed., Saunders College Publishing, Philadelphia, 1985, sections 12.1-12.4.

EXPERIMENTAL PROCEDURE

(Study this section and the PRE-LABORATORY QUESTIONS before coming to the laboratory. **Wear safety goggles when performing this experiment.**)

A. Determination of the Freezing point of Pure Naphthalene

Obtain a 25 x 200-mm test tube and a wire-loop stirrer in a slit, #4 2-hole rubber stopper from the equipment bench. Weigh your clean, dry, 25 x 200-mm test tube on a beam balance to the nearest 0.1 g (Laboratory Methods **C**). Add about 10. g of naphthalene to the test tube, and weigh it again. Record all masses in TABLE 20.1.

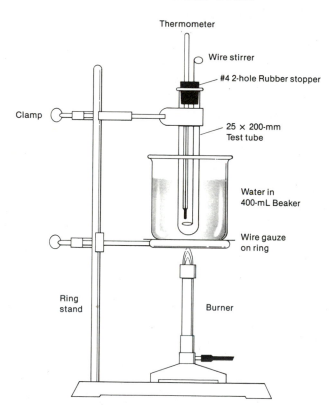

FIGURE 20.1. Apparatus for determining freezing point.

258

Assemble the apparatus shown in FIGURE 20.1. Carefully place the thermometer in the slit rubber stopper. The top of the stopper should be at the 74° mark on the thermometer scale to enable you to make the necessary readings easily.

Heat the water in the beaker sufficiently to cause all of the naphthalene to melt. Be sure that the bulb of the thermometer dips completely below the surface of the molten naphthalene and that all of the naphthalene is below the level of the water in the beaker. Stop heating the water, and allow the apparatus to cool slowly while you stir the naphthalene continuously with a slow up-and-down motion of the wire stirrer. The stirrer must not touch the thermometer. When the thermometer reading has dropped to 82°C, start reading and recording the temperature to the nearest 0.1° every half-minute. Continue to take readings until it becomes impossible to stir. Record your temperature-time data in TABLE 20.1.

Save the test tube and pure naphthalene for your experimental determination of the freezing point of the solution of your unknown solid in naphthalene.

B. Determination of the Freezing Point of the Solution of Your Unknown Solid in Naphthalene

Weigh using an analytical balance approximately 1 g of unknown solid onto a piece of smooth, preweighed weighing paper that has been creased along its diagonals (Laboratory Methods C). Record the masses in TABLE 20.1.

Partially melt the naphthalene in the 25 x 200-mm test tube from part A so that you can remove the stopper with the thermometer and stirrer. Transfer the weighed sample of unknown solid to the test tube containing the naphthalene, being careful not to lose any. Tap the test tube so that all of the unknown solid is down in the naphthalene. Replace the test tube in the beaker of water, and heat the water sufficiently to cause all the naphthalene to melt. When all of the naphthalene is melted, stop heating the water. Insert the thermometer into the test tube to the same depth as in part A. When the temperature has dropped to 82°C, record the temperature of the solution of unknown in naphthalene every half-minute in TABLE 20.1, until stirring becomes impossible.

As soon as you have completed the experiment, warm the test tube gently in your water bath until the naphthalene just melts, and remove and wipe clean the stirrer and the thermometer. Pour the molten naphthalene solution into a jar on the reagent bench marked USED NAPHTHALENE. **Do not pour the naphthalene into the sink!** Return the clean rubber stopper and stirrer to your instructor.

The melting point of naphthalene is lowered 6.85°C per mole of solute in 1000. g of solvent. Perform the calculations in TABLE 20.5 including the sample calculations on the back of TABLE 20.5.

Enthalpy Change of a Chemical Reaction

PURPOSE OF EXPERIMENT: Determine the enthalpy change for the reaction of magnesium metal with hydrochloric acid.

A chemical reaction involves breaking, making, or changing chemical bonds. These alterations in bonding lead to **a change in enthalpy**, ΔH^O, between the products and reactants. This enthalpy change takes into account the change in internal energy in the system and the work done by the system on its surroundings during the reaction. If the reaction occurs at a constant pressure and in an **adiabatic** apparatus (no heat can leave or enter), the change in enthalpy will be equal to the heat absorbed or evolved, reflected by the temperature of the system decreasing or increasing, respectively. The heat of a reaction is determined in the laboratory by measuring this temperature change in a **calorimeter**. If you know the amount of heat required to change the temperature of the calorimeter, reactants, and products, you can calculate the heat of reaction or the enthalpy change.

In this experiment you will measure the heat evolved in the reaction of magnesium with hydrochloric acid.

$$Mg(s) \ + \ 2 \ H^+(aq) \ ---> \ Mg^{2+}(aq) \ + \ H_2(g)$$

In order to calculate the heat of this reaction very precisely, you would need to know the **heat capacities** (heat required to change the temperature of one gram one degree Celsius) of everything that changes temperature during the reaction, and the magnitudes of corrections for the amount of heat lost to the surroundings and for the heat produced by the mechanical stirring of the contents of the calorimeter. To simplify the calculations in this experiment, we will assume that the reactants will not change temperature during the reaction, as they will be consumed quickly, and that the heat absorbed by the hydrogen gas formed will be too small to add appreciably to the other terms. The corrections for heat loss and the other factors related to the apparatus will be included in one term, which we will call the **heat capacity of the calorimeter**.

You will determine a value for the heat capacity of your calorimeter by studying a system for which the heat of reaction is known: the formation of water from its ions in dilute solution.

$$H^+(aq) \ + \ OH^-(aq) \ ---> \ H_2O(aq) \qquad \Delta H^O \ = \ -57,320 \ Joules$$

REFERENCES

(1) Kotz, J. C., and Purcell, K. F., _Chemistry and Chemical Reactivity_, Saunders College Publishing, Philadelphia, 1987, sections 2.5, 3.1-3.4, 4.4, 5.1-5.8, 17.1, 20.1, and 20.4.

(2) Masterton, W. L., Slowinski, E. J., and Stanitski, C. L., _Chemical Principles_, 6th ed., Saunders College Publishing, Philadelphia, 1985, sections 2.5, 2.6, 3.6, 4.3, 5.1-5.5, 8.5, and 19.6.

EXPERIMENTAL PROCEDURE

(Study this section and the PRE-LABORATORY QUESTIONS before coming to the laboratory. **Wear safety goggles when performing this experiment.**)

A. Heat Capacity of the Calorimeter

The calorimeter in this experiment will consist of two nested styrofoam cups. Add to your calorimeter about 50. mL of approximately 2 $\underline{M}$ hydrochloric acid, HCl, solution, measured accurately in your graduated cylinder (Laboratory Methods **B**). Record the volume and actual molarity of the HCl solution in TABLE 21.1**A**. Calculate the volume of sodium hydroxide, NaOH, solution (approximately 2 $\underline{M}$) you need to neutralize your HCl solution.

Measure this volume plus an additional 0.5 mL in your graduated cylinder. Record the volume and the actual molarity of this NaOH solution in TABLE 21.1**A**. Put it into another beaker. Measure the temperatures of both the HCl and the NaOH solutions, and adjust these temperatures using cool tap water surrounding the containers so that they are the same and preferably a few degrees below room temperature. **Be careful to rinse and wipe your thermometer clean before shifting from acid to base or vice versa.** Record these temperatures in TABLE 21.1**A**. Add 2 drops of phenolphthalein to your HCl solution.

Pour the NaOH solution into the calorimeter, and stir **very gently** with your thermometer. Your mixture should be pink if you have added enough base to neutralize the acid. At intervals of 15 seconds from the time of mixing, read your thermometer to the nearest 0.1°C, and record the values in TABLE 21.1**A**. Continue stirring and recording the temperatures <u>until the temperature of the solution has reached a maximum</u> and has decreased for at least two consecutive readings. Do not miss the maximum temperature. It may occur between time increments.

Clean and dry your apparatus, and repeat the experiment once or twice more as indicated by your instructor.

B. Enthalpy Change for Reaction of Magnesium with Hydrochloric Acid

Using an analytical balance, weigh 0.5 g of magnesium, Mg, turnings onto weighing paper, and record your masses in TABLE 21.1**B**. (Do not handle the magnesium with your fingers because grease from them will inhibit the reaction.) Place the magnesium in your clean, dry calorimeter. Add to a 250-mL or larger beaker about 50. mL of the approximately 2 $\underline{M}$ HCl, measured accurately in your graduated cylinder. Record both the volume and the actual molarity of HCl in TABLE 21.1**B**. Then dilute with 50. mL of distilled water, recording the volume in TABLE 21.1**B**. Adjust the temperature of the HCl solution so it is equal to the initial temperature in your previous calibration experiments in part **A**, and record it in TABLE 21.1**B**.

Add the HCl solution to the calorimeter and stir **very gently** with your thermometer. At intervals of 30 seconds from the time of addition of the HCl at first, and at 15-second intervals later, read your thermometer to the nearest 0.1°C, and record the values in TABLE 21.1**B**. Continue stirring and recording the temperatures <u>until the temperature of the solution has reached a maximum</u> and has decreased for at least two consecutive readings. Do not miss the maximum temperature. It may occur between time increments.

Clean and dry your apparatus, and repeat the experiment once or twice more as indicated by your instructor.

Enthalpy, Entropy, and Free Energy Changes for a Chemical Reaction

PURPOSE OF EXPERIMENT: Determine the standard enthalpy, entropy, and free energy changes for dissolving lead(II) chloride, $PbCl_2$, by measuring its change in solubility with temperature.

The prediction of whether or not a given chemical reaction or physical change will occur spontaneously is an important application of thermodynamics. For a process to be spontaneous at constant temperature and pressure, the difference between the **Gibbs' free energy** of products and reactants, ΔG, must be less than zero.

$$\Delta G = \Delta H - T\Delta S < 0$$

The change in free energy, ΔG, takes into account the energy required as heat, the change in **enthalpy**, ΔH, and the ratio of the probabilities of the final and initial states, the change in **entropy**, ΔS. Thus, if the initial and final states do not differ in enthalpy, that is, ΔH is essentially zero, the probability term, $T\Delta S$, determines whether the change occurs. On the other hand, if the probabilities of initial and final states are the same, that is, ΔS is essentially zero, the difference in enthalpy, ΔH, determines whether the reaction or process goes as written.

We will apply these thermodynamic concepts to a solubility equilibrium, one involving the dissolving of a sparingly soluble salt. The extent to which a given salt dissolves in water at a particular temperature is given by an equilibrium constant called the solubility product constant for that temperature. Dissolving of a sparingly soluble salt such as lead(II) chloride, $PbCl_2$, can be represented by the following equation

$$PbCl_2(s) \rightleftharpoons Pb^{2+}(aq) + 2\ Cl^-(aq)$$

for which the solubility product constant is given by

$$K_{sp} = [Pb^{2+}][Cl^-]^2$$

if we assume very dilute solutions in which molar concentrations in the square brackets approximate the activities of the ions.

The extent to which a given salt dissolves in water at a particular temperature also depends on the standard free energy change between products and reactants, ΔG^O, for that temperature. For the dissolving of $PbCl_2$ at constant temperature and pressure, including nonstandard conditions, the following equations are valid.

$$\Delta G = \Delta H - T\Delta S = \Delta G^O + RT \ln Q$$

$Q = [Pb^{2+}][Cl^-]^2$, and the molar concentrations to be used are those giving rise to the free energy change, ΔG. ΔG must be negative for a chemical reaction to be spontaneous at a given temperature and pressure. However, for a chemical reaction at equilibrium as in this experiment, $Q = K_{sp}$, and $\Delta G = 0$ because there is no net tendency toward further change in the system. Thus,

$$\Delta G = \Delta G^O + RT \ln K_{sp} = 0$$

Experiment 22

Rearranging,

$$\Delta G^O = -RT \ln K_{sp}$$

On the other hand, the standard free energy change between products and reactants, ΔG^O, is not zero, but is a measure of the driving force for $PbCl_2$ to dissolve in water under standard conditions. ΔG^O is also related to the standard enthalpy and entropy changes, ΔH^O and ΔS^O, between reactants and products.

$$\Delta G^O = \Delta H^O - T\Delta S^O$$

Setting the right sides of the last two equations equal to each other, and rearranging, we get

$$\ln K_{sp} = -\frac{\Delta H^O}{R}\left(\frac{1}{T}\right) + \frac{\Delta S^O}{R}$$

Over the temperature range of this experiment, ΔH^O and ΔS^O are essentially constant. Therefore, a plot of $\ln K_{sp}$ versus $1/T$ is a straight line having a slope equal to $-\Delta H^O/R$ and an intercept equal to $\Delta S^O/R$. ΔH^O and ΔS^O can thus be calculated from the slope and intercept, respectively, and ΔG^O can be determined either from $\Delta G^O = \Delta H^O - T\Delta S^O$ or $\Delta G^O = -RT \ln K_{sp}$.

In this experiment you will determine the molar solubility of $PbCl_2$ as a function of temperature by means of a precipitation titration to determine the molar concentration of chloride ion at each temperature. From this data you will calculate K_{sp} at each temperature, plot $\ln K_{sp}$ versus $1/T$, determine ΔH^O and ΔS^O from the slope and intercept, and then calculate ΔG^O.

REFERENCES

(1) Kotz, J. C., and Purcell, K. F., <u>Chemistry and Chemical Reactivity</u>, Saunders College Publishing, Philadelphia, 1987, sections 4.4, 5.4, 5.7, 15.2-15.4, and 18.1-18.7.

(2) Masterton, W. L., Slowinski, E. J., and Stanitski, C. L., <u>Chemical Principles</u>, 6th ed., Saunders College Publishing, Philadelphia, 1985, sections 2.6, 5.1, 5.2, 5.5, 12.2, 14.1-14.4, and 18.1-18.3.

EXPERIMENTAL PROCEDURE

(Study this section and the PRE-LABORATORY QUESTIONS before coming to the laboratory. **Wear safety goggles when performing this experiment.**)

Add 35 mL of 0.1 <u>M</u> sodium chromate, Na_2CrO_4, solution to each of three 250-mL Erlenmeyer flasks and to a 250-mL or larger beaker labeled 1, 2, 3, and 4, respectively. **CAUTION: <u>Chromates are recognized carcinogens to the lungs, nasal cavity, and sinuses. Rinse thoroughly with water if there is a spill or spatter</u>.**

Using a beam balance (Laboratory Methods C), weigh 9.0 g of lead(II) chloride, $PbCl_2$, onto weighing paper. **CAUTION: <u>$PbCl_2$ is poisonous; wash your hands thoroughly after use</u>.** Put the $PbCl_2$ sample into a 400-mL beaker, and add 250-mL of distilled water. Heat the mixture to boiling (Laboratory Methods D), stirring to insure the formation of a saturated solution. **CAUTION: <u>Do not stir the solution with your thermometer</u>.** As soon as the mixture has come to a boil, stop heating, and allow the undissolved $PbCl_2$ to settle before removing 25.0-mL portions.

Warm a 100-mL graduated cylinder by pouring hot distilled water into it. After the graduated cylinder is warm, drain the water completely. When the temperature of your $PbCl_2$ solution is down to about 90.°C (Laboratory Methods **E**), record the actual temperature in TABLE 22.1, handle the beaker carefully with a cloth or "hot hand," measure 25.0 mL of the clear $PbCl_2$ solution into your warm graduated cylinder, and quickly transfer it to flask 1. Rinse your graduated cylinder with 5 mL of distilled water, and add it to flask 1. [If there is a large amount of solid $PbCl_2$ floating on the surface when you remove the 25.0-mL portion, you may have to use vacuum filtration (Laboratory Methods **I**) to remove the $PbCl_2$ before transfer to the Na_2CrO_4 solution.]

The original saturated solution cools rapidly at first and then more slowly later. Repeat the procedure of the previous paragraph to transfer 25.0-mL portions at about 75°C to flask 2, at about 50.°C to flask 3, and at about 25°C to beaker 4. Be sure to record your actual temperatures in TABLE 22.1.

Set up a ring stand and buret clamp, clean a buret carefully with soap solution, and rinse it thoroughly with distilled water. Practice reading the buret and manipulating the stopcock or pinch clamp (Laboratory Methods **B**) before starting the following titration. Finally, drain the buret.

Obtain in a graduated cylinder 55 mL of approximately 0.12 **M** silver nitrate, $AgNO_3$, solution. **CAUTION: AgNO_3 is corrosive and stains your skin, and is toxic. Wash your hands thoroughly after use or if there is any spill**. Record the exact molarity of the $AgNO_3$ solution in TABLE 22.1. Rinse your buret with several milliliters of $AgNO_3$ solution. Then add the remainder of the $AgNO_3$ solution, making certain that there are no air bubbles in the buret tip. Record the initial volume of $AgNO_3$ solution in TABLE 22.1. Swirl the mixture in flask 1 slowly as you add the $AgNO_3$ solution to it. Titrate the Cl^- in flask 1 to the first permanent appearance of a faint pink or dull red color of the solution. The endpoint requires some care in discerning it. The indicator is excess $CrO_4{}^{2-}$ which causes the solution to be yellow and leads to the formation of yellow $PbCrO_4(s)$, but the endpoint is the first permanent appearance of a reddish color due to the formation of the first trace of $Ag_2CrO_4(s)$. This solid is red and forms only after all the chloride ion has reacted to form the less soluble $AgCl(s)$. Near the endpoint the mixture of precipitates becomes very granular and settles more rapidly. When the reddish color forming at the liquid surface persists for a longer time, begin adding the $AgNO_3$ solution dropwise, and stop swirling occasionally to observe the color of the supernatant liquid. The solution actually turns red before the solid; therefore, allow the solid to settle at intervals during the latter part of the titration so that you can see the solution clearly. If you titrate until you obtain a red-brown solid, the solution is already clear, and you are as much as 50.% past the endpoint. When you have reached the endpoint, record the final volume of $AgNO_3$ solution in TABLE 22.1. Repeat the procedure in this paragraph to titrate solutions in flasks 2 and 3, and beaker 4, recording initial and final volumes of $AgNO_3$ solution in TABLE 22.1, and obtaining more $AgNO_3$ solution, if necessary. Finally, empty any remaining $AgNO_3$ solution into a bottle marked USED $AgNO_3$.

Perform the calculations in TABLE 22.2 including sample calculations for one run on the back of TABLE 22.2. After calculating K_{sp} for each temperature, you will need to plot $\ln K_{sp}$ on the y-axis versus the reciprocal of the absolute temperature, $1/T$, on the x-axis, draw the best straight line, and determine the slope and intercept before you can calculate ΔH^o, ΔS^o, and ΔG^o. Use good quality graph paper with at least 10. divisions per inch, and choose labels for your axes so that your actual plotting encompasses at least half of your graph paper in each direction. Turn in your graph with your report.

TABLE 22.1. Temperature and volume data.

	Flask 1	Flask 2	Flask 3	Beaker 4
Temperature of PbCl$_2$ solution ($^{\circ}$C)				
Volume of PbCl$_2$ solution (mL)				
Molarity of AgNO$_3$ solution ($\underline{M}$)				
Initial volume of AgNO$_3$ solution (mL)				
Final volume of AgNO$_3$ solution (mL)				

Instructor's initials _____

Name _____

Student ID number _____

Section _____ Date _____

Instructor _____

TABLE 22.2. Solubilities and thermodynamic functions for dissolving $PbCl_2$.

	Flask 1	Flask 2	Flask 3	Beaker 4
Temperature of saturated $PbCl_2$ solution (K)				
Volume of $AgNO_3$ solution used (mL)				
Moles of $AgNO_3$ used (mol)				
Moles of Cl^- in saturated $PbCl_2$ solution (mol)				
Molarity of Cl^- in saturated $PbCl_2$ solution (M)				
Molarity of Pb^{2+} in saturated $PbCl_2$ solution (M)				
K_{sp} for $PbCl_2$				
ln K_{sp} (for $PbCl_2$)				
$1/T$ (K^{-1})				
Slope (K)				
Intercept				
ΔH^o (kJ/mol)				
ΔS^o (J/K mol)				
ΔG^o, 298 K (kJ/mol)				

Experiment 22

Sample calculations for Run __

   Moles of $AgNO_3$ used

   Molarity of Cl^- in saturated $PbCl_2$ solution

   K_{sp} for $PbCl_2$

   Slope

   Intercept

   ΔH^o

   ΔS^o

   ΔG^o

Name ______________________ **Experiment 22**

Student ID number ______________

Section __________ Date __________

Instructor ______________________

PRE-LABORATORY QUESTIONS

1. Show in detail how the equation

$$\ln K_{sp} = -\frac{\Delta H^\circ}{R}\left(\frac{1}{T}\right) + \frac{\Delta S^\circ}{R}$$

can be derived from

$$\Delta G = \Delta G^\circ + RT \ln Q$$

for a chemical reaction at equilibrium.

2. What is the general form of an equation of a straight line? Is the equation that you derived in PRE-LABORATORY QUESTION 1 the equation of a straight line? Justify your answer.

3. If the slope and intercept from a plot of $\ln K_{sp}$ versus $1/T$ for a given salt are -6.0×10^3 K and +3.8, respectively, calculate ΔH°, ΔS°, and ΔG°. Show clearly your calculations.

4. Would your experimental data be of no use if you had started with 12 g of lead(II) chloride, $PbCl_2$, rather than the suggested 9 g? Why, or why not?

5. Why do you warm your graduated cylinder for the transfer of a 25.0-mL portion of saturated solution at the 90.°C temperature?

6. What is the purpose of the sodium chromate, Na_2CrO_4, in this experiment? What additional chemical reaction occurs that requires you to add so much Na_2CrO_4?

7. Why can't you look for a red solid rather than a red solution as the endpoint of your $AgNO_3$ titration?

1. Which factor, ΔH^o or $T\Delta S^o$, is more important for determining the extent of the solubility of $PbCl_2$ in water at a given temperature? Justify your answer.

2. The solubility of $PbCl_2$ in water at $0.^oC$ is 6.73 g/liter and at $100.^oC$ is 33.4 g/liter. Plot these values on the graph of your data. Do they fall on the line you drew through your experimental points? What changes can you suggest in your experimental procedure that might provide better data?

3. a. In your experiments you measured 25.0 mL of $PbCl_2$ solution. To this you added 5.0 mL of rinse water. You then calculated the molarity of the Pb^{2+} and Cl^- based on 25.0 mL of solution. Why didn't you use 30.0 mL, the sum of the two volumes?

    b. Would your K_{sp} for $PbCl_2$ at any given temperature have been higher than, lower than, or unchanged from the correct value if you had based your molarity calculations on a 30.0-mL rather than a 25.0-mL portion? Indicate clearly your reasoning.

4. Write balanced complete equations describing all possible precipitation reactions that occurred during titration with your $AgNO_3$ solution.

5. Would your values of ΔH^o and ΔS^o be higher than, lower than, or unchanged from the correct value for each of the following mistakes? Indicate clearly your calculations and reasoning.

   a. You misread your thermometer and recorded your highest temperature as 89.7°C when it was actually 79.7°C.

   b. You went well past the endpoint before recording your final volume of $AgNO_3$ solution for your 25°C portion.

Chemical Kinetics

PURPOSE OF EXPERIMENT: Determine the orders of reagents and a specific rate constant by measuring the rate of oxygen evolution from the iodide-catalyzed decomposition of hydrogen peroxide, H_2O_2.

The rate of a chemical reaction can be determined by following the rate at which one of the products is formed or the rate at which one of the reactants is consumed. The rate of a given reaction depends on the concentration of reagents and the temperature. The specific dependence of the rate of reaction on the concentration of the reagents is summarized by the **rate law**, which has the general form

$$\text{rate of reaction} = k[A]^m[B]^n...$$

where m and n are the appropriate powers to which the concentrations of reagents A and B, respectively, are raised in order to summarize the experimental data. The proportionality constant, k, is called the **specific rate constant** and is characteristic of a given reaction and the temperature. The rate law must be determined by experiment. It cannot be deduced from the balanced equation of the overall reaction.

In this experiment you will study the rate of decomposition of hydrogen peroxide catalyzed by iodide ion, I^-, or really by I_3^- produced at the very beginning of the reaction by the oxidation of I^- by H_2O_2. The following equation summarizes the reaction

$$2 H_2O_2(aq) \quad ---> \quad 2 H_2O(l) \quad + \quad O_2(g)$$

and an appropriate rate expression is

$$\text{rate} = k[H_2O_2]^m[I^-]^n$$

You will follow the rate of reaction by monitoring the rate of oxygen evolution, and investigate how changes in the concentration of H_2O_2 and I^- affect the rate of oxygen evolution. You will summarize your kinetic data with an appropriate rate law. Your problem is to determine the numerical values of the specific rate constant, k, and exponents of the concentration of reagent terms, m and n.

The exponents of the concentration of reagent terms (the order of the reaction with respect to each reagent) can be determined in several ways. First, you will plot the volume of oxygen evolved versus elapsed time. Such a plot should give a straight line for a zero order reaction. Therefore, the shape of the plot will immediately give you information about whether the reaction is zero order overall or is behaving as a pseudo-zero order reaction, that is, whether m and/or n are zero. You will then draw a line tangent to each curve on your plots and determine the slope of each line which corresponds to the initial rates of reaction for different concentrations of reagents. Qualitative examination of these initial rates will give you an estimate about the order with respect to each reagent. For example, if the rate is unchanged when a reagent concentration is doubled, the reaction is zero order with respect to that reagent; if the rate is doubled when a reagent concentration is doubled, the reaction is first order with respect to that reagent; and if the rate is quadrupled when a reagent concentration is doubled, the reaction is second order with respect to that reagent. Of course, more complex orders including fractional orders are also possible. Quantitative use of the initial rates allows calculation of the order with respect to each reagent and the specific rate constant by the following procedure. The rate expression

$$rate = k[H_2O_2]^m[I^-]^n$$

can be written in a natural logarithmic form.

$$\ln rate = \ln k + m \ln [H_2O_2] + n \ln [I^-]$$

Initial rates will have been determined from three experiments using molar concentrations of H_2O_2 and I^- which can be calculated. Then two pairs of simultaneous equations that have identical terms for the molar concentration of either H_2O_2 or I^- can be written. The exponents, m and n, can be calculated by solving these two pairs of simultaneous equations. Finally, the specific rate constant, k, can be determined after substituting m and n and appropriate reactant concentrations into the original rate expression. For example, suppose that the initial rates shown in TABLE 23.1 are determined for the general reaction

$$A + B \longrightarrow C$$

for which an appropriate rate expression is

$$rate = k[A]^m[B]^n$$

TABLE 23.1. Initial rates as a function of concentrations.

Experiment	[A] (M)	[B] (M)	Initial rate (M/s)
1	0.13	0.31	6.2×10^{-4}
2	0.26	0.31	1.3×10^{-3}
3	0.13	0.61	2.4×10^{-3}

Qualitatively, doubling [A] doubles the rate; therefore, the reaction must be first order with respect to A. Moreover, doubling [B] quadruples the rate; therefore, the reaction must be second order with respect to B. These same results can be determined quantitatively by setting up and solving pairs of simultaneous equations, one pair of which is

$$\ln (6.2 \times 10^{-4}) = \ln k + m \ln (0.13) + n \ln (0.31)$$

$$\ln (1.3 \times 10^{-3}) = \ln k + m \ln (0.26) + n \ln (0.31)$$

Solving from this pair

$$\ln (6.2 \times 10^{-4}) - \ln (1.3 \times 10^{-3}) = m [\ln (0.13) - \ln (0.26)]$$

m = 1.1, or m = 1 assuming it is an integer. It can be determined that n = 2 from the other pair of simultaneous equations. Finally,

$$rate = k[A][B]^2$$

from which

$$k = \frac{rate}{[A][B]^2} = \frac{6.2 \times 10^{-4} \text{ M/s}}{(0.13 \text{ M})(0.31 \text{ M})^2} = 5.0 \times 10^{-2} \text{ M}^{-2} \text{ s}^{-1}$$

Two additional values of k can be calculated from the data from Experiments 2 and 3, and an average can be determined.

REFERENCES

(1) Kotz, J. C., and Purcell, K. F., <u>Chemistry and Chemical Reactivity</u>, Saunders College Publishing, Philadelphia, 1987, sections 4.4, 6.5, 12.1, 13.1-13.8, and 22.2b.

(2) Masterton, W. L., Slowinski, E. J., and Stanitski, C. L., <u>Chemical Principles</u>, 6th ed., Saunders College Publishing, Philadelphia, 1985, sections 2.6, 6.4, 12.2, 16.1-16.7, and 26.1.

EXPERIMENTAL PROCEDURE

(Study this section and the PRE-LABORATORY QUESTIONS before coming to the laboratory. **Wear safety goggles when performing this experiment.**)

You will work in pairs in this experiment unless your instructor states otherwise.

<u>**Following rigorously the procedures given in Laboratory Methods F for inserting glass tubing into rubber stoppers**</u>, carefully insert one medicine dropper into a #5 1-hole rubber stopper (A) and another medicine dropper into a #0 1-hole rubber stopper (B) if you are using a measuring tube as shown in FIGURE 23.1 or into a #00 1-hole rubber stopper if you are using a buret in place of the measuring tube.

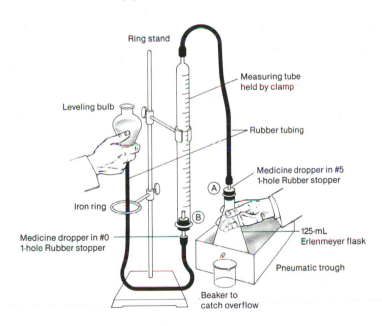

FIGURE 23.1. Apparatus for decomposition of H_2O_2.

Construct the apparatus in FIGURE 23.1. Note that your instructor may designate that a buret be used as the measuring tube. Nearly fill the pneumatic trough with water that is exactly at room temperature. A carboy of room-temperature water may be available for your use. Add hot or cold tap water to adjust the temperature, if necessary. Make certain that the stopcock is open if a buret with a stopcock is used as the measuring tube. Add room-temperature water to the leveling bulb until the measuring tube is filled to the top calibration mark (0.00 mL for the measuring tube or 50.00 mL for a buret) when the water level in the leveling bulb is the same as in the measuring tube. Check the apparatus for leaks by lowering and raising the leveling bulb. If all the joints are tight, the level of the water in the measuring tube (or buret) will return to its original level when the leveling bulb is raised to the same original level.

For volumetric measurements in the following experiments, use your 10-mL graduated cylinder for the H_2O_2 solution and a 25-mL graduated cylinder for water and for the KI solution.

Experiment 1. Record in TABLE 23.2 the concentration of the stock solution of hydrogen peroxide, H_2O_2.

Clean the 125-mL Erlenmeyer flask. Rinse the flask with distilled water, and drain it thoroughly. Add 10.0 mL of 0.10 $\underline{M}$ potassium iodide, KI, solution and 15.0 mL of distilled water to your flask. Swirl the flask in the water in the pneumatic trough for several minutes so that the solution comes to the temperature of the bath. Record the bath temperature in TABLE 23.2. Add 5.0 mL of the stock solution of H_2O_2, and quickly stopper the flask. One student should keep swirling the flask in the bath as vigorously and <u>uniformly</u> as possible throughout the experiment. **It is very important that the swirling be done at a constant rate.** The other student should observe the volume of oxygen evolved at various intervals. The first reading should be taken when 2.0 mL of oxygen has been evolved. To take a reading, one student matches up the water levels by manipulating the leveling bulb and then reads the volume to the nearest 0.1 mL. The other student, still swirling the flask, records the time at the instant the volume is read. Take readings of volume and time at <u>exactly</u> 2-mL intervals until no more oxygen is evolved. Record volumes and times in TABLE 23.2.

Experiment 2. Clean the 125-mL Erlenmeyer flask. Rinse the flask with distilled water, and drain it thoroughly. Add 10.0 mL of 0.10 $\underline{M}$ KI and 10.0 mL of distilled water to your flask. Swirl your flask in the pneumatic trough to bring your reagents to the temperature of the bath. Record the bath temperature in TABLE 23.2. Add 10.0 mL of the stock solution of H_2O_2. Quickly stopper the flask, and take readings at <u>exactly</u> 2-mL intervals until 14 mL of oxygen has accumulated. Record volumes and times in TABLE 23.2.

Experiment 3. Clean the 125-mL Erlenmeyer flask thoroughly again. Add 20.0 mL of 0.10 $\underline{M}$ KI and 5.0 mL of distilled water. Swirl your flask in the pneumatic trough to bring your reagents to the temperature of your bath. Record the bath temperature in TABLE 23.2. Add 5.0 mL of the stock solution of H_2O_2. Quickly stopper the flask, and take readings at <u>exactly</u> 2-mL intervals until 14 mL of oxygen has accumulated. Record volumes and times in TABLE 23.2.

Perform the calculations in TABLE 23.3 including sample calculations on the back of TABLE 23.3. Note that since there is an induction period in the iodide-catalyzed decomposition of H_2O_2, the calculated initial concentrations after dilution closely approximate those for the solution at the beginning of each experiment. Note also that you must plot on the same graph for all three experiments the volume of O_2 evolved (on the y-axis) versus time (on the x-axis) before you can determine each initial rate of reaction (slope of the tangent at the beginning of the reaction). Zero time corresponds to the first volume reading at 2.0 mL. Use good quality graph paper with at least 10. divisions per inch, and choose labels for your axes so that your actual plotting encompasses at least half of your graph paper in each direction. Turn in your graph with your report.

Name _____

Student ID number _____

Section _____ Date _____

Instructor _____

DATA

TABLE 23.2. Volume of O_2 versus time for decomposition of H_2O_2.

Molarity of stock solution of H_2O_2 ($\underline{M}$)			
Temperature of water bath ($^\circ$C)			
Times at 2-mL intervals of O_2 formed (min and s)	Experiment 1	Experiment 2	Experiment 3
2.0 mL			
4.0			
6.0			
8.0			
10.0			
12.0			
14.0			
16.0			
18.0			
20.0			
22.0			
24.0			
26.0			
28.0			
30.0			
32.0			
34.0			
36.0			
38.0			
40.0			
42.0			
44.0			
46.0			
48.0			

Instructor's initials _____

Name ______________________________

Student ID number ______________

Section __________ Date __________

Instructor ________________________

Experiment 23

CALCULATIONS

TABLE 23.3. Orders of reaction and specific rate constant.

	Experiment 1	Experiment 2	Experiment 3
Initial molarity of H_2O_2 after dilution ($\underline{M}$)			
Initial molarity of I^- after dilution ($\underline{M}$)			
Initial rate of reaction (slope of tangent) ($\underline{M}$ H_2O_2/s)			
Order with respect to H_2O_2 (estimated)			
Order with respect to I^- (estimated)			
Order with respect to H_2O_2 (calculated)			
Order with respect to I^- (calculated)			
Specific rate constant (specify the units)			
Average specific rate constant (specify the units)			

Experiment 23

Sample calculations for Experiment __

Initial molarity of H_2O_2 after dilution

Initial rate of reaction (slope of tangent)

Order with respect to H_2O_2 (calculated)

Order with respect to I^- (calculated)

Specific rate constant (specify the units)

1. Why is it important that room-temperature water be used in both the pneumatic trough and the measuring tube (or buret) and that temperature equilibration occurs before adding the H_2O_2 solution in each experiment?

2. When raising the water to the top calibration mark of the measuring tube at the beginning of the experiment, why is it important to adjust the leveling bulb so that water levels in the leveling bulb and the measuring tube are exactly the same, since there is only air present at that point anyway?

3. a. Since there is already oxygen from the air present in the Erlenmeyer flask and rubber tubing before each experiment is started, how can you possibly get an accurate measure of the volume of additional oxygen evolved from the decomposition of H_2O_2?

    b. What law relating to the properties of ideal gases comes to your rescue from the dilemma in a.? How does it help you?

4. Calculate the initial molar concentrations of potassium iodide, KI, after dilution, for each of the three experiments.

5. a. Use the data from Experiments 1 and 3 in TABLE 23.1 to set up two simultaneous equations in the natural logarithmic form of the rate expression, rate = $k[A]^m[B]^n$.

    b. Solve the pair of simultaneous equations in <u>a</u>. in order to calculate n, the order of the reaction with respect to reactant B.

    c. Calculate a value for the specific rate constant, k, using m = 1, your value of n from <u>b</u>., and the data from Experiment 3 in TABLE 23.1. Compare your result with the value given in the Introduction.

Name _____

Student ID number _____

Section _____ Date _____

Instructor _____

POST-LABORATORY QUESTIONS

1. Would the exponents, m and n, in your rate expression be larger than, smaller than, or unchanged from the correct values as a result of each of the following mistakes? Indicate clearly your reasoning.

    a. The water in the pneumatic trough warmed up between Experiment 1 and Experiments 2 and 3.

    b. All of the rinse water was not drained from the reaction flask in Experiments 1 and 2.

    c. A leak occurred around stopper Ⓐ in the mouth of the Erlenmeyer flask (FIGURE 23.1) during Experiment 3.

    d. You used tap water instead of distilled water to dilute your KI solution before adding H_2O_2 solution.

2. Would your specific rate constant, k, be larger than, smaller than, or unchanged from the correct value as a result of each of the following mistakes? Justify your answers.

    a. Your water bath in the pneumatic trough was about 5°C warmer than room temperature for all three of your experiments.

    b. You accidentally substituted a value for n twice as great as it should have been when calculating k.

3. a. Can the iodide-catalyzed decomposition of H_2O_2 occur by a single-step mechanism? Why, or why not?

    b. Propose a two-step mechanism for the iodide-catalyzed decomposition of H_2O_2 that is consistent with your kinetic data.

Chemical Equilibrium

PURPOSE OF EXPERIMENT: Study the properties of a system at chemical equilibrium, and determine the value of the equilibrium constant.

When reactants are first mixed and held at a given temperature, the concentrations of the reactants decrease and the concentrations of the products increase. However, since the products can also be converted back to the reactants, you have opposing changes. With time and under a given set of conditions, you reach a point where the concentrations of all species remain constant, even though the reactions in both directions continue. The chemical system is then said to have attained a state of **chemical equilibrium.** This state persists as long as the conditions remain constant.

In this experiment you will study qualitatively a complex ion equilibrium, and you will determine the equilibrium constant for the formation of $FeNCS^{2+}$ and see if the value stays the same as you change the concentration. The net ionic equation that describes the reaction is

$$Fe^{3+}(aq) \ + \ NCS^-(aq) \ \rightleftharpoons \ FeNCS^{2+}(aq)$$

The following expression for the equilibrium constant describes the state of equilibrium,

$$K \ = \ \frac{[FeNCS^{2+}]}{[Fe^{3+}][NCS^-]}$$

where brackets indicate molar concentrations. Such equilibrium constants are dependent only on temperature. As long as all of the components of the reaction are present, K will be satisfied. To calculate K, it is necessary to determine the molar concentration of Fe^{3+}, NCS^-, and $FeNCS^{2+}$ at equilibrium. Determining quantities of reactants and products present at equilibrium is a problem that often plagues researchers.

Solutions of the $FeNCS^{2+}$ complex ion appear blood-red (red-orange) because they absorb the blue and blue-green wavelengths (centered around 447 nm) of visible light, and therefore you see red and orange wavelengths. The $FeNCS^{2+}$ ion will be the only highly colored species present in the equilibrium mixture if the starting $Fe(NO_3)_3$ solution is acidified to minimize the yellow-brown color of hydrolyzed Fe^{3+} ions.

You will prepare an $FeNCS^{2+}$ solution of known concentration by reacting a given quantity of NCS^- with a very large excess of Fe^{3+}. The very large excess of Fe^{3+} accomplishes two functions. First, it insures that essentially all of the NCS^- is converted to $FeNCS^{2+}$, since the association equilibrium lies very far to the right. Therefore, the equilibrium concentration of NCS^- is nearly zero, and the final concentration of $FeNCS^{2+}$ nearly equals the initial concentration of NCS^- adjusted to the new volume. Second, the very high $[Fe^{3+}]/[NCS^-]$ ratio essentially prevents the formation of higher complexes such as $Fe(NCS)_2^+$.

In this experiment you will study the properties of the system at chemical equilibrium, and you will determine the equilibrium constant for the reaction by using either a <u>Visual</u> <u>Method</u> or an <u>Instrumental</u> <u>Method</u>. These methods are based on the following general ideas. The intensity of the color of a solution of $FeNCS^{2+}$ will depend on the concentration of this ion in the solution and the depth of the solution through which you look. In the <u>Visual</u> <u>Method</u> you will compare a solution of known concentration with a

305

solution of unknown concentration. Your goal will be to change the depth of an $FeNCS^{2+}$ solution of known concentration so that its color intensity matches that of a fixed depth of an $FeNCS^{2+}$ solution of unknown concentration. Then the following relationship will permit you to calculate the concentration of the unknown solution.

$[FeNCS^{2+}]$unknown concentration x depth of unknown solution =

$[FeNCS^{2+}]$known concentration x depth of known solution

For example, if it takes 35 mm of depth of an 0.0010 M $FeNCS^{2+}$ solution to match the color intensity of 70. mm of depth of an unknown $FeNCS^{2+}$ solution, the unknown solution must have a concentration of 0.00050 M $FeNCS^{2+}$. Since you know the initial concentrations of Fe^{3+} and NCS^- and the final equilibrium concentration of $FeNCS^{2+}$, you can calculate the equilibrium concentrations of Fe^{3+} and NCS^- by difference.

The Instrumental Method uses a Bausch and Lomb Spectronic 20. This spectrophotometer is more reliable than the human eye and brain for determining the light aborbance of solutions. It is also more sensitive and thus must operate at lower concentrations. The major components of this instrument are a white-light source, a monochromator (which separates and chooses light of a narrow wavelength range), a sample compartment, a measuring phototube connected to a meter that presents the results visually, as well as associated lenses, slits, filters, and a shutter. The instrument measures the amount of light of a chosen wavelength absorbed by a sample solution compared to the amount of light of the same initial intensity and wavelength absorbed by a reference or "blank" solution. Distilled water is used as a reference in this experiment because it is the solvent and it does not absorb any light at 447 nm, the wavelength of interest. The instrument is constructed in such a way that the absorbance reading on the meter is directly proportional to the concentration of the absorbing species, $FeNCS^{2+}$, in this experiment. This can be expresssed mathematically as

$$A = ab[FeNCS^{2+}]$$

where A is the absorbance reading on the meter, a is an absorptivity constant that depends on the nature of the absorbing species but that is constant for any given species, and b is the pathlength of the sample cell. Since a and b are constant throughout the experiment, A is proportional to $[FeNCS^{2+}]$, and the following equation can be obtained.

$$\frac{A_1}{[FeNCS^{2+}]_1} = \frac{A_2}{[FeNCS^{2+}]_2}$$

Thus, if the absorbance A_1 of a solution of known $[FeNCS^{2+}]_1$ is determined, it can be used along with another observed absorbance A_2 to calculate the unknown $[FeNCS^{2+}]_2$. For example, if the absorbance of a 1.50×10^{-4} M $FeNCS^{2+}$ solution is found to be 0.325, and the absorbance of a second $FeNCS^{2+}$ solution with different concentration but in the same cell and at the same wavelength and temperature is 0.0705, then the concentration of the second $FeNCS^{2+}$ solution must be $(1.50 \times 10^{-4}$ M$)(0.0705/0.325)$ or 3.25×10^{-5} M. Since you know the initial concentrations of Fe^{3+} and NCS^-, as well as how much is required to form $FeNCS^{2+}$ of known concentration, you can determine the final concentrations of Fe^{3+} and NCS^- by difference.

REFERENCES

(1) Kotz, J. C., and Purcell, K. F., <u>Chemistry and Chemical Reactivity</u>, Saunders College Publishing, Philadelphia, 1987, sections 4.4, 12.1, 14.1-14.5, 25.3, and 25.6.

(2) Masterton, W. L., Slowinski, E. J., and Stanitski, C. L., <u>Chemical Principles</u>, 6th ed., Saunders College Publishing, Philadelphia, 1985, sections 2.6, 12.2, 15.2-15.4, 21.1, and 21.3.

EXPERIMENTAL PROCEDURE

(Study this section and the PRE-LABORATORY QUESTIONS before coming to the laboratory. **Wear safety goggles when performing this experiment.**)

A. Properties of a System at Equilibrium

In dilute solution, iron(III) nitrate, $Fe(NO_3)_3$(aq), and sodium thiocyanate, NaNCS(aq), are completely dissociated. When these two solutions are mixed, the following equilibrium is established.

$$Fe^{3+}(aq) + NCS^-(aq) \rightleftharpoons FeNCS^{2+}(aq)$$

Of the five ions in solution, Na^+(aq), NO_3^-(aq), and NCS^-(aq) are colorless, Fe^{3+}(aq) is almost colorless, but $FeNCS^{2+}$(aq) is deep red. Changes in the concentration of $FeNCS^{2+}$ are indicated by changes in the intensity of the color of the solution.

Add 1 mL of 0.20 $\underline{M}$ iron(III) nitrate, $Fe(NO_3)_3$, and 2 mL of 0.20 $\underline{M}$ sodium thiocyanate, NaNCS, to 50. mL of distilled water in a 100-mL beaker, and stir the mixture. Divide this solution equally among four 16 x 150-mm test tubes. To one test tube add 2 mL of 0.20 $\underline{M}$ $Fe(NO_3)_3$, to the second add 2 mL of 0.20 $\underline{M}$ NaNCS, and to the third add a few crystals of mercury(II) nitrate, $Hg(NO_3)_2$. Shake the contents of each tube, and compare the resulting color of each tube with that of the fourth tube, which has been reserved as a standard. (The colors may be seen best when the tubes are held in front of a sheet of white paper.) Record your observations and account for the changes in the color that you observe in TABLE 24.1A.

B. Preparation of Solutions for Determination of Equilibrium Constant

Work in pairs as assigned by your instructor. Record all volumes of reagents used in TABLE 24.1C and D.

1. Visual method. Label five clean, dry 16 x 150-mm test tubes 1 through 5. Obtain from the reagent bench about 40. mL of 2.00 x 10^{-3} $\underline{M}$ iron(III) nitrate, $Fe(NO_3)_3$, solution in 1 $\underline{M}$ HNO_3 in a clean, dry 150-mL or larger beaker. Using a buret (Laboratory Methods B), measure 5.0 mL of the $Fe(NO_3)_3$ solution into each test tube. Next obtain 30. mL of 2.00 x 10^{-3} $\underline{M}$ sodium thiocyanate, NaNCS, solution in another clean, dry 150-mL or larger beaker. Using a second buret, measure 1.0, 2.0, 3.0, 4.0, and 5.0 mL of the NaNCS solution into the tubes labeled 1 through 5, respectively. (To tube 1 add 1.0 mL; to tube 2 add 2.0 mL; etc.) Empty and thoroughly clean the buret containing $Fe(NO_3)_3$ solution, and then fill the buret with distilled water. Add 4.0 mL of distilled water to tube 1; 3.0 mL to tube 2; 2.0 mL to tube 3; and 1.0 mL to tube 4. No distilled water is added to tube 5. Mix each solu- tion thoroughly with a glass stirring rod. Be sure to rinse and dry the stirring rod after mixing each solution.

Prepare a solution of $FeNCS^{2+}$ of "known concentration" by combining 10.0 mL of 2.00×10^{-1} $\underline{M}$ $Fe(NO_3)_3$ in 1 $\underline{M}$ HNO_3 measured with your 10-mL graduated cylinder, 2.00 mL of 2.00×10^{-3} $\underline{M}$ NaNCS measured by buret, and 8.00 mL of distilled water measured by buret. Mix the solution thoroughly with a clean stirring rod. In this solution, the concentration of Fe^{3+} is much, much greater than that of NCS^-. Consequently, formation of $FeNCS^{2+}$ is driven to the right, essentially to completion. You can assume without serious error that all the NCS^- is converted to $FeNCS^{2+}$.

2. Instrumental method. Label five clean, dry 16 x 150-mm test tubes 1 through 5. Obtain from the reagent bench about 40. mL of 5.00×10^{-2} $\underline{M}$ iron(III) nitrate, $Fe(NO_3)_3$, solution in 1 $\underline{M}$ HNO_3 in a clean, dry 150-mL or larger beaker. Using a buret (Laboratory Methods B), measure 5.0 mL of the $Fe(NO_3)_3$ solution into each test tube. Next obtain 30. mL of 5.00×10^{-4} $\underline{M}$ sodium thiocyanate, NaNCS, solution in another clean, dry 150-mL or larger beaker. Using a second buret, measure 1.0, 2.0, 3.0, 4.0, and 5.0 mL of the NaNCS solution into the tubes labeled 1 through 5, respectively. (To tube 1 add 1.0 mL; to tube 2 add 2.0 mL; etc.) Empty and thoroughly clean the buret containing $Fe(NO_3)_3$ solution, and then fill the buret with distilled water. Add 4.0 mL of distilled water to tube 1; 3.0 mL to tube 2; 2.0 mL to tube 3; and 1.0 mL to tube 4. No distilled water is added to tube 5. Mix each solution thoroughly with a glass stirring rod. Be sure to rinse and dry the stirring rod after mixing each solution.

Prepare a solution of $FeNCS^{2+}$ of "known concentration" by combining 10.0 mL of 2.00×10^{-1} $\underline{M}$ $Fe(NO_3)_3$ in 1 $\underline{M}$ HNO_3 measured with your 10-mL graduated cylinder, 2.00 mL of 2.00×10^{-3} $\underline{M}$ NaNCS measured by buret, and 8.00 mL of distilled water measured by buret. Mix the solution thoroughly with a clean stirring rod. In this solution, the concentration of Fe^{3+} is much, much greater than that of NCS^-. Consequently, formation of $FeNCS^{2+}$ is driven to the right, essentially to completion. You can assume without serious error that all the NCS^- is converted to $FeNCS^{2+}$.

C. Determination of Equilibrium Constant by the Visual Method

The concentration of $FeNCS^{2+}$ in test tubes 1 through 5 will be determined by comparing the intensity of the red color in each tube with that of the standard solution of "known concentration" by using the human eye. This can be done by placing the test tube labeled 1 next to a test tube of <u>identical diameter</u> which contains your solution of "known concentration" and by looking down both tubes toward a well illuminated piece of white paper as the background. To facilitate viewing, wrap each test tube with a piece of white paper as provided on the reagent bench. Adjust the depth of the solution of "known concentration" by removing or adding solution by using a disposable pipet. **Do not pour the solution out of the test tube. Do not discard any of the solution of "known concentration."** When the intensities of the colors in the two tubes match, measure the depths of each solution with a ruler (Laboratory Methods **A**) to the nearest millimeter, and record the depths in TABLE 24.1C. Repeat the procedure for matching of color intensities of tubes 2, 3, 4, and 5 with the solution of "known concentration."

D. Determination of Equilibrium Constant by the Instrumental Method

The concentration of $FeNCS^{2+}$ in test tubes 1 through 5 will be determined by measuring the absorbance of each solution and then by comparing the absorbance with that measured for the solution of "known concentration." The absorbances will be measured at a wavelength of 447 nm by using a Bausch and Lomb Spectronic 20 spectrophotometer (Laboratory Methods **N**).

Turn on the instrument, if necessary, and allow it to warm up for 5-10 minutes. Adjust the wavelength control so that the dial reads 447 nm, a wavelength in the region of the visible spectrum that is strongly absorbed by the $FeNCS^{2+}$ ion. **Do not move the wavelength control again during the experiment.**

With no cuvette (or test tube) in the sample holder (so that a shutter automatically falls into the light beam) **and with the sample compartment cover closed**, use the amplifier control to set the meter needle at infinite absorbance (optical density). This adjustment defines a point on the scale where no light from the source is transmitted to the measuring phototube. This adjustment should be checked before each measurement.

Each of you will be provided with two cuvettes. (Alternatively, your instructor may have you index and use two 13 x 100-mm test tubes.) These cuvettes are rather costly and must be handled with utmost care and only on the upper one-quarter of the outside surface. Fill both cuvettes three-quarters full of distilled water. Make certain that there are no air bubbles inside the cuvettes and that there are no fingerprints or moisture on the lower three-quarters of the outside of the cuvettes. Place one of the cuvettes in the sample holder, aligned so that the trademark is facing you, and push the tube down firmly until you hear a click. This click means that the shutter has moved out of the light path and light is now being transmitted through the cuvette and is striking the measuring phototube. With the sample compartment closed, use the light control to set the meter needle at zero absorbance (optical density). This adjustment must also be checked before each measurement.

Recheck the infinite-absorbance and zero-absorbance settings for the same cuvette until the needle returns to each setting without adjustment of either knob. Then replace the first cuvette with the second cuvette and recheck the infinite-absorbance and zero-absorbance settings. If transmittance meter readings on the upper scale for both cuvettes do not agree within 1 percent, obtain assistance from your instructor.

Leave distilled water in one of the cuvettes. Using a solution of known concentration, rinse the other cuvette with two small portions of the solution, discard the rinsings, and fill the cuvette three-quarters full of the solution. Recheck the infinite-absorbance and zero-absorbance meter readings, using the cuvette filled with distilled water. Then insert the cuvette containing the $FeNCS^{2+}$ solution and read the meter to the nearest 0.1 division on the absorbance (optical density) scale. Repeat the checks with the distilled-water cuvette and the absorbance readings with the $FeNCS^{2+}$ solution cuvette several times. Record the absorbance readings in TABLE 24.1D. Repeat the absorbance measurements on solutions from test tubes 1 through 5, and record your absorbance readings in TABLE 24.1D.

Perform the calculations in TABLE 24.2 including sample calculations for one test tube on the back of TABLE 24.2.

Name _____

Student ID number _____

Section _____ Date _____

Instructor _____

TABLE 24.1. Observations and related questions for comparing color intensities of Fe^{3+}-NCS^- equilibrium mixtures.

A **Observation** Change in color intensity of the Fe^{3+}-NCS^--$FeNCS^{2+}$ equilibrium mixture when more Fe^{3+} was added

QUESTION Account for the effect of adding Fe^{3+}.

Observation Change in color intensity of the Fe^{3+}-NCS^--$FeNCS^{2+}$ equilibrium mixture when more NCS^- was added

QUESTION Account for the effect of adding NCS^-.

Observation Change in color intensity of the Fe^{3+}-NCS^--$FeNCS^{2+}$ equilibrium mixture when $Hg(NO_3)_2$ was added

QUESTION Account for the effect of adding $Hg(NO_3)_2$.

TABLE 24.1 (continued)

C	Volume (mL) 2.00 x 10^{-3} $\underline{M}$ Fe(NO$_3$)$_3$	Volume (mL) 2.00 x 10^{-3} $\underline{M}$ NaNCS	Volume (mL) H$_2$O	Depth (mm) Standard	Unknown
Test tube					
1					
2					
3					
4					
5					

D	Volume (mL) 5.00 x 10^{-2} $\underline{M}$ Fe(NO$_3$)$_3$	Volume (mL) 5.00 x 10^{-4} $\underline{M}$ NaNCS	Volume (mL) H$_2$O	Absorbance Readings
Test tube				
1				
2				
3				
4				
5				
Standard solution of "known concentration"				

Instructor's initials _____

Name ________________________________ **Experiment 24**

Student ID number ________________

Section __________ Date __________

Instructor ________________________

CALCULATIONS

TABLE 24.2. Equilibrium concentrations and equilibrium constant.

Experimental Method	
Molarity of $FeNCS^{2+}$ in solution of "known concentration" (M)	

Test tube	Initial number of moles		Total volume (mL)	Number of moles at equilibrium		
	Fe^{3+}	NCS^-		Fe^{3+}	NCS^-	$FeNCS^{2+}$
1						
2						
3						
4						
5						

Test tube	Equilibrium concentrations (M)			Equilibrium constant, K
	Fe^{3+}	NCS^-	$FeNCS^{2+}$	
1				
2				
3				
4				
5				

Average K

Experiment 24

Sample calculations for test tube #___

    Molarity of $FeNCS^{2+}$ in solution of "known concentration"

    Equilibrium concentration of $FeNCS^{2+}$

    Initial number of moles of Fe^{3+}

    Initial number of moles of NCS^-

    Total volume of solution

    Moles of $FeNCS^{2+}$ at equilibrium

    Moles of Fe^{3+} at equilibrium

    Moles of SCN^- at equilibrium

    Equilibrium concentration of Fe^{3+}

    Equilibrium concentration of NCS^-

    Equilibrium constant, K

Name _____   **Experiment 24**

Student ID number _____

Section _____ Date _____

Instructor _____   **PRE-LABORATORY QUESTIONS**

1. Complete the following table for the experimental method which you will
   use in the laboratory.

Test tube	1	2	3	4	5
Volume of $Fe(NO_3)_3$ solution (mL)					
Volume of NaNCS solution (mL)					
Volume of H_2O (mL)					

2. Calculate the concentration of $FeNCS^{2+}$ in the solution of "known concentration" using the quantities of reagents as described in part B.

3. Explain fully why the method used to determine the concentration of
   $FeNCS^{2+}$ in the solution of "known concentration" is valid.

4. What assumption must be made in order to estimate the concentration of
   $FeNCS^{2+}$ in the solution of "known concentration"?

5. Would it be possible to determine the concentration of $FeNCS^{2+}$ in a solution which was prepared from Fe^{3+} and a large excess of NCS^-? Why, or why not?

6. Why are the depths of solutions not used in the Instrumental Method to calculate the concentration of $FeNCS^{2+}$?

7. Why is it important for each of your test tubes to be clean but especially dry?

8. Why is HNO_3 added to the $Fe(NO_3)_3$ solution used in this experiment? How does the HNO_3 accomplish its function?

Instructor _____   **POST-LABORATORY QUESTIONS**

1. For which test tube should your value for the equilibrium constant be most reliable? Justify your selection.

2. For which test tube should your value for the equilibrium constant be least reliable? Justify your selection.

3. a. Using your average value for the equilibrium constant, calculate the equilibrium NCS^- concentration in your solution of "known concentration."

    b. Was the assumption of complete reaction to form $FeNCS^{2+}$ in this solution valid? Why, or why not?

4. Show clearly your reasoning in describing what would be the effect on the calculated equilibrium constant for each of the following mistakes.

   a. Your buret was not rinsed with NaNCS solution before use and still contained traces of distilled water.

   b. A bubble of air remained in the glass jet below your buret throughout your metering of 5-mL portions of NaNCS solutions.

   c. Unknown to you, your ruler was broken off and there was a constant error of 3 mm in all your depth measurements (answer only if you use the visual method).

5. How might the use of tap water in this experiment affect your results?

Acid-Base Titration and Molecular Weight of an Unknown Acid

PURPOSE OF EXPERIMENT: Determine the molecular weight of an unknown acid by reacting the acid with standardized sodium hydroxide.

An acid reacts with a base to form a salt plus water. Many acids have one proton available for reaction with a base such as sodium hydroxide. An acid with one reactive proton per molecule or formula unit is called a monoprotic acid or a monobasic acid. A diprotic acid or a dibasic acid has a maximum of two reactive protons per molecule. A triprotic acid or a tribasic acid has a maximum of three reactive protons per molecule. There are also examples of acids which have four and even more reactive protons per molecule. In contrast, most useful bases have one or two available hydroxide ions per formula unit.

Titration is the process for ascertaining the exact volume of a solution that reacts stoichiometrically according to a balanced chemical equation with a given volume of a second solution. One reagent is added by means of a buret until the endpoint is reached. The endpoint occurs when stoichiometric quantities of reagents have been mixed. The endpoint of a titration for reactions of acids and bases is usually indicated by a third reagent, the indicator, which has an abrupt and distinctive color change at the hydrogen ion concentration which is present after the stoichiometric reaction has occurred. The typical indicator for titrations of strong acids and bases is phenolphthalein. Phenolphthalein is colorless in acidic solution and red (pink in dilute solution) in basic solution. Since it is much easier and distinctive to see a color change from colorless to pink rather than from red to pink to colorless, sodium hydroxide is added by means of the buret to the acid, usually contained in an Erlenmeyer flask.

In this experiment, you will standardize (determine precisely the concentration) a solution of sodium hydroxide, NaOH, using oxalic acid dihydrate, $H_2C_2O_4 \cdot 2H_2O$, as a primary standard acid. A primary standard acid is a solid acid whose mass is an accurate measure of the number of moles of protons the acid will furnish. Oxalic acid, $H_2C_2O_4$, is a diprotic acid and provides two reactive protons per molecule according to the following net ionic equation for the neutralization reaction.

$$H_2C_2O_4(aq) \ + \ 2 \ OH^-(aq) \ \longrightarrow \ C_2O_4^{2-}(aq) \ + \ 2 \ H_2O(l)$$

In the second part of the experiment you will titrate an unknown acid with your standardized NaOH solution using phenolphthalein as the indicator. Your goal will be to calculate the molecular weight of your acid. Your instructor will tell you the number of protons your acid furnishes for reaction with base.

REFERENCES

(1) Kotz, J. C., and Purcell, K. F., *Chemistry and Chemical Reactivity*, Saunders College Publishing, Philadelphia, 1987, sections 2.5, 4.4, 12.1, 16.4, 16.5a, 17.1a, 17.3, and 17.4.

(2) Masterton, W. L., Slowinski, E. J., and Stanitski, C. L., *Chemical Principles*, 6th ed., Saunders College Publishing, Philadelphia, 1985, sections 2.5, 2.6, 12.2, 19.3, 19.4, 19.6, and 19.7.

EXPERIMENTAL PROCEDURE

(Study this section and the PRE-LABORATORY QUESTIONS before coming to the laboratory. **Wear safety goggles when performing this experiment.**)

A. Standardization of a Solution of Sodium Hydroxide

Thoroughly clean, using cleanser and buret brush if necessary, a buret, a graduated cylinder, a 500-mL Florence flask, and three 250-mL Erlenmeyer flasks so that water will drain well from them. Fill your wash bottle with distilled water to use at your desk throughout this experiment. Rinse the apparatus you have cleaned with distilled water. **Do this at the sink, not at the distilled-water tap, and do not waste distilled water.**

Prepare a dilute solution of sodium hydroxide by adding about 50. mL of the stock solution of sodium hydroxide, NaOH, provided on the reagent table, to 250. mL of distilled water contained in a 500-mL Florence flask. Shake the flask to mix the solution. Keep the flask stoppered except when transferring the solution to your buret. Assume the concentration of the NaOH solution you have just prepared is 0.12 $\underline{M}$.

Using an analytical balance (Laboratory Methods **C**), weigh onto weighing paper three separate samples of about your calculated number of grams of the acid (PRE-LABORATORY QUESTION 5). Record your masses in TABLE 25.1A. It is not necessary to weigh exactly the amount calculated, but it is imperative that the mass of each sample be known precisely. After each sample has been weighed, transfer it quantitatively to a clean, clearly labeled 250-mL Erlenmeyer flask, and add 25 mL of distilled water and two drops of phenolphthalein indicator. You will titrate the solution of NaOH you have prepared against each of these solutions of standard acid.

Set up a ring stand and buret clamp, clean a buret carefully with soap solution, and rinse it thoroughly with distilled water. Practice reading the buret and manipulating the stopcock or pinch clamp (Laboratory Methods **B**) before starting the following titration. Rinse your buret once with distilled water and then twice with 5-mL portions of the solution of sodium hydroxide you have prepared, draining the solution through the stopcock into a beaker for waste liquid. Fill the buret, using your funnel, nearly to the top of the graduated portion with the solution of sodium hydroxide, making sure that the stopcock and the glass tip are completely filled with the solution. Touch the inner wall of the beaker to the tip of the buret to remove any hanging drop of solution. Make sure you remove your funnel from the top of your buret.

Make a preliminary titration, using one of your solutions of oxalic acid to see how the neutralization proceeds. Place a sheet of white paper under the flask so that the color of the solution is easily observed. Read and record in TABLE 25.1A the position on the buret of the lowest point of the meniscus of the solution of sodium hydroxide. Swirl the sample in the flask with your right hand, and add sodium hydroxide rather rapidly from the buret until, finally, one drop of the alkaline solution changes the colorless sample to a permanent **pink**. The stopcock of the buret is controlled with your left hand. (The directions given are for a right-handed person.) No drop should be left hanging on the tip, and the walls of the flask should be rinsed with distilled water from the wash bottle to insure mixing of all the base with the acid. You probably will overrun the endpoint in this first titration, but it will provide a useful rough measure of the volume of the sodium hydroxide solution needed to neutralize the acid solution. Read and record the level of the meniscus in the buret in TABLE 25.1A and compute the volume of basic solution used in the titration.

Now titrate the remaining samples of standard acid, being certain each time to refill the buret to nearly the top graduation with the sodium hydroxide solution and to record the buret reading in TABLE 25.1A. In these runs, add the sodium hydroxide from the buret very rapidly up to about 2 mL of the volume that you estimate will be needed on the basis of your first titration. Then carefully add the rest of the base drop by drop so that you can determine the endpoint accurately. Record the data for your titrations in TABLE 25.1A. This solution which you have just standardized is used in part B, so do not waste it or discard it.

B. Determination of Molecular Weight of an Unknown Acid

Your instructor will give you an unknown acid and tell you the mass of sample of unknown acid to weigh and the number of reactive protons per molecule. Record the proposed mass and the number of reactive protons per molecule in TABLE 25.1B. Carefully clean three 250-mL Erlenmeyer flasks. Weigh on an analytical balance, using weighing paper, three separate samples of your unknown, and record your masses in TABLE 25.1B. Quantitatively transfer each precisely weighed sample to the appropriately labeled Erlenmeyer flask. Add about 50. mL of distilled water and two drops of phenolphthalein to each flask, and then titrate each solution with your standardized NaOH solution. (Use the identical procedure described for the standardization of NaOH solution in part A.) The unknown sample may not all go into solution when you add distilled water, but dissolution should be completed as you add NaOH solution. Record your titration data in TABLE 25.1B.

Perform the calculations in TABLE 25.2 including the sample calculations for run 2 on the back of TABLE 25.2.

TABLE 25.1. Mass and volume data for titration of primary standard acid and unknown acid with sodium hydroxide.

		Run 1	Run 2	Run 3
A	Mass of weighing paper (g)			
	Mass of weighing paper and $H_2C_2O_4 \cdot 2H_2O$ (g)			
	Initial reading of buret (mL)			
	Final reading of buret (mL)			
B	Unknown number ___	Run 1	Run 2	Run 3
	Mass of unknown acid suggested for titration (g)			
	Number of reactive protons per molecule of unknown acid			
	Mass of weighing paper (g)			
	Mass of weighing paper and unknown (g)			
	Initial reading of buret (mL)			
	Final reading of buret (mL)			

Instructor's initials _____

CALCULATIONS

TABLE 25.2. Molarity of NaOH and molecular weight of the unknown acid.

	Run 1	Run 2	Run 3
A Mass of $H_2C_2O_4 \cdot 2H_2O$ used (g)			
Moles of $H_2C_2O_4 \cdot 2H_2O$ (mol)			
Number of protons available for reaction with OH^-			
Moles of OH^- which reacted (mol)			
Volume of NaOH solution used (mL)			
Molarity of NaOH solution ($\underline{M}$)			
Average molarity of NaOH ($\underline{M}$)			
B Mass of unknown acid used for titration (g)			
Volume of NaOH solution used (mL)			
Moles of NaOH which reacted (mol)			
Moles of unknown acid (mol)			
Molecular weight of unknown acid (g/mol)			
Average molecular weight of unknown acid (g/mol)			

Experiment 25

Sample calculations for Run 2

A Moles of $H_2C_2O_4 \cdot 2H_2O$ used

   Protons available for reaction with hydroxide ion

   Moles of hydroxide ion which reacted with protons

   Molarity of NaOH solution

B Moles of NaOH which reacted with your unknown acid

   Moles of unknown acid which reacted with hydroxide ion

   Molecular weight of your unknown acid

326

Name ______________________________

Student ID number _____________

Section __________ Date __________

Instructor ________________________

Experiment 25

PRE-LABORATORY QUESTIONS

1. Explain fully why it is good technique to wash the sides of the Erlenmeyer flask during the addition of NaOH solution.

2. What size graduated cylinder, 10-mL or 100-mL, will you need to clean for this experiment? Justify your answer.

3. Why is sodium hydroxide rather than the acid placed in the buret for acid-base titrations which use phenolphthalein as the indicator?

4. If the student is right-handed, which hand holds the flask for swirling
    and which hand controls the stopcock of the buret?

5. If you assume that the NaOH which you will prepare for standardization
    is approximately 0.12 $\underline{M}$, calculate the number of moles and the number of
    grams of $H_2C_2O_4 \cdot 2H_2O$ required to neutralize 35 mL of this NaOH solution.

6. What are the units of molecular weight as used in this experiment?

Name ______________________________

Student ID number ________________

Section __________ Date __________

Instructor _______________________

Experiment 25

POST-LABORATORY QUESTIONS

1. Criticize each of the following techniques.

   a. A student rushes through part **A** but takes great care with part **B**.

   b. A student weighs only enough oxalic acid dihydrate to require about 8 mL of NaOH solution in part **A**.

   c. A student weighs twice the amount of oxalic acid dihydrate needed for one titration. He dissolves the sample in 50.00 mL of water and then titrates 25.00-mL portions of this solution to standardize his base.

2. How would each of the following experimental errors change the experimental molecular weight of the unknown acid? Justify your answers of increase, decrease, or no change.

   a. The oxalic acid contained chemicals which did not react with NaOH.

   b. The NaOH solution evaporated slightly between the standardization and the titration of the unknown acid.

   c. The unknown acid contained an impurity which did not react with NaOH.

3. Calculate the grams of NaOH per milliliter of your standardized solution.

Analysis of an Unknown Mixture
by Acid-Base Titration

PURPOSE OF EXPERIMENT: Determine the percent composition of a two-component mixture of tartaric acid and a component which is unreactive with acids and bases.

Titration is the process for ascertaining the exact volume of a solution, such as a base, that reacts stoichiometrically according to a balanced chemical equation with a given quantity of a second reagent, such as an acid. The goal of this experiment is to analyze a mixture of a known acid and a second component, which is unreactive with base, by using the chemical technique of titration. In this experiment, the known acid is tartaric acid, $C_4H_4O_6H_2$, a diprotic acid or a dibasic acid. A diprotic acid has two protons which can react with sodium hydroxide, NaOH. (You must read the Introduction to Experiment 25 for a more complete description of acid-base titrations, indicators, and the meaning of standardization.) Tartaric acid occurs naturally in many fruits. The potassium salt of tartaric acid, $K_2O_6C_4H_4$, has been known to man since ancient times because this salt forms a fine crystalline crust on the sediment obtained from the fermentation of grape juice. Today, tartaric acid and its salts are commonly used in soft drinks, in bakery products, and in gelatin desserts.

In this experiment as in Experiment 25, you will standardize a solution of sodium hydroxide using oxalic acid dihydrate, $H_2C_2O_4 \cdot 2H_2O$, as the primary standard. This solution of base will then be used to titrate the unknown mixture. Phenolphthalein will be used as the indicator of the endpoint in all titrations. Your goal is to determine the percent composition by mass of the mixture.

REFERENCES

(1) Kotz, J. C., and Purcell, K. F., Chemistry and Chemical Reactivity, Saunders College Publishing, Philadelphia, 1987, sections 2.5, 4.4, 12.1, 16.4, 16.5a, 17.1a, 17.3, and 17.4.

(2) Masterton, W. L., Slowinski, E. J., and Stanitski, C. L., Chemical Principles, 6th ed., Saunders College Publishing, Philadelphia, 1985, sections 2.5, 2.6, 12.2, 19.3, 19.4, 19.6, and 19.7.

EXPERIMENTAL PROCEDURE

(Study this section and the PRE-LABORATORY QUESTIONS before coming to the laboratory. **Wear safety goggles when performing this experiment.**)

A. Standardization of a Solution of Sodium Hydroxide

Thoroughly clean, using cleanser and buret brush if necessary, a buret, a graduated cylinder, a 500-mL Florence flask, and three 250-mL Erlenmeyer flasks so that water will drain well from them. Fill your wash bottle with distilled water to use at your desk throughout this experiment. Rinse the apparatus you have cleaned with distilled water. **Do this at the sink, not at the distilled-water tap, and do not waste distilled water.**

Prepare a dilute solution of sodium hydroxide by adding about 50. mL of the stock solution of sodium hydroxide, NaOH, provided on the reagent table, to 250. mL of distilled water contained in a 500-mL Florence flask. Stopper and shake the flask to mix the solution. Keep the flask stoppered except when transferring the solution to your buret. Assume the concentration of the NaOH solution you have just prepared is 0.12 $\underline{M}$.

Using an analytical balance (Laboratory Methods **C**), weigh onto weighing paper three separate samples of about your calculated number of grams of the acid (PRE-LABORATORY QUESTION 2). Record your masses in TABLE 26.1A. It is not necessary to weigh exactly the amount calculated, but it is imperative that the mass of each sample be known precisely. After each sample has been weighed, transfer it quantitatively to a clean, clearly labeled 250-mL Erlenmeyer flask, and add 25 mL of distilled water and two drops of phenolphthalein indicator. You will titrate the solution of NaOH you have prepared against each of these solutions of standard acid.

Set up a ring stand and buret clamp, clean a buret carefully with soap solution, and rinse it thoroughly with distilled water. Practice reading the buret and manipulating the stopcock or pinch clamp (Laboratory Methods **B**) before starting the following titration. Rinse your buret once with distilled water and then twice with 5-mL portions of the solution of sodium hydroxide you have prepared, draining the solution through the stopcock into a beaker for waste liquid. Fill the buret, using your funnel, nearly to the top of the graduated portion with the solution of sodium hydroxide, making sure that the stopcock and the glass tip are completely filled with the solution. Touch the inner wall of the beaker to the tip of the buret to remove any hanging drop of solution. Make sure you remove your funnel from the top of your buret.

Make a preliminary titration, using one of your solutions of oxalic acid to see how the neutralization proceeds. Place a sheet of white paper under the flask so that the color of the solution is easily observed. Read and record in TABLE 26.1A the position on the buret of the lowest point of the meniscus of the solution of sodium hydroxide. Swirl the sample in the flask with your right hand, and add sodium hydroxide rather rapidly from the buret until, finally, one drop of the alkaline solution changes the colorless sample to a permanent **pink**. The stopcock of the buret is controlled with your left hand. (The directions given are for a <u>right</u>-handed person.) No drop should be left hanging on the tip, and the walls of the flask should be rinsed with distilled water from the wash bottle to insure mixing of all the base with the acid. You probably will overrun the endpoint in this first titration, but it will provide a useful rough measure of the volume of the sodium hydroxide solution needed to neutralize the acid solution. Read and record the level of the meniscus in the buret in TABLE 26.1A, and compute the volume of basic solution used in the titration.

Now titrate the remaining samples of standard acid, being certain each time to refill the buret to nearly the top graduation with the sodium hydroxide solution and to record the buret reading in TABLE 26.1A. In these runs, add the sodium hydroxide from the buret very rapidly up to about 2 mL of the volume that you estimate will be needed on the basis of your first titration. Then carefully add the rest of the base drop by drop so that you can determine the endpoint accurately. Record the data for your titrations in TABLE 26.1A. This solution which you have just standardized is used in part **B**, so do not waste it or discard it.

B.   Determination of Percent Composition of an Unknown Mixture

Your instructor will give you an unknown mixture of tartaric acid and a second unreactive component. Record your "Unknown Number" in TABLE 26.1B. Carefully clean three 250-mL Erlenmeyer flasks. Using an analytical balance and weighing paper, weigh three separate samples of unknown, and record your masses in TABLE 26.1B. Each sample of unknown should weigh in the range of 1.2-1.4 g.   Transfer each precisely weighed sample to the appropriately labeled Erlenmeyer flask.   Add about 50. mL of distilled water and two drops of phenolphthalein to each flask, and then titrate each solution with your standardized NaOH solution.   (Use the identical procedure described for the standardization of NaOH solution in part A.) Record your titration data in TABLE 26.1B.

Perform the calculations in TABLE 26.2 including sample calculations for run 2 on the back of TABLE 26.2.

Name _____
Student ID number _____
Section _____ Date _____
Instructor _____   D A T A

TABLE 26.1. Mass and volume data for titration of primary standard acid and unknown acid with sodium hydroxide.

		Run 1	Run 2	Run 3
A	Mass of weighing paper (g)			
	Mass of weighing paper and $H_2C_2O_4 \cdot 2H_2O$ (g)			
	Initial reading of buret (mL)			
	Final reading of buret (mL)			
B	Unknown number ___	Run 1	Run 2	Run 3
	Mass of unknown acid suggested for titration (g)			
	Number of reactive protons per molecule of unknown acid			
	Mass of weighing paper (g)			
	Mass of weighing paper and unknown (g)			
	Initial reading of buret (mL)			
	Final reading of buret (mL)			

Instructor's initials _____

Name ______________________________

Student ID number ________________

Section __________ Date __________

Instructor ______________________

Experiment 26

CALCULATIONS

TABLE 26.2. Molarity of NaOH and percent composition of unknown mixture.

		Run 1	Run 2	Run 3
A	Mass of $H_2C_2O_4 \cdot 2H_2O$ used (g)			
	Moles of $H_2C_2O_4 \cdot 2H_2O$ used (mol)			
	Number of moles of protons available for reaction with OH^- (mol)			
	Moles of OH^- which reacted (mol)			
	Volume of NaOH solution used (mL)			
	Molarity of NaOH solution ($\underline{M}$)			
	Average molarity of NaOH ($\underline{M}$)			
		Run 1	Run 2	Run 3
B	Mass of unknown mixture used for titration (g)			
	Volume of NaOH solution used (mL)			
	Moles of NaOH which reacted (mol)			
	Moles of protons which reacted with OH^- (mol)			
	Moles of tartaric acid which reacted with OH^- (mol)			
	Mass of tartaric acid present in weighed sample of unknown (g)			
	Percent tartaric acid by mass in unknown (%)			
	Average percent tartaric acid (%)			
	Average percent unreactive component (%)			

Experiment 26

Sample calculations for Run 2

A   Moles of $H_2C_2O_4 \cdot 2H_2O$ used

Moles of protons available for reaction with hydroxide ion

Molarity of NaOH solution

B   Moles of NaOH which reacted with unknown

Moles of tartaric acid which reacted with NaOH

Mass of tartaric acid which reacted with NaOH

Percent tartaric acid in unknown

Name ______________________________ **Experiment 26**

Student ID number ________________

Section __________ Date __________

Instructor ________________________

PRE-LABORATORY QUESTIONS

(In addition to the following questions, you should answer the PRE-LABORATORY QUESTIONS in Experiment 25.)

1. Explain why you can add distilled water to the sample of primary standard for the purposes of washing the walls of the flask and not change the calculated concentration of NaOH.

2. Calculate the number of moles and number of grams of $H_2C_2O_4 \cdot 2H_2O$ required to neutralize 35 mL of 0.12 M NaOH.

3. How many protons will 3.156 g of tartaric acid provide for reaction with NaOH?

4. During the standardization of NaOH, what happens to the water molecules in the primary standard, $H_2C_2O_4 \cdot 2H_2O$? Why don't these water molecules change the concentration of NaOH?

5. What is the meaning of the term, primary standard?

6. Calculate the number of milliliters of 0.12 $\underline{M}$ NaOH required to neutralize 1.3 g of pure tartaric acid.

Name _____

Student ID number _____

Section _____ Date _____

Instructor _____

POST−LABORATORY QUESTIONS

1. Criticize each of the following techniques.

    a. A student fails to rinse the walls of the Erlenmeyer flask with distilled water near the end of a titration.

    b. A student fails to swirl the Erlenmeyer flask during the addition of NaOH solution.

    c. A student fails to have a piece of white paper under the Erlenmeyer flask during the titration.

2. An unknown contains citric acid, $C_6H_8O_7$, a tribasic acid, and sodium sulfate, Na_2SO_4. If 1.731 g of unknown requires 29.32 mL of 0.487 $\underline{M}$ NaOH for complete neutralization of all reactive protons, calculate the percent by mass of citric acid in the sample. Indicate clearly your calculations and reasoning.

3. How would each of the following errors change the experimentally determined percent of tartaric acid in your unknown? Justify your answers of increase, decrease, or no change.

   a. The unknown contained a component which reacted with NaOH but was not tartaric acid.

   b. The NaOH solution was more concentrated than the solution was believed to be according to the standardization experiment.

Ionic Equilibria, pH,

Indicators, and Buffers

PURPOSE OF EXPERIMENT: Use wide-range acid-base indicator paper, solutions of acid-base indicators, and a pH meter to measure the pH of solutions of electrolytes so as to calculate equilibrium constants and/or describe the net reactions that occur as equilibrium is established.

Dissociation of a substance into its ions in aqueous solution is a reversible change. For example, HCN dissociates into H^+ and CN^-, and at the same time, H^+ and CN^- recombine to form HCN. These two reactions are summarized by the following equation.

$$HCN(aq) \rightleftharpoons H^+(aq) + CN^-(aq)$$

In time a state of equilibrium will be established at which point the concentrations of each species show no further change. This chemical equilibrium at a given temperature is described by an equilibrium constant, K, which is defined by its mass-action expression, the product of the concentrations (conventionally expressed in moles/liter) of the products, raised to appropriate powers, divided by the product of the concentrations of the reactants, raised to appropriate powers. Thus,

$$K = \frac{[H^+][CN^-]}{[HCN]}$$

Increasing the concentration of HCN will change the equilibrium concentrations; $[H^+]$ and $[CN^-]$ will increase, and [HCN] will decrease until the equilibrium constant, K, is again satisfied. Likewise, removal of H^+ also changes the equilibrium concentrations; HCN will dissociate to form more CN^- and H^+ until K is again satisfied. Any set of concentrations, when substituted into the mass-action expression, must equal the equilibrium constant.

Equilibria are very rapidly established in aqueous solutions where reversible ionic reactions are taking place. These reactions frequently involve water molecules and the hydrogen and hydroxide ions provided by the ionization of water. Much can be learned about these equilibria and the net changes that occur to establish them from a quantitative study of the hydrogen-ion and hydroxide-ion concentrations. Weak and strong electrolytes can be distinguished by the extent to which they ionize when placed in solution.

In any aqueous solution, the partial dissociation of water must be considered

$$H_2O(l) \rightleftharpoons H^+(aq) + OH^-(aq)$$

for which the equilibrium constant expression is

$$K = \frac{[H^+][OH^-]}{[H_2O]}$$

However, in dilute aqueous solutions the concentration of water, $[H_2O]$, is so large (55.5 $\underline{M}$) compared to the concentration of the H^+ and OH^- ions that it does not change appreciably with changes in the concentration of H^+ or OH^-. Therefore, we can let $[H_2O]$ be effectively constant and define a new constant, K_W, called the water ion-product constant, as being equal to the product of K and $[H_2O]$.

$$K[H_2O] = K_W = [H^+][OH^-] = 1.0 \times 10^{-14} \text{ at } 25^oC$$

In any dilute aqueous solution at 25^oC, the product of the molar concentrations of H^+ and OH^- must always have this value when equilibrium has been established. In acidic solutions $[H^+]$ is greater then 1.0×10^{-7} $\underline{M}$ and the $[OH^-]$ is correspondingly smaller, according to the requirements of the equation for K_W. In basic solutions $[H^+]$ is less than 1.0×10^{-7} $\underline{M}$ and $[OH^-]$ is greater. In neutral solutions both $[H^+]$ and $[OH^-]$ are 1.0×10^{-7} $\underline{M}$.

Concentrations of acids and bases are conventionally expressed in terms of molarity when the concentrations are greater then 0.1 $\underline{M}$. However, for very dilute solutions of acids and bases where exponential numbers are required to describe $[H^+]$ and $[OH^-]$ in a solution, it is more convenient to use a compressed logarithmic scale for $[H^+]$ and to express pH mathematically as

$$pH = -\text{logarithm(base 10)} [H^+]$$

Thus, when $[H^+] = 1.0 \times 10^{-7}$ $\underline{M}$, the pH is 7.0. Note that the pH scale is actually defined more precisely by pH $= -\log A_{H^+}$, where A_{H^+} is the "activity" of hydrogen ion in the given solution. For dilute aqueous solutions, where the total concentration of ions is less than about 0.01 $\underline{M}$, A_{H^+} is approximately equal to $[H^+]$. For concentrated aqueous solutions and for nonaqueous solutions, this approximation is less useful because of the effects of electrical forces between ions. A pOH scale can be similarly defined: pOH $= -\log [OH^-]$. Since the relation $[H^+][OH^-] = K_W = 1.0 \times 10^{-14}$ holds approximately for dilute aqueous solutions (i.e., for solutions dilute enough that activities are approximated by concentrations), one finds that pH + pOH = 14.

Strong electrolytes, for example, hydrochloric acid, HCl, and sodium hydroxide, NaOH, dissociate essentially completely.

$$HCl(aq) \quad ---> \quad H^+(aq) + Cl^-(aq)$$

$$NaOH(aq) \quad ---> \quad Na^+(aq) + OH^-(aq)$$

Thus, $[H^+]$ or $[OH^-]$ is essentially the original concentration of the strong monoprotic acid or the strong monohydroxy base. Weak electrolytes dissociate only partially as was described earlier for HCN,

$$HCN(aq) \quad \rightleftharpoons \quad H^+(aq) + CN^-(aq)$$

and $[H^+]$ is much less than the original concentration of HCN.

Certain cations and anions exhibit acidic or basic behavior in aqueous solution by virtue of hydrolysis. **Hydrolysis** involves the reaction of a cation or anion with water to liberate small amounts of H^+ or OH^-. Hydrolysis is easiest to understand in terms of the Brønsted-Lowry approach to acid-base reactions. This theory defines acids as proton donors and bases as proton acceptors. To every acid, A, there is a corresponding conjugate base, B; that is to say, the conjugate base of an acid is the species formed by the removal of the acidic proton of the acid. For example, the conjugate base of the weak acid, acetic acid, $HC_2H_3O_2$, is the acetate ion, $C_2H_3O_2^-$; the conjugate base of water, H_2O, is the hydroxide ion, OH^-; the conjugate base of ammonium ion, NH_4^+, is ammonia, NH_3; the conjugate base of hydronium ion, H_3O^+, or for short, H^+, is water, H_2O. The hydrolysis reaction which occurs when the salt of a weak acid and strong base (e.g., sodium acetate, $NaC_2H_3O_2$, the salt of acetic acid and sodium hydroxide) is dissolved in water, can be viewed as a reaction between conjugate acids and bases.

$$C_2H_3O_2^-(aq) \; + \; H_2O(l) \; \rightleftharpoons \; HC_2H_3O_2(aq) \; + \; OH^-(aq)$$

$$B_1 \qquad\qquad A_2 \qquad\qquad A_1 \qquad\qquad B_2$$

       Conjugate   Conjugate
       base of A_1,   acid of B_2,
       acetic acid   hydroxide ion

Since acetate ion is a weak base compared to OH^-, the reaction does not go very far to the right; however it still goes far enough to the right that a solution of sodium acetate is basic (pH > 7). Similarly, when the salt of a weak base and a stong acid (e.g., ammonium chloride, NH_4Cl, the salt of the weak base, NH_3, and the strong acid, HCl) is dissolved in water, one has

$$NH_4^+(aq) \; + \; H_2O(l) \; \rightleftharpoons \; NH_3(aq) \; + \; H_3O^+(aq)$$

$$A_1 \qquad\qquad B_2 \qquad\qquad B_1 \qquad\qquad A_2$$

     Conjugate acid   Conjugate base
     of B_1, ammonia,   of A_2, hydronium
     NH_3 ion, H_3O^+

Since NH_4^+ is a weaker acid than H_3O^+, the reaction will go only a little to the right, but still far enough that a solution of ammonium chloride is acidic (pH < 7). Note that in the above reactions H_2O can act as either an acid or a base. Note also that $Na^+(aq)$ and $Cl^-(aq)$ do not affect the above equilibria and are omitted from the net ionic equations for the hydrolysis reactions; such ions are termed "spectator" ions since they do not change the acid-base equilibrium, even though they are present in solution.

Mixing a weak acid and a soluble salt of that acid or mixing a weak base and a soluble salt of that base produces a buffer solution. Within a limited capacity a **buffer** solution is capable of maintaining nearly constant pH, even upon the addition of small amounts of strong acids or bases. The pH characteristics that give a buffer solution this capability are interesting and easy to understand. If a weak acid such as acetic acid, $HC_2H_3O_2$, and a salt of that weak acid, sodium acetate, $NaC_2H_3O_2$, are dissolved in water at roughly the same concentrations (within a factor of 10), the amount of anion produced by dissociation of the acid

$$HC_2H_3O_2(aq) \; \rightleftharpoons \; H^+(aq) \; + \; C_2H_3O_2^-(aq)$$

and the amount of acid produced by hydrolysis of the anion of the salt

$$C_2H_3O_2^-(aq) \; + \; H_2O(l) \; \rightleftharpoons \; HC_2H_3O_2(aq) \; + \; OH^-(aq)$$

are negligible compared to the original concentrations of the undissociated acid and the anion of the salt put into solution. (This is synonymous with neglecting x in the calculation that follows.) This is so because both of these equilibria are suppressed by the common ion effect according to Le Chatelier's principle. The result is that both $[H^+]$ and the pH can be calculated easily as illustrated by the following example. Suppose that 0.010 mole of $HC_2H_3O_2$ and 0.0020 mole of $NaC_2H_3O_2$ are dissolved in enough water to make 1.00 L of solution. Initial concentrations, the change in concentrations to achieve equilibrium, and equilibrium concentrations can be expressed and calculated.

$$HC_2H_3O_2(aq) \rightleftharpoons H^+(aq) + C_2H_3O_2^-(aq)$$

0.010 $\underline{M}$	0	0.0020 $\underline{M}$	initial
$- x\ \underline{M}$	$+ x\ \underline{M}$	$+ x\ \underline{M}$	change
$(0.010 - x)\ \underline{M}$	$x\ \underline{M}$	$(0.0020 + x)\ \underline{M}$	equilibrium
0.010 $\underline{M}$	$9.0 \times 10^{-5}\ \underline{M}$	0.0020 $\underline{M}$	

The pH of the buffered solution can then be determined.

$$pH = -\log [H^+] = -\log (9.0 \times 10^{-5}) = 4.05$$

Then suppose that 0.0010 mole of strong base is added to the buffer solution in order to see whether pH remains nearly constant. The following neutralization reaction goes to completion

$$HC_2H_3O_2(aq) + OH^-(aq) \longrightarrow H_2O(l) + C_2H_3O_2^-(aq)$$

0.010 $\underline{M}$	0.0010 $\underline{M}$	0.0020 $\underline{M}$	initial
$-0.0010\ \underline{M}$	$-0.0010\ \underline{M}$	$+0.0010\ \underline{M}$	change
0.009 $\underline{M}$	0	0.0030 $\underline{M}$	completion

after which a new equilibrium problem can be solved and the pH can be determined.

$$HC_2H_3O_2(aq) \rightleftharpoons H^+(aq) + C_2H_3O_2^-(aq)$$

0.009 $\underline{M}$	0	0.0030 $\underline{M}$	initial
$- y\ \underline{M}$	$+ y\ \underline{M}$	$+ y\ \underline{M}$	change
$(0.009 - y)\ \underline{M}$	$y\ \underline{M}$	$(0.0030 + y)\ \underline{M}$	equilibrium
0.009 $\underline{M}$	$5 \times 10^{-5}\ \underline{M}$	0.0030 $\underline{M}$	

$$pH = -\log (5 \times 10^{-5}\ \underline{M}) = 4.3$$

The pH clearly went up as would be expected upon the addition of strong base; nevertheless, it did not go up very much, and pH remained <u>nearly</u> constant. (Note that if 0.0010 mole of strong base had been added to water, the pH would have risen sharply to 11.0 rather than to the 4.3 value for the buffered solution.) Very stringent buffering is required in biological fluids like blood where the pH must be kept within very narrow limits.

Acid-base indicator solutions (Laboratory Methods P) will be used in this experiment as one means of measuring pH. Most acid-base indicator solutions are solutions of weak organic acids for which the following equilibrium and mass-action expression are appropriate.

$$HIn(aq) \rightleftharpoons H^+(aq) + In^-(aq)$$

acid form   conjugate base form

$$K = \frac{[H^+][In^-]}{[HIn]} \quad \text{and} \quad \frac{[In^-]}{[HIn]} = \frac{K}{[H^+]}$$

For $[H^+] \ll K$, the indicator will be present predominantly in the In^- or conjugate base form, and its color will dominate. For $[H^+] \gg K$, the indicator will be present predominantly in the HIn or acid form, and its color will dominate. For $[H^+] \sim K$, both forms will be present in comparable concentrations. The color changes and the pH range in which they occur are shown in TABLE 27.1 for the indicators used in this experiment. Indicator solutions at various pH values will be available in the laboratory for comparison as standards.

TABLE 27.1.   Indicators used in this experiment.

Indicator	Approximate pH range for color change	Color change
Crystal Violet	0.0 to 1.8	yellow to blue-green
Thymol Blue	1.2 to 2.8	red to yellow
Methyl Orange	3.2 to 4.4	red to yellow
Methyl Red	4.8 to 6.0	red to yellow
Neutral Red	6.8 to 8.0	red to amber
Thymol Blue	8.0 to 9.0	yellow to blue
Alizarin Yellow R	10.1 to 12.0	yellow to red
1,3,5-Trinitrobenzene	12.0 to 14.0	colorless to yellow
		(pH = 13) yellow to orange

In this experiment you will use wide-range acid-base indicator paper, acid-base indicator solutions, and a pH meter to measure the pH of solutions of electrolytes, and you will interpret such pH data in terms of the molar concentrations of H^+ and OH^- present. From these concentrations you will calculate equilibrium constants for some of the equilibria.

REFERENCES

(1)   Kotz, J. C., and Purcell, K. F., _Chemistry and Chemical Reactivity_, Saunders College Publishing, Philadelphia, 1987, sections 4.1, 4.4, 12.1, 16.1-16.5, 17.2, and 17.4.

(2)   Masterton, W. L., Slowinski, E. J., and Stanitski, C. L., _Chemical Principles_, 6th ed., Saunders College Publishing, Philadelphia, 1985, sections 2.6, 3.7, 12.2, 15.4, 19.1-19.6, and 20.1-20.5.

EXPERIMENTAL PROCEDURE

(Study this section and the PRE-LABORATORY QUESTIONS before coming to the laboratory. **Wear safety goggles when performing this experiment.**)

A.   Determining pH with Indicator Solutions

1. Acids, bases, and hydrolysis. Record in TABLE 27.2A1 the colors of indicator solution standards at various pH values that are provided in the laboratory. These will serve as references for all use of indicator solutions in this experiment.

Determine the pH of 0.10 M solutions of each of the following

Acetic acid, $HC_2H_3O_2$	Citric acid, $C_6H_8O_7$
Aluminum chloride, $AlCl_3$	Hydrochloric acid, HCl
Ammonium chloride, NH_4Cl	Sodium acetate, $NaC_2H_3O_2$
Aqueous ammonia, NH_3	Sodium carbonate, Na_2CO_3
Boric acid, H_3BO_3	Sodium hydrogen carbonate, $NaHCO_3$
Borax, $Na_2B_4O_7$	Sodium hydroxide, NaOH

using indicator solutions, and record all observations of color and pH data in TABLE 27.2A1. Into a clean 13 x 100-mm or larger test tube, pour approximately 1 mL of the solution to be tested. Add 1 drop of Neutral Red indicator solution. Mix and determine by the resulting color whether the pH of the solution is greater than 8 (i.e., an amber color) or less than 6.8 (i.e., a red color). If the pH is greater than 8, to another different 1 mL sample of the solution to be tested, add 1 drop of Thymol Blue indicator solution, and compare the color produced with the color standards of Thymol Blue for pH values 8, 9, and 10. If the pH of this last sample appears to be greater than 10, carry out a similar test on a new 1-mL sample of the solution, using 1 drop of the Alizarin Yellow R indicator solution. If this last test indicates a pH greater than 12, try a new 1-mL sample of the solution with 1,3,5-Trinitrobenzene as the indicator. If your first test with the solution using Neutral Red as an indicator showed that the pH was less than 6.8, carry out a similar procedure on different 1-mL samples of the solution, using Methyl Red, Methyl Orange, and Crystal Violet indicator solution, as required.

2. **Buffer solution.** Into a clean 100-mL beaker, add 10. mL of distilled water and 3 drops of Methyl Orange indicator solution. Record the color you observe in TABLE 27.2**A2**. Then add 0.10 $\underline{M}$ hydrochloric acid, HCl, solution dropwise until the indicator turns red, counting the drops required and recording the number of drops in TABLE 27.2**A2**. Then clean the beaker, and measure into it 1.0 mL of 0.10 $\underline{M}$ acetic acid, $HC_2H_3O_2$, solution, 1.0 mL of 0.10 $\underline{M}$ sodium acetate, $NaC_2H_3O_2$, solution, and 8.0 mL of distilled water. Add 3 drops of Methyl Orange indicator solution, and record the color you observe in TABLE 27.2**A2**. Then add 0.10 $\underline{M}$ HCl solution dropwise until the indicator solution turns red, counting the drops required and recording the number of drops in TABLE 27.2**A2**.

B. Determining pH with Wide-Range pH Paper

Determine the pH of the following solutions

Aspirin	Non-cola carbonated drink
Buffered aspirin	Orange juice
0.10 $\underline{M}$ Citric acid	Soap solution
Cola carbonated drink	Tomato juice
Lemon juice	Vinegar

using wide-range pH paper, and record all observations of color and pH data in TABLE 27.2**B**. Into a clean 13 x 100-mm or larger test tube, pour approximately 1 mL of the solution to be tested. Dip a <u>clean</u> stirring rod into the solution, and touch it to a small piece of wide-range pH paper placed on a clean watch glass. Compare the color produced with the colors on the chart supplied with the paper, and estimate the pH of the solution. The stirring rod must be cleaned after each use by rinsing it with distilled water and drying it.

Determine the pH of a soil sample by the following procedure. Pour a sample of soil into a clean 13 x 100-mm test tube to a depth of about 2 cm. Add twice the volume of <u>freshly boiled</u> distilled water, and stir thoroughly. Use either gravity filtration or vacuum filtration (Laboratory Methods **I**) to filter the mixture. Then test the pH of the filtrate with wide-range pH paper, and record the result in TABLE 27.2**B**.

C. Determining pH with a pH Meter

Determine the pH of 20-mL portions of the following solutions contained in a 50-mL beaker

0.10 $\underline{M}$ Acetic acid, $HC_2H_3O_2$
0.010 $\underline{M}$ $HC_2H_3O_2$
0.0010 $\underline{M}$ $HC_2H_3O_2$

using a pH meter (Laboratory Methods **Q**), if available, and record your pH data in TABLE 27.2**C**. Note that you must rinse the electrodes before and after use and that the electrodes must be handled very carefully. The 0.010 $\underline{M}$ solution is prepared by diluting 2.0 mL of 0.10 $\underline{M}$ solution to 20.0 mL, each volume being measured in the smallest possible graduated cylinder. The 0.0010 $\underline{M}$ solution is prepared by diluting 0.2 mL of 0.10 $\underline{M}$ solution, measured in a pipet (Laboratory Methods **B**), to 20.0 mL, measured in the smallest possible graduated cylinder.

TABLE 27.2. Indicator colors, pH data, and related questions.

(Answer all QUESTIONS before handing in your report.)

A1	Indicator standard solution	pH	Color
	Crystal Violet		
	Thymol Blue		
	Methyl Orange		
	Methyl Red		
	Neutral Red		
	Thymol Blue		
	Alizarin Yellow R		
	1,3,5-Trinitrobenzene		

TABLE 27.2 (continued)

0.10 M Solutions tested	Indicator solution(s) used	Color	pH
Acetic acid, $HC_2H_3O_2$			
Aluminum chloride, $AlCl_3$			
Ammonium chloride, NH_4Cl			
Aqueous ammonia, NH_3			
Boric acid, H_3BO_3			
Borax, $Na_2B_4O_7$			
Citric acid, $C_6H_8O_7$			
Hydrochloric acid, HCl			

TABLE 27.2 (continued)

0.10 M Solutions tested	Indicator solution(s) used	Color	pH
Sodium acetate, $NaC_2H_3O_2$			
Sodium carbonate, Na_2CO_3			
Sodium hydrogen carbonate, $NaHCO_3$			
Sodium hydroxide, NaOH			

QUESTION Write a balanced net ionic equation to explain the observed pH for each of the solutions tested.

TABLE 27.2   (continued)

A2   **Observation**   Color   of   Methyl   Orange   indicator   solution in distilled water

Observation   Number of drops of 0.10 $\underline{M}$ HCl to turn indicator red

Observation   Color of Methyl Orange indicator solution in buffer solution of $HC_2H_3O_2$ and $NaC_2H_3O_2$

Observation   Number of drops of 0.10 $\underline{M}$ HCl to turn indicator red

QUESTION   How do your observations indicate that the buffer solution maintains near constancy of pH?

QUESTION   How do your observations indicate that a buffer solution has a finite capacity?

QUESTION   Why did the color of the Methyl Orange change at all in the presence of the buffer components?

TABLE 27.2   (continued)

B	Solutions tested	pH
	Aspirin	
	Buffered aspirin	
	0.10 M Citric acid	
	Cola carbonated drink	
	Lemon juice	
	Non-cola carbonated drink	
	Orange juice	
	Soap solution	
	Tomato juice	
	Vinegar	

QUESTION   What can you conclude about the pH of most solutions that we
            ingest? When do we ingest solutions in the other half of the
            pH range?

Observation   pH of soil sample

TABLE 27.2   (continued)

C	Solution tested	pH	$[H^+]$ $(\underline{M})$	$[C_2H_3O_2^-]$ $(\underline{M})$	K	Percent dissociation
	0.10 $\underline{M}$ $HC_2H_3O_2$					
	0.010 $\underline{M}$ $HC_2H_3O_2$					
	0.0010 $\underline{M}$ $HC_2H_3O_2$					

QUESTION   Write the net ionic equation describing the dissociation of acetic acid in aqueous solution.

QUESTION   Calculate $[H^+]$ and $[C_2H_3O_2^-]$ for each concentration above, showing a sample calculation here.

QUESTION   Write the equilibrium constant expression for acetic acid, and calculate a K value and an apparent percent dissociation for each concentration, showing sample calculations here.

QUESTION   What is the effect of dilution on the percent dissociation? Why should this behavior be expected?

Instructor's initials _____

Name ______________________________ **Experiment 27**

Student ID number ________________

Section __________ Date __________

Instructor ________________________

PRE-LABORATORY QUESTIONS

1. Define clearly each of the following terms, using a specific example wherever possible.

   a. Acid.

   b. Base.

   c. Equilibrium.

   d. Mass-action or equilibrium constant expression.

   e. pH.

   f. Dissociation.

   g. Hydrolysis.

   h. Strong electrolyte.

   i. Weak electrolyte.

   j. Buffer solution.

   k. Acid-base indicator.

2. Why should _freshly_ _boiled_ distilled water be mixed with the soil sample before determining the pH in part **B**?

3. Describe clearly how a buffer composed of aqueous ammonia, NH_3, and ammonium chloride, NH_4Cl, could neutralize either strong acid or strong base in order to maintain nearly constant pH.

4. Write the formula and name for each of the following.

   a. The conjugate base of NH_4^+.

   b. The conjugate base of H_2O.

   c. The conjugate base of HNO_2.

   d. The conjugate acid of H_2O.

   e. The conjugate acid of ClO^-.

   f. The conjugate acid of HCO_3^-.

Name _____

Student ID number _____

Section _____ Date _____

Instructor _____

POST-LABORATORY QUESTIONS

1. Arrange the solutions for which the pH was tested in part **A1** in order of increasing <u>acid</u> strength. Indicate clearly your reasoning.

2. Using your pH data from part **A1**, calculate an approximate equilibrium constant, K, and percent ionization of aqueous ammonia, NH_3. Compare your result with the accepted value, and account for any difference.

3. Describe clearly why a single indicator solution is only useful for measuring pH over a rather narrow range. Use a specific indicator solution to illustrate your arguments.

4. Why is a dilute solution of boric acid, H_3BO_3, used as an eyewash?

5. a. Write a mass-action or equilibrium constant expression for the hydrolysis of sodium acetate, $NaC_2H_3O_2$.

   b. Derive a relationship between the equilibrium constant for hydrolysis in <u>a</u>. and the equilibrium constant for the dissociation of acetic acid, $HC_2H_3O_2$.

   c. Evaluate the equilibrium constant for hydrolysis in <u>b</u>. using K_w from the Introduction and the equilibrium constant for the dissociation of $HC_2H_3O_2$ calculated in part **c**.

   d. Compare your calculated equilibrium constant for hydrolysis in <u>c</u>. with the value that can be calculated from your pH data in part **A1**.

Solubility Product of CdC_2O_4 and Formation Constant of $Cd(NH_3)_4^{2+}$

PURPOSE OF EXPERIMENT: Determine the solubility product, K_{sp}, of CdC_2O_4 and the formation constant, K_f, of $[Cd(NH_3)_4]^{2+}$.

When an excess of a sparingly soluble ionic solid is placed in water, the solid will dissolve until the solution phase is saturated. An equilibrium will then exist between the solid phase and the ions in the saturated solution. For example, when solid cadmium oxalate, CdC_2O_4, is in contact with its saturated solution, the following equilibrium equation and mass-action expression can be written.

$$CdC_2O_4(s) \rightleftharpoons Cd^{2+}(aq) + C_2O_4{}^{2-}(aq)$$

$$K = \frac{[Cd^{2+}][C_2O_4{}^{2-}]}{[CdC_2O_4(s)]}$$

The concentration of solid CdC_2O_4 is a constant. Therefore, if the equilibrium constant, K, is multiplied by $[CdC_2O_4(s)]$, a new constant is obtained, the **solubility product, K_{sp}**.

$$K_{sp} = K[CdC_2O_4(s)] = [Cd^{2+}][C_2O_4{}^{2-}]$$

The product of the concentrations of the cadmium ion and the oxalate ion at equilibrium with the solid phase must always equal the solubility product. If the product of the concentrations of the ions is greater than K_{sp}, CdC_2O_4 will precipitate until the K_{sp} is satisfied. If the product of the concentrations of the ions is less than K_{sp}, CdC_2O_4 will dissolve. The only requirement for the system to be in equilibrium with the solid is that the product of the concentrations of the ions equals the solubility product. It is not necessary for the cadmium ion concentration to be equal to the oxalate ion concentration. Any set of concentrations that when multiplied together equal K_{sp} is possible.

Consider the following system. A saturated solution of CdC_2O_4 in equilibrium with solid CdC_2O_4 is reacted with aqueous ammonia, NH_3. The concentration of Cd^{2+} will decrease because $[Cd(NH_3)_4]^{2+}$ is formed. Now a new equilibrium will be established, and a new equilibrium constant, K_f, must be satisfied.

$$Cd^{2+}(aq) + 4 NH_3(aq) \rightleftharpoons [Cd(NH_3)_4]^{2+}(aq)$$

$$K_f = \frac{[Cd(NH_3)_4{}^{2+}]}{[Cd^{2+}][NH_3]^4}$$

The solubility product must also be satisfied simultaneously. The total equation and the total equilibrium constant for the system are obtained by adding the two equations and by multiplying the equilibrium constants.

$$CdC_2O_4(s) \rightleftharpoons Cd^{2+}(aq) + C_2O_4{}^{2-}(aq)$$

$$Cd^{2+}(aq) + 4 NH_3(aq) \rightleftharpoons [Cd(NH_3)_4]^{2+}(aq)$$

$$CdC_2O_4(s) + 4 NH_3(aq) \rightleftharpoons [Cd(NH_3)_4]^{2+}(aq) + C_2O_4{}^{2-}(aq)$$

$$K = K_f \cdot K_{sp} = \frac{[Cd(NH_3)_4^{2+}][C_2O_4^{2-}]}{[NH_3]^4}$$

In this experiment you will determine the solubility product of CdC_2O_4 by reacting equimolar quantities of $Cd(NO_3)_2$ and $Na_2C_2O_4$ to form solid CdC_2O_4. The equilibrium concentration of $C_2O_4^{2-}$ in the solution phase will be determined by titration with $KMnO_4$. The Cd^{2+} concentration will be equal to the $C_2O_4^{2-}$ concentration, owing to the nature of the experiment and the reaction. The formation constant of $[Cd(NH_3)_4]^{2+}$ will be determined by using titration data from the reaction of a $CdC_2O_4(s)-Cd^{2+}(aq)-C_2O_4^{2-}(aq)$ equilibrium mixture with aqueous ammonia, NH_3.

REFERENCES

(1) Kotz, J. C., and Purcell, K. F., _Chemistry and Chemical Reactivity_, Saunders College Publishing, Philadelphia, 1987, sections 3.4, 4.4, 12.1, 15.2, 15.4, 15.6, and 25.3.

(2) Masterton, W. L., Slowinski, E. J., and Stanitski, C. L., _Chemical Principles_, 6th ed., Saunders College Publishing, Philadelphia, 1985, sections 2.6, 12.2, 18.1-18.3, 20.5, 21.1, and 23.2.

EXPERIMENTAL PROCEDURE

(Study this section and the PRE-LABORATORY QUESTIONS before coming to the laboratory. **Wear safety goggles when performing this experiment.**)

A. Preparation of Saturated Solutions of Cadmium Oxalate

You need to prepare **three** saturated solutions of CdC_2O_4. Mark three small beakers 1, 2, and 3. Clean, and properly rinse your buret (Laboratory Methods B). Fill the buret with 0.100 $\underline{M}$ cadmium nitrate, $Cd(NO_3)_2$, solution. To beaker 1 add 200. mL of distilled water and 25.00 mL of $Cd(NO_3)_2$ solution. Add 10.00-mL portions of your $Cd(NO_3)_2$ solution to beakers 2 and 3. Clean your buret again, and fill it this time with 0.100 $\underline{M}$ sodium oxalate, $Na_2C_2O_4$, solution. Add 25.00 mL of this $Na_2C_2O_4$ solution to beaker 1, and 10.00 mL each to beakers 2 and 3. Stir the mixtures for 10. minutes to allow the particle size to grow, and then permit the precipitates to settle.

B. Determination of the Solubility Product of CdC_2O_4

Using either gravity filtration or suction filtration (Laboratory Methods I), filter the mixture in beaker 1. Clean your buret, and rinse it with a few milliliters of the filtrate. Using your buret, measure 100. mL of filtrate into each of two Erlenmeyer flasks. You will need to fill your buret twice in order to measure the required 100. mL of filtrate for each of the two flasks. Add 20. mL of 3 $\underline{M}$ sulfuric acid, H_2SO_4, solution to each flask. CAUTION: Dilute H_2SO_4 is corrosive and causes burns on your skin or holes in your clothing. If there are any spills or spatters onto your skin or clothing, rinse the affected area thoroughly with water.

Clean, rinse, and fill your buret with approximately 0.01 $\underline{M}$ potassium permanganate, $KMnO_4$, solution. Record in TABLE 28.1B the $KMnO_4$ molarity shown on the bottle. Heat one 100-mL portion of the filtrate to 55°C-60.°C, and titrate the hot solution with the $KMnO_4$ solution. The endpoint will be the point in the titration at which the pink color of $KMnO_4$ first persists for about 30. seconds. Record your titration data in TABLE 28.1B. Repeat the titration with the second 100-mL portion of CdC_2O_4 solution, and record your data in TABLE 28.1B.

C. Determination of the Equilibrium Constant for the Formation of $[Cd(NH_3)_4]^{2+}$

Clean, rinse, and fill your buret with approximately 5 $\underline{M}$ aqueous ammonia, NH_3. Record in TABLE 28.1C the molarity of the NH_3 shown on the bottle. Titrate the mixture in beaker 2. The endpoint will be the point at which the solid CdC_2O_4 completely disappears. In order to determine this endpoint easily, put a <u>dark</u> <u>cross</u> on a piece of white paper that is placed under the beaker of mixture you are titrating. When all solid CdC_2O_4 is gone, you will be able to see the cross clearly. Record your titration data in TABLE 28.1C.

Then repeat the titration of the mixture in beaker 3, and record your data in TABLE 28.1C.

Perform the calculations in TABLE 28.2 including sample calculations for one run following TABLE 28.2.

TABLE 28.1. Volumes and molarities for titration of CdC_2O_4 solutions.

		Run 1	Run 2
B	Initial reading of buret ($KMnO_4$) (mL)		
	Final reading of buret ($KMnO_4$) (mL)		
	Molarity of $KMnO_4$ solution ($\underline{M}$)		
		Beaker 2	Beaker 3
C	Initial reading of buret (NH_3) (mL)		
	Final reading of buret (NH_3) (mL)		
	Molarity of NH_3 solution ($\underline{M}$)		

Instructor's initials _____

TABLE 28.2. Solubility product of CdC_2O_4 and formation constant of $[Cd(NH_3)_4]^{2+}$.

		Run 1	Run 2
B	Volume of $KMnO_4$ solution (mL)		
	Moles of MnO_4^- used for titration (mol)		
	Moles of $C_2O_4^{2-}$ in 100.0 mL of solution (mol)		
	Molarity of $C_2O_4^{2-}$ ($\underline{M}$)		
	Molarity of Cd^{2+} ($\underline{M}$)		
	K_{sp} of CdC_2O_4		
	Average K_{sp} of CdC_2O_4		
		Beaker 2	Beaker 3
C	Total volume of solution after titration (mL)		
	Total moles of $C_2O_4^{2-}$ (mol)		
	Molarity of $C_2O_4^{2-}$ ($\underline{M}$)		
	Total moles of Cd^{2+} (mol)		
	Moles of $[Cd(NH_3)_4]^{2+}$ (mol)		
	Molarity of $[Cd(NH_3)_4]^{2+}$ ($\underline{M}$)		
	Moles of NH_3 added by titration (mol)		
	Moles of NH_3 that did not react with Cd^{2+} (mol)		
	Molarity of NH_3 that did not react with Cd^{2+} ($\underline{M}$)		
	K_f for $[Cd(NH_3)_4]^{2+}$		
	Average K_f for $[Cd(NH_3)_4]^{2+}$		

Name _____

Student ID number _____

Section _____ Date _____

Instructor _____

Sample calculations for Run 2 or Beaker 2

B   Moles of MnO_4^- used for titration of saturated solution of CdC_2O_4

Moles of $C_2O_4^{2-}$ in 100.0 mL of saturated solution of CdC_2O_4

Molarity of $C_2O_4^{2-}$ in saturated solution of CdC_2O_4

Molarity of Cd^{2+} in saturated solution of CdC_2O_4

Solubility product, K_{sp}, of CdC_2O_4

C   Total volume of solution after titration with NH_3

Total moles of $C_2O_4^{2-}$

Experiment 28

Molarity of $C_2O_4{}^{2-}$

Total moles of Cd^{2+}

Moles of $[Cd(NH_3)_4]^{2+}$ after titration

Molarity of $[Cd(NH_3)_4]^{2+}$ after titration

Moles of NH_3 added by titration

Moles of NH_3 that did not react with Cd^{2+}

Molarity of NH_3 that did not react with Cd^{2+}

Formation constant for $[Cd(NH_3)_4]^{2+}$

Name ______________________________ **Experiment 28**

Student ID number ________________

Section __________ Date __________

Instructor ________________________

PRE-LABORATORY QUESTIONS

1. Write a balanced net ionic equation for the reaction of $KMnO_4$ with the saturated solution of CdC_2O_4 in dilute H_2SO_4. The major products are Mn^{2+} and CO_2.

2. Why is the concentration of solid CdC_2O_4 considered a constant?

3. How many total mL of 0.100 M $Cd(NO_3)_2$ will you need to take from the reagent shelf in order to prepare the solutions in beakers 1 plus 2 plus 3? Justify your answer by showing how you will use this total volume of solution.

4. Why is the saturated solution of CdC_2O_4 heated to 55°C-60.°C for the $KMnO_4$ solution? (HINT: Read Experiment 15 or 30.)

5. What is the indicator for the $KMnO_4$ titration? (HINT: Read Experiment 15 or 30.)

6. Write the electron configuration including the population of all partially filled subshells for Cd^{2+}.

7. Predict the structure of (1) $C_2O_4^{2-}$ and (2) $[Cd(NH_3)_4]^{2+}$.

POST-LABORATORY QUESTIONS

1. For each of the following possible errors, determine whether the calcu-
    lated solubility product of CdC_2O_4 in part **B** will increase, decrease, or
    not change. Justify your answers.

    a. The measured volume of $Cd(NO_3)_2$ solution was assumed to be 10.00 mL,
        but the actual volume was 12.00 mL.

    b. The molarity of the $KMnO_4$ solution was actually larger than that in-
        dicated on the reagent bottle.

    c. The solution for titration by $KMnO_4$ solution was heated to $80.^\circ C$.

2. How would the formation constant for $[Cd(NH_3)_4]^{2+}$ be affected (increase, decrease, or no change) if not all of the CdC_2O_4 precipitate reacted with NH_3? Justify your answer.

3. Use your calculated values of K_{sp} and K_f to explain why CdC_2O_4 is dissolved by ammonia.

4. What are the significant chemical species which are present in a solution of aqueous ammonia?

Galvanic and Electrolytic Cells

PURPOSE OF EXPERIMENT: Construct several galvanic cells, and measure their voltages; construct an electrolytic cell, and determine the copper oxidized and hydrogen liberated during electrolysis of a dilute H_2SO_4 solution.

Oxidation-reduction reactions are the basis of the branch of chemistry called **electrochemistry**. Such a reaction may occur spontaneously and produce electrical energy, as in a galvanic cell. If the reaction does not occur spontaneously, the addition of electrical energy may initiate a chemical change, a process called electrolysis.

All electrochemical cells involve two half-reactions: an oxidation half-reaction in which electrons are released, and a reduction half-reaction in which electrons are taken up. The net voltage of the cell, the only quantity that can be measured experimentally, is the algebraic sum of the potentials for the two half-reactions. Each potential is a measure of the relative ability of a given half-reaction to occur. However, since the potential for a half-reaction cannot be measured directly, numerical values for half-reaction potentials are arbitrary and must be based on a reference potential. The hydrogen half-reaction serves as the reference for all electrochemical potentials.

Oxidation-reduction reactions that occur spontaneously and liberate energy may be used to push electrons through an external circuit and to generate an electric current. Galvanic cells operate on this principle. A common lead storage battery consists of a number of such cells connected in series. In a galvanic cell, the oxidation half-reaction occurs at one electrode while the reduction half-reaction occurs at the other electrode. The system is so arranged that the electrons released in the oxidation half-reaction taking place at one electrode must flow through a metallic conductor to get to the other electrode of the cell where the reduction half-reaction occurs. The stream of electrons flowing through the external metallic conductor constitutes an electric current. The net reaction for the cell as a whole is one of oxidation-reduction.

Oxidation-reduction reactions that do not occur spontaneously require the addition of electrical energy to make them take place. When electrolysis is carried out, an electric current (a stream of electrons) acts on a solution of an electrolyte, and chemical reactions occur at the two electrodes. Electrons pumped through the external circuit are taken up at the cathode by a reduction half-reaction. At the anode an oxidation half-reaction releases electrons, which travel through the external circuit. Transfer of charge through the solution in the electrolytic cell is due to the migration of ions toward the electrode of opposite charge. The number of electrons entering the cell at the cathode for a given period of operation is just equal to the number of electrons leaving the cell at the anode.

In this experiment you will first construct several galvanic cells and measure their voltages. One half-reaction will be the same for all these cells, so that your measurements will enable you to set up a qualitative order showing the relative tendencies of the half-reactions to occur. Then you will set up an electrolytic cell having copper electrodes and will electrolyze a dilute solution of sulfuric acid, running the cell for a measured time at a measured amperage. You will determine the amount of copper oxidized at the anode, and collect and measure the amount of hydrogen that is liberated simultaneously at the cathode. The quantities of material that have undergone chemical change at the electrodes can then be related to the current that has flowed through the circuit.

REFERENCES

(1) Kotz, J. C., and Purcell, K. F., <u>Chemistry and Chemical Reactivity</u>, Saunders College Publishing, Philadelphia, 1987, sections 2.5, 3.4, 4.1, 4.4, 6.1, 6.3-6.5, 19.1, 19.2, 19.5, 20.4, and 25.2.

(2) Masterton, W. L., Slowinski, E. J., and Stanitski, C. L., <u>Chemical Principles</u>, 6th ed., Saunders College Publishing, Philadelphia, 1985, sections 2.5, 2.6, 3.7, 4.3, 6.1-6.4, 8.5, 23.1-23.4, and 25.2.

EXPERIMENTAL PROCEDURE

(Study this section and the PRE-LABORATORY QUESTIONS before coming to the laboratory. **Wear safety goggles when performing this experiment.**)

A. Galvanic Cells

Obtain a zinc electrode and a copper electrode from the reagent table, and construct the cell shown in FIGURE 29.1. Place 25 mL of a 0.5 $\underline{M}$ sodium sulfate, Na_2SO_4, solution in the U-tube. Drop in several small crystals of copper(II) sulfate pentahydrate, $CuSO_4 \cdot 5H_2O$. Place the copper electrode so that it extends through the lower portion of the U-tube and is in contact with the crystals, **being careful not to stir up the solution.** Insert a stopper to hold the wire in place. Adjust the zinc electrode in the other arm of the U-tube, fixing it with a stopper in the proper position.

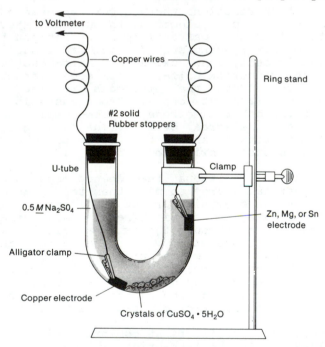

FIGURE 29.1. A galvanic cell.

As soon as you have made up the cell, use the following procedure to determine which electrode is releasing electrons to the external circuit and which is taking up electrons. Pour approximately 2 mL of a starch-potassium iodide, KI, solution into a small beaker. Add 2 drops of 3 $\underline{M}$ hydrochloric acid, HCl, solution and stir. Thoroughly wet a piece of filter paper in the beaker, and place it on a watch glass. With the <u>very clean</u> electrode wires from the cell about 0.5-1.0 cm apart, touch the wires to the filter paper. A dark blue color (the starch-iodine complex) will appear after about 1-2 minutes (sometimes longer) around the wire at which the half-reaction taking up electrons is occurring. Record your observations in TABLE 29.1A.

Immediately determine the voltage of your galvanic cell by connecting the wires to the proper terminals of the voltmeter as indicated by your starch-potassium iodide test. Continue to read the voltmeter for several minutes to make sure that you observe the maximum voltage of your cell. Record the maximum voltage in TABLE 29.1A. Remove the zinc electrode, clean it, and return it to the reagent shelf.

Construct galvanic cells by replacing the zinc electrode first with a tin foil electrode and then with magnesium ribbon. Determine which electrode is taking up electrons in each case, and record the voltage generated by each of these two cells in TABLE 29.1A. Clean and return all electrodes and equipment to the reagent shelf.

B. Electrolysis

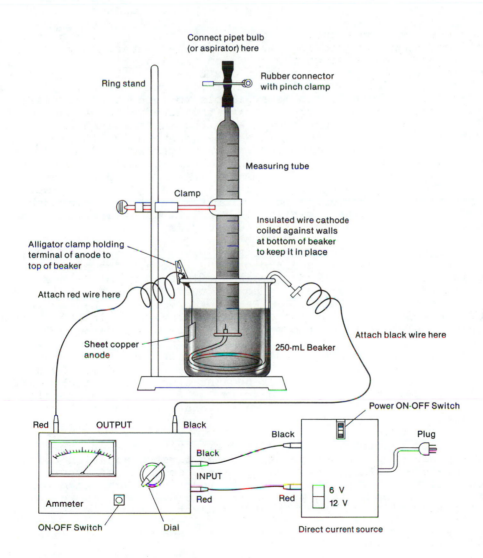

FIGURE 29.2. An electrolytic cell.

Experiment 29

Construct the electrolytic cell shown in FIGURE 29.2 as follows. Invert a measuring tube into a 250-mL beaker, and clamp the measuring tube securely. (Alternatively, your instructor may designate that a buret be used as the gas measuring tube in FIGURE 29.2.) Place an insulated copper wire cathode as indicated in the figure. All of the end of the exposed copper wire must be <u>inside</u> the measuring tube so that all of the hydrogen gas which is formed is collected by the measuring tube. Prepare a 1 $\underline{M}$ solution of sulfuric acid, H_2SO_4, by diluting 50. mL of 3 $\underline{M}$ H_2SO_4 with the proper amount of distilled water (PRE-LABORATORY QUESTION 1). **CAUTION: <u>Dilute sulfuric acid is corrosive and causes burns on your skin and holes in your clothing. If any spills or spatters occur onto your skin or clothing, rinse the affected area thoroughly with water</u>**. Add the 1 $\underline{M}$ H_2SO_4 to the beaker of the electrolysis cell. Fill the measuring tube with the H_2SO_4 solution by first squeezing the air out of your pipet bulb (Laboratory Methods B) and then by attaching it to the tip of the measuring tube. With the pinch clamp (or the stopcock) of the measuring tube open, draw up the solution into the measuring tube, then close the pinch clamp (or stopcock). Repeat the procedure with the pipet bulb until the measuring tube is completely filled with H_2SO_4. (Alternatively, the 1 $\underline{M}$ H_2SO_4 solution can be drawn up into the measuring tube using an aspirator connected to the tip.)

Clean a sheet copper anode by dipping it into 3 $\underline{M}$ nitric acid, HNO_3, solution and then by rinsing it in distilled water. Dry the anode with pieces of filter paper. Weigh the anode on the analytical balance (Laboratory Methods C). Record the mass in TABLE 29.1B. Place this anode in the cell as shown in FIGURE 29.2. A source of direct current, with ammeter, switch, and connecting wires is available to connect to your electrolytic cell. The direct current source converts the line alternating current to direct current of low amperage. With the power switch OFF, connect the direct current source and ammeter in series as shown in FIGURE 29.2. Connect the two output leads from the ammeter together to "short" the system, plug in the direct current source to the laboratory electricity, and switch the electrical power ON. With the ammeter switch ON, adjust the dial on the ammeter so that the amperage output is 0.7-0.8 A. Turn the power switch to the OFF position. Separate the leads from the ammeter, and attach them to your electrolytic cell as shown in FIGURE 29.2. Your instructor must check your apparatus before you begin the electrolysis.

<u>Take care that the electrodes are not moved during electrolysis</u>. Turn ON the electrical power, and record the time and the ammeter reading in TABLE 29.1B. Continue the electrolysis until 30-40 mL of H_2 has been collected. Record the ammeter reading again, turn OFF the electrical power and record the time in TABLE 29.1B. Record the volume of H_2, and measure the height of the solution in the measuring tube above that in the beaker, the temperature, and the atmospheric pressure. Record all data in TABLE 29.1B.

Disconnect the copper anode, and dip it into a beaker of distilled water to remove the acid. Permit the electrode to air dry. Do not dry the electrode by rubbing it with paper because copper metal flakes off. Weigh the dry electrode, and record the mass in TABLE 29.1B. Remove all acid from the apparatus, rinse with distilled water, and return both electrodes to the reagent table.

If you used a buret as the gas measuring tube, measure the volume between the pinch clamp (or stopcock) and the first graduation using the following procedure. Fill the uncalibrated section of your buret with distilled water. Empty the water into your 10-mL graduated cylinder, and note the volume. Repeat this process twice more, and record the average volume of the uncalibrated section of the buret in TABLE 29.1B.

Perform the calculations in TABLE 29.2 and the sample calculations on the back of TABLE 29.2.

Name _____

Student ID number _____

Section _____ Date _____

Instructor _____

DATA

TABLE 29.1. Data for galvanic cells and for the electrolysis of H_2SO_4 using copper electrodes.

A Electrode (copper or zinc) at which blue color appears	
Voltage of copper-zinc galvanic cell (V)	
Electrode (copper or tin) at which blue color appears	
Voltage of copper-tin galvanic cell (V)	
Electrode (copper or magnesium) at which blue color appears	
Voltage of copper-magnesium galvanic cell (V)	
B Mass of copper anode before electrolysis (g)	
Mass of dry copper anode after electrolysis (g)	
Time electrolysis started (hr, min, and sec)	
Time electrolysis stopped (hr, min, and sec)	
Amperage at start of electrolysis (A)	
Amperage at end of electrolysis (A)	
Observed volume of hydrogen collected (mL)	
Volume of uncalibrated section of buret (mL)	
Height of solution above that in beaker (mm)	
Atmospheric pressure (mmHg)	
Room temperature (°C)	

Instructor's initials _____

TABLE 29.2. Moles of H_2 formed, electrons transferred, and copper oxidized during electrolysis of H_2SO_4 using copper electrodes.

A	Reduction potential for $Cu^{2+}(aq) + 2\ e^- \longrightarrow Cu(s)$ assuming $E^o = -0.76$ V for $Zn^{2+}(aq) + 2\ e^- \longrightarrow Zn(s)$	
	Reduction potential for $Sn^{2+}(aq) + 2\ e^- \longrightarrow Sn(s)$	
	Reduction potential for $Mg^{2+}(aq) + 2\ e^- \longrightarrow Mg(s)$	
B	Total volume of H_2 collected (mL) (add uncalibrated volume if buret used)	
	Pressure equivalent of height of solution above that in beaker (mmHg)	
	Total pressure of dry H_2 (mmHg)	
	Moles of H_2 formed during electrolysis (mol)	
	Moles of electrons transferred during electrolysis based on H^+ reduced (mol)	
	Average amperage during the electrolysis (A)	
	Number of Faradays of electrons used during electrolysis based on time of electrolysis (F)	
	Mass of copper oxidized during electrolysis (g)	
	Moles of copper oxidized (mol)	
	Moles of electrons transferred during electrolysis based on copper oxidized (mol)	
	Oxidation state of copper ions formed in solution	
	Balanced chemical equation for overall electrolysis reaction	

Experiment 29

Sample calculations

A Reduction potential for $Cu^{2+}(aq) + 2e^- \longrightarrow Cu(s)$ assuming $E^o = -0.76$ V for $Zn^{2+}(aq) + 2e^- \longrightarrow Zn(s)$

Reduction potential for $Sn^{2+}(aq) + 2e^- \longrightarrow Sn(s)$ using answer from above

B Pressure equivalent of the height of solution in the measuring tube above the beaker (The density of mercury is 13.53 g/mL.)

Moles of H_2 formed during electrolysis

Moles of electrons transferred during electrolysis based on moles of H_2 formed

Number of Faradays of electrons transferred during electrolysis based on time of electrolysis

Moles of copper oxidized during electrolysis

Moles of electrons transferred during electrolysis based on moles of copper oxidized

Name ______________________________ **Experiment 29**

Student ID number ________________

Section __________ Date __________

Instructor ________________________

PRE-LABORATORY QUESTIONS

1. Calculate the volume of water which must be added to 50. mL of 3 $\underline{M}$ H_2SO_4 to prepare a 1 $\underline{M}$ solution.

2. How is the starch-potassium iodide used to determine the direction of electron flow in a galvanic cell?

3. How is the uncalibrated volume of the buret measured in this experiment?

4. Suggest a second experimental method by which you could measure the un-
    calibrated volume of the buret, and explain the necessary calculations.

5. Suggest a possible reason why the electrodes must not be moved during
    electrolysis.

6. Why is the copper anode in the electrolysis experiment air dried before
    weighing?

1. Write balanced net ionic equations for each of the overall reactions which occurred in the three galvanic cells in part **A**.

2. Compare the potentials, as measured in this experiment, for the oxidation-reduction reactions which occurred in the three galvanic cells with the standard voltages calculated from standard reduction potentials. Suggest possible reasons for any differences.

3. Compare the moles of electrons transferred during electrolysis as based on moles of H^+ reduced, moles of copper oxidized, and the amperage and time of electrolysis. Arrange these three in terms of decreasing accuracy. Justify your arrangement.

4. Which metallic element and which nonmetallic element are commonly prepared in commercial quantities by electrolysis? Give a possible reason
   why each element which you selected is best prepared electrochemically.

5. Electrolysis resulting in the deposition of a metal on an object is
   called electroplating. Such a process is often used for metal objects
   to protect them against corrosion or to enhance their attractiveness.
   Thus, eating utensils are silver-plated in an electrolytic cell by
   making the clean cutlery the cathode and a silver bar the anode in an
   aqueous potassium silver cyanide, $KAg(CN)_2$, solution. If 0.0137 g of
   silver is required to electroplate one spoon, how long must 0.93 amp of
   current be applied to electroplate 75 spoons?

Oxidation-Reduction Titration and
Analysis of an Unknown Mixture

PURPOSE OF EXPERIMENT: Standardize a solution of $KMnO_4$, and determine the percent by mass of $Na_2C_2O_4$ in an unknown mixture.

The process of titration may be used for the standardization of solutions of oxidizing and/or reducing agents, provided a suitable method for observing the endpoint of the reaction is available. When potassium permanganate, $KMnO_4$, is used as a titrant, the endpoint is easily apparent. The intensely purple colored MnO_4^- in acidic solution produces the manganese(II) ion, which is very pale pink in color, but dilute solutions are practically colorless. Thus, as MnO_4^- is initially added to a solution of a colorless reactant, the resulting solution will remain essentially colorless. After excess MnO_4^- has been added, the solution will have a pink color.

In this experiment you will prepare a dilute solution of MnO_4^- and standardize it by titration with an acidified solution of ferrous ammonium sulfate hexahydrate, $Fe(NH_4)_2(SO_4)_2 \cdot 6H_2O$. During standardization, MnO_4^- reacts very rapidly at room temperature with iron(II) in acidic solution to produce solutions of manganese(II) and iron(III). You will then use your standardized solution of MnO_4^- to analyze a mixture of sodium oxalate, $Na_2C_2O_4$, and a second component, potassium sulfate, K_2SO_4, which does not react with MnO_4^-. The oxalate ion reacts slowly in dilute acidic solutions with MnO_4^- to produce carbon dioxide and manganese(II) ion. Since this reaction is slow at room temperature, the solution containing the oxalate ion will be heated and titrated at 55°C. In addition, the rate of reaction will be enhanced by the formation of manganese(II) because manganese(II) is an autocatalyst. In autocatalysis, a product of the reaction acts as a catalyst of the reaction. By knowing the balanced chemical equation and the number of moles of MnO_4^- which react, the number of moles of oxalate ion and the number of moles of sodium oxalate can be calculated. The number of moles of sodium oxalate can, in turn, be used to calculate the percent composition by mass of sodium oxalate in the unknown mixture.

REFERENCES

(1) Kotz, J. C., and Purcell, K. F., <u>Chemistry and Chemical Reactivity</u>, Saunders College Publishing, Philadelphia, 1987, sections 1.3, 2.5, 2.12, 3.1, 3.2, 3.4, 4.4, and 25.1.

(2) Masterton, W. L., Slowinski, E. J., and Stanitski, C. L., <u>Chemical Principles</u>, 6th ed., Saunders College Publishing, Philadelphia, 1985, sections 1.6, 2.5, 2.6, 3.6, 3.7, 12.2, 23.2, 25.2, and 25.3.

EXPERIMENTAL PROCEDURE

(Study this section and the PRE-LABORATORY QUESTIONS before coming to the laboratory. **Wear safety goggles when performing this experiment.**)

A. Standardization of a Potassium Permanganate Solution

Clean your buret, a 100-mL graduated cylinder, three 250-mL Erlenmeyer flasks, a 100-mL beaker, and a 500-mL Florence flask.

Prepare a dilute solution of potassium permanganate, $KMnO_4$, by mixing in your 500-mL Florence flask about 25 mL of the concentrated $KMnO_4$ solution, provided on the reagent table, and 275 mL of distilled water. Shake or swirl the flask thoroughly to mix the solution. Determine the concentration of this solution by titrating it against a known quantity of iron(II) by the following procedure.

Label the three Erlenmeyer flasks 1, 2, and 3. Using your analytical balance (Laboratory Methods C), weigh onto weighing paper the ferrous ammonium sulfate hexahydrate, $Fe(NH_4)_2(SO_4)_2 \cdot 6H_2O$, as calculated in PRE-LABORATORY QUESTION 3. (It is not necessary to have exactly the calculated mass, but you must know the mass precisely.) Record the masses in TABLE 30.1A. Place this sample in flask 1. Weigh two more samples, put them into flasks 2 and 3, and record the masses in TABLE 30.1A. Dissolve each sample of the salt in 30. mL of distilled water (Laboratory Methods K), and add 10. mL of 3 $\underline{M}$ sulfuric acid, H_2SO_4, solution.

Make a rapid preliminary titration of sample 1 with your $KMnO_4$ solution to gain experience with the procedure. Place a white sheet of paper under the flask so that the color of the solution may be seen clearly. Add the $KMnO_4$ solution from the buret, swirling the sample constantly, until the last drop leaves a permanent pink color. Rinse the walls of the flask with distilled water from your wash bottle to make sure all the Fe^{2+} reacts. The first appearance of a permanent pink color from MnO_4^- indicates the endpoint of the titration. Record the initial and final volumes in TABLE 30.1A.

Titrate samples 2 and 3, adding the first 20. mL of $KMnO_4$ rapidly and then approaching the endpoint with care. Record your initial and final volumes in TABLE 30.1A.

B. Analysis of $Na_2C_2O_4$ in an Unknown Solid Mixture

Obtain an unknown from your instructor. Clean two 250-mL Erlenmeyer flasks, and rinse them with distilled water. Using your analytical balance, weigh onto weighing paper two samples of unknown of approximately 0.25 g each, record your masses in TABLE 30.1B, and put the samples into the flasks, labeling them 1 and 2.

Add 25 mL of distilled water and 25 mL of 3 $\underline{M}$ sulfuric acid, H_2SO_4, solution to flask 1. **CAUTION:** <u>**Dilute sulfuric acid is corrosive and causes burns on your skin or holes in your clothing. If any spills or spatters occur onto your skin or clothing, wash the affected area thoroughly with water**</u>. Swirl the mixture to hasten dissolution. (You should remember that H^+ reacts with $C_2O_4^{2-}$ to form $H_2C_2O_4$.) Fill your buret with your standard $KMnO_4$ solution, and record the initial reading of the buret in TABLE 30.1B. Add rapidly about 15 mL of $KMnO_4$, swirl the mixture, and let the solution stand until the pink color of MnO_4^- disappears. This should occur in less than five minutes. Heat the solution to between 55°C and 60.°C (Laboratory Methods D), and complete the titration by adding MnO_4^- solution slowly. (At temperatures below 55°C the reaction between MnO_4^- and $H_2C_2O_4$ is too slow to give a good endpoint, but above 60.°C the oxalic acid decomposes.) Therefore, **temperature control is very important.** The pink color of MnO_4^- should persist for 30. seconds at the endpoint. Record the final buret reading in TABLE 30.1B.

Dissolve your sample 2 in 25 mL of distilled water and 25 mL of 3 $\underline{M}$ H_2SO_4. Heat the solution to between 55°C and 60.°C. Then titrate, adding the $KMnO_4$ very rapidly up to about 2 mL of the volume you estimate will be needed on the basis of your titration of sample 1. Then slowly add more MnO_4^- solution until you reach the endpoint. Record your initial and final volumes in TABLE 30.1B.

Perform the calculations in TABLE 30.2 including the sample calculations for one run on the back of TABLE 30.2.

Student ID number _____

Section _____ Date _____

Instructor _____

DATA

TABLE 30.1. Masses and volumes for oxidation-reduction titrations.

	Run 1	Run 2	Run 3
A Mass of weighing paper (g)			
Mass of $Fe(NH_4)_2(SO_4)_2 \cdot 6H_2O$ and weighing paper (g)			
Initial reading of buret (mL)			
Final reading of buret (mL)			

	Run 1	Run 2
B Mass of weighing paper (g)		
Mass of unknown and weighing paper (g)		
Initial reading of buret (mL)		
Final reading of buret (mL)		

Instructor's initials _____

Name _____

Student ID number _____

Section _____ Date _____

Instructor _____

CALCULATIONS

TABLE 30.2. Molarity of $KMnO_4$ solution and percent $Na_2C_2O_4$ by mass.

	Run 2	Run 3
A Mass of $Fe(NH_4)_2(SO_4)_2 \cdot 6H_2O$ which reacted with MnO_4^- (g)		
Moles of iron(II) which reacted with MnO_4^- (mol)		
Moles of MnO_4^- which reacted with iron(II) (mol)		
Volume of MnO_4^- solution which reacted with iron(II) (mL)		
Molarity of $KMnO_4$ solution (M)		
Average molarity of $KMnO_4$ (M)		

	Run 1	Run 2
B Volume of MnO_4^- which reacted with oxalic acid (mL)		
Moles of MnO_4^- which reacted with oxalic acid (mol)		
Moles of oxalic acid which reacted with MnO_4^- (mol)		
Moles of $Na_2C_2O_4$ present in unknown sample (mol)		
Mass of $Na_2C_2O_4$ present in unknown sample (g)		
Mass of unknown sample (g)		
Percent by mass of $Na_2C_2O_4$ in unknown sample (%)		
Average percent by mass of $Na_2C_2O_4$ (%)		

Experiment 30

Sample calculations for Run 2

A   Moles of iron(II) which reacted with MnO_4^-

   Moles of MnO_4^- which reacted with iron(II)

   Molarity of $KMnO_4$ solution

B   Moles of MnO_4^- which reacted with oxalic acid

   Moles of oxalic acid which reacted with MnO_4^-

   Moles of $Na_2C_2O_4$ present in unknown sample

   Mass of $Na_2C_2O_4$ present in unknown sample

   Percent $Na_2C_2O_4$ by mass in unknown sample

Name ______________________________ **Experiment 30**

Student ID number ________________

Section __________ Date __________

Instructor ________________________

PRE-LABORATORY QUESTIONS

1. What is the indicator in $KMnO_4$-iron(II) titrations? Describe how it functions.

2. Write a balanced net ionic equation for the reaction of MnO_4^- with iron(II) to produce manganese(II) and iron(III) in aqueous acidic solution.

3. Calculate the mass of ferrous ammonium sulfate hexahydrate, $Fe(NH_4)_2(SO_4)_2 \cdot 6H_2O$, required to react with 25 ml of 0.020 $\underline{M}$ potassium permanganate, $KMnO_4$, solution.

4. What is the oxidation state of iron in $Fe(NH_4)_2(SO_4)_2 \cdot 6H_2O$? Justify your answer.

5. Write a balanced net ionic equation for the reaction which occurs when MnO_4^- reacts with a solution prepared from $Na_2C_2O_4$ in acidic solution. (Note: The ion, $C_2O_4^{2-}$, is the anion of oxalic acid, $H_2C_2O_4$, a very weak acid.)

6. If you heat the acidified solution of your unknown above 55°C, will problems occur? If so, identify the problem, and state how the problem will affect your analysis of $Na_2C_2O_4$ in the unknown.

Name ______________________________

Student ID number ______________

Section __________ Date __________

Instructor ________________________

Experiment 30

POST-LABORATORY QUESTIONS

1. How will the following experimental errors influence the calculated percent by mass of $Na_2C_2O_4$ in your unknown? Justify your answer.

   a. The $Fe(NH_4)_2(SO_4)_2 \cdot 6H_2O$ was wet, i.e., the sample had more water than the six moles of water of hydration.

   b. You heated the $Na_2C_2O_4$ solution to 75°C for 15 minutes before titration with the $KMnO_4$ solution.

   c. Your instructor made your unknown from $Na_2C_2O_4$ and Rb_2SO_4 rather than K_2SO_4.

d. Your instructor made your unknown from $Na_2C_2O_4$ and KCl rather than K_2SO_4.

e. You added 50. mL of 3 $\underline{M}$ H_2SO_4 rather than the required 25 mL of 3 $\underline{M}$ H_2SO_4.

2. What is the meaning of the term, autocatalytic reaction? How could you prove experimentally that a reaction was "autocatalytic"?

3. What is the color of each of the following?

a. $KMnO_4(s)$.

b. $Na_2C_2O_4(s)$.

c. $Mn^{2+}(aq)$.

Preparation of Aspirin,

Determination of Its Molecular Weight,

and the Assay of Commercial Aspirin

PURPOSE OF EXPERIMENT: Prepare aspirin, determine its molecular weight by freezing point depression, and assay commercial aspirin by titration with NaOH.

Aspirin is one of three salicylic acid derivatives that find use in medicine as antipyretics (substances that reduce or prevent fever) and analgesics (substances that reduce or prevent pain). Salicylic acid is the substance that produces beneficial responses upon absorption through the intestinal membrane. However, it has the disadvantage of being so acidic that it irritates the mouth. This difficulty is overcome by using the less acidic acetylsalicylic acid (aspirin). This derivative is not hydrolyzed by the weakly acidic digestive juices of the mouth or the stomach and passes through with no irritating action. It is, nevertheless, readily hydrolyzed by the alkaline fluids of the intestinal tract to salicylic acid, which can then carry out its beneficial action.

Aspirin is synthesized commercially by a series of organic reactions. Phenol, C_6H_5OH, is converted in several steps to salicylic acid, which is then converted to acetylsalicylic acid. This last conversion uses acetic anhydride according to the following reaction scheme.

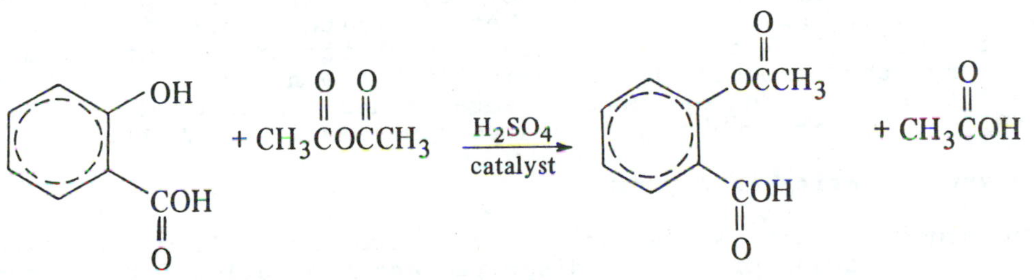

| salicylic acid | acetic anhydride | | aspirin | acetic acid |

Several different acetylating agents may be used, but acetic anhydride is cheap and forms a by-product, acetic acid, which is noncorrosive and can be recovered to make more acetic anhydride. As with many esterification reactions, a trace of concentrated acid, such as sulfuric acid, acts as a catalyst to increase the rate of the reaction.

In this experiment you will prepare aspirin by the above reaction. You will further characterize your product by determination of its melting point and its molecular weight. Finally, you will assay (determine the amount of) aspirin in commercial aspirin tablets by titration with NaOH solution. Aspirin tablets have a binder which helps to hold the tablet together. The binder does not react with NaOH.

REFERENCES

(1) Kotz, J. C., and Purcell, K. F., <u>Chemistry and Chemical Reactivity</u>, Saunders College Publishing, Philadelphia, 1987, sections 2.5, 2.10, 4.1-4.4, 12.1, 12.3, 17.1, 23.5, and 23.10.

(2) Masterton, W. L., Slowinski, E. J., and Stanitski, C. L., <u>Chemical Principles</u>, 6th ed., Saunders College Publishing, Philadelphia, 1985, sections 2.5, 3.7, 3.8, 12.4, 19.6, 19.7, and 28.1.

EXPERIMENTAL PROCEDURE

(Study this section and the PRE-LABORATORY QUESTIONS before coming to the laboratory. **Wear safety goggles when performing this experiment.**)

A. Preparation of Aspirin

Using a beam balance (Laboratory Methods C), weigh 3.0 g of salicylic acid, $C_7H_6O_3$, into an evaporating dish, recording your masses in TABLE 31.1A. Working in the hood, cover the crystals with 5 mL of acetic anhydride, $C_4H_6O_3$. CAUTION: <u>These substances are extremely corrosive. Avoid contact with the skin; do not breathe the vapors</u>. Add only 2 or 3 drops of concentrated (18 $\underline{M}$) sulfuric acid, H_2SO_4, solution as a catalyst. CAUTION: <u>Concentrated sulfuric acid is very corrosive and causes burns on your skin and holes in your clothing. If any spills or spatters occur onto your skin or clothing, rinse the affected area thoroughly with water</u>. Heat the dish on a steam bath (Laboratory Methods M), and stir the contents. As soon as the mixture reaches 80°C-90°C, remove the dish from the steam bath, and allow the mixture to cool for 10-15 minutes.

Break up the crystalline mass with a stirring rod, and add, with stirring, about 10. mL of toluene, C_7H_8. Filter the product by vacuum filtration (Laboratory Methods I). (Wet the filter paper with toluene.) As soon as the product is free from toluene, wash the product with 2 mL of ice water. Dry the product by pressing it between several layers of filter paper and then spreading it out on several other layers of dry filter paper. Weigh the dry, crude product. Record your masses in TABLE 31.1A.

B. Recrystallization of Aspirin

The crude product may be purified by a process called recrystallliza-tion. If the crude product is dissolved and then allowed to recrystallize slowly, most of the impurities will remain in solution. A solvent convenient for this recrystallization process is a mixture of ethyl alcohol and water.

Dissolve the crude product, contained in a 50-mL beaker, in 6 to 9 mL of ethyl alcohol, C_2H_5OH. Add to the solution 18 mL of hot water. If a solid separates at this point, warm until solution is complete. Cover the beaker with a watch glass, and set it aside to cool slowly. Filter the crystals using gravity filtration or vacuum filtration, and dry them by pressing them between several layers of filter paper and then spreading them out on several layers of dry filter paper. Weigh the recrystallized product when it is dry. Record the mass in TABLE 31.1B.

If melting-point apparatus is available, your instructor will demon-strate its use. Grind a very small portion of your recrystallized product, place 2 mm of sample in the base of a capillary tube, and determine the melting point. If the melting point is significantly below 137°C, repeat the recrystallization and drying procedure on a small portion of your sample, and take another melting point. Record the melting points in TABLE 31.1B.

C. Molecular Weight of Aspirin by Freezing Point Depression

By referring to Experiment 20 (Freezing-Point Depression and Molecular Weight of an Unknown Solid), devise a reasonable method for determining the molecular weight of aspirin. Your instructor must approve an outline of your method and the quantities you plan to use before you begin work. Record all experimental data in TABLE 31.1C.

Experiment 31

D. Molecular Weight of Aspirin by Acid-Base Titration

By referring to Experiment 25 (Acid-Base Titration and Molecular Weight of an Unknown Acid), devise a reasonable method for determining the molecular weight of aspirin. Your instructor must approve an outline of your method and the quantities you plan to use before you begin work. The pK_a for aspirin is 3.48, which lies between the two pK_a's of oxalic acid used in Experiment 25. An approximately 0.1 $\underline{M}$ standardized NaOH solution will be provided. Record all data in TABLE 31.1D.

E. Assay of Commercial Aspirin by Sodium Hydroxide Titration

By referring to Experiment 25 (Acid-Base Titration and Molecular Weight of an Unknown Acid), devise a reasonable method for determining the percent of aspirin in commercial aspirin tablets. The pK_a for aspirin is 3.48, which lies between the two pK_a's of oxalic acid used in Experiment 25. Your **instructor must approve an outline of your method and the quantities you plan to use before you begin work.** An approximately 0.1 $\underline{M}$ standardized NaOH solution will be provided. Record all experimental data in TABLE 31.1E.

Perform the calculations in TABLE 31.2, showing all aspects of all calculations.

TABLE 31.1. Masses and data from your devised experiments.

A	Mass of evaporating dish	
	Mass of evaporating dish and salicylic acid	
	Mass of watch glass	
	Mass of watch glass and crude aspirin	
B	Mass of watch glass	
	Mass of watch glass and recrystallized aspirin	
	Melting point of recrystallized aspirin	

TABLE 31.1 (continued)

C Determination of the Molecular Weight of Aspirin by Freezing-Point Depression

(Clearly record and identify all experimental data.)

TABLE 31.1 (continued)

D   Determination of Molecular Weight of Aspirin by Acid–Base Titration

(Clearly record and identify all experimental data.)

Experiment 31

TABLE 31.1 (continued)

E Assay of Commercial Aspirin

   (Clearly record and identify all experimental data.)

Instructor's initials _____

CALCULATIONS

TABLE 31.2. **Percent yield and molecular weight of aspirin and percent aspirin in commercial aspirin.**

A Percent yield of both crude and pure, recrystallized aspirin

&

B

C Molecular weight of aspirin by freezing-point depression

TABLE 31.2 (continued)

D Molecular weight of aspirin by acid-base titration

E Percent aspirin by mass in commercial aspirin

Name ______________________________ **Experiment 31**

Student ID number ______________

Section __________ Date __________

Instructor ______________________

PRE-LABORATORY QUESTIONS

1. Which chemical provides the beneficial effects to the body when aspirin is taken orally? Write a balanced chemical equation to show how this chemical is formed in the body.

2. Calculate the theoretical yield of aspirin expected from 3.0 g of salicylic acid.

3. Account for the observation that one usually smells acetic acid (as in vinegar) when a bottle of older aspirin is newly opened.

4. What is the purpose of adding water to an ethanol solution of aspirin for recrystallization?

5. What is formed when a solution for a freezing-point depression measurement freezes?

6. Explain why a melting point below 137°C is indicative of an impure sample of aspirin.

Section _____ Date _____

Instructor _____   POST-LABORATORY QUESTIONS

1. Suggest three reasons why the percent yield of aspirin is less than 100%.

2. Compare the molecular weight of aspirin as calculated from your experimental data with the molecular weight as calculated from the molecular formula. Explain the difference. Draw structures if necessary.

3. Compare the experimental molecular weight of aspirin as calculated from freezing-point depression measurements with that calculated from NaOH titration data. Explain the difference.

4. Using your experimental value for the percent by mass of aspirin in commercial aspirin, calculate the cost of the aspirin in each tablet and the cost of the total aspirin in the bottle. Assume a bottle of 100. aspirin tablets costs $2.91 and that the cost of the bottle and binder is negligible.

5. Write a balanced chemical equation, and draw all structures for the reaction of aspirin with <u>boiling</u> aqueous sodium hydroxide (excess). Assume that all appropriate functional groups react completely.

Polymers

PURPOSE OF EXPERIMENT: Prepare polymers by both addition polymerization and condensation polymerization, and break down a polymer.

Polymers are long-chain compounds built up from small repeating structural units, called monomers, which are derived from small molecules. Fundamental differences between small molecules and polymers include molecular weight, structure, and bonding, and they usually exhibit different physical or chemical properties, by which they can often be distinguished. The small molecules tend to be volatile liquids of low viscosity, or even gases, at room temperature. Polymers, on the other hand, tend to be nonvolatile, highly viscous liquids, glasses, or solids that soften only at high temperatures. Artificial materials such as fibers, films, plastics, semisolid resins, synthetic rubbers, and silicate glasses, and natural materials such as proteins, natural rubber, cellulose, starch, and complex silicate minerals are examples of polymers that have great practical importance because of the useful properties they exhibit.

Polymers are formed by two general methods: addition polymerization and condensation polymerization.

Addition polymerization is a process in which small molecules join together to form polymers having the same percent composition as the original monomers. A double or triple bond is frequently required in the original small molecule, and atoms that can increase their coordination numbers must be present. Conversion of a double bond in each reacting molecule to two single bonds, with coincident increase in coordination number and bond energy, overcomes the decrease in entropy and is the driving force for the reaction. Thus the polymerization is exothermic, and heat is evolved.

A very important synthetic polymer formed by addition polymerization is polyethylene. Under appropriate conditions, one molecule of ethylene adds to another, and the addition process continues to produce a very long chain of $-CH_2-CH_2-$ units.

$$n[CH_2{=}CH_2] \xrightarrow{\text{catalyst}} -CH_2-CH_2\underbrace{+CH_2-CH_2+}_{\text{repeating unit}}CH_2-CH_2-$$

ethylene   polyethylene

If mixtures of two different reactants are used, **copolymers** containing two kinds of monomers in every polymer molecule are formed.

Condensation polymerization is a process in which polymers are produced as a result of elimination of small, volatile molecules in each step of the process, leaving an extensive polymeric molecule. Organic or inorganic reactants containing more than one functional group are required so that simple coupling reactions can take place at several sites on each reactant. The condensation reaction is often endothermic and is forced to completion by removing the volatile condensation product from the reaction mixture.

A very important synthetic polymer formed by condensation polymerization is marketed under the name Dacron. Under appropriate conditions, one molecule of terephthalic acid and one molecule of ethylene glycol, both bifunctional reactants, condense to form a molecule of water and a species that is still bifunctional and that can react repeatedly in the same manner to form a very long chain polymer.

$$HO-\underset{O}{\overset{O}{C}}-\langle\bigcirc\rangle-\underset{O}{\overset{O}{C}}-OH \ + \ HO-CH_2CH_2-OH \ \rightarrow$$

terephthalic acid   ethylene glycol

$$H_2O \ + \ HO-\underset{O}{\overset{O}{C}}-\langle\bigcirc\rangle-\underset{O}{\overset{O}{C}}-O-CH_2CH_2-OH$$

bifunctional species

$$n\left[HO-\underset{O}{\overset{O}{C}}-\langle\bigcirc\rangle-\underset{O}{\overset{O}{C}}-OH\right] \ + \ n[HO-CH_2CH_2-OH] \ \rightarrow$$

$$2nH_2O \ + \ \left[-\underset{O}{\overset{O}{C}}-\langle\bigcirc\rangle-\underset{O}{\overset{O}{C}}-O-CH_2CH_2-O-\right]_n$$

Dacron

Polymers are frequently classified in terms of bonding in one dimension versus bonding in two or three dimensions. Bonding in one dimension results in **linear polymers** with single-strand chains. Bonding in two or three dimensions results in **cross-linked polymers** having infinite sheets or three-dimensional networks. Linear polymers are produced by addition polymerization if the reactant has only one double bond or by condensation polymerization if the reactant or reactants each have two reactive sites. Such polymers are usually soluble in suitable solvents. Since they also tend to soften when heated, they are called **thermoplastic polymers.** Cross-linked polymers may be produced by addition polymerization if the reactant has more than one double bond, or by condensation polymerization if the reactant or reactants each have more than two reactive sites. Such network polymers are usually insoluble and infusible and are called **thermosetting polymers.**

Some polymers can be **depolymerized** or broken down to small monomer units by appropriate chemical reactants that cleave the linkages between repeating units. Many biological polymers can be depolymerized. Starch, a carbohydrate stored in seeds and roots of many vegetables, including corn and potatoes, exhibits this behavior. Starch is about a 20:80 mixture of two types of condensation polymers of the sugar, glucose. One type called amylose is insoluble in water and is a linear polymer with nearly 2000 glucose units in a chain whereas the other called amylopectin is soluble in water and consists of highly cross-linked chains of 20 to 25 glucose units. Both types can be broken down by the action of either acids or enzymes, freeing the monomer glucose molecules.

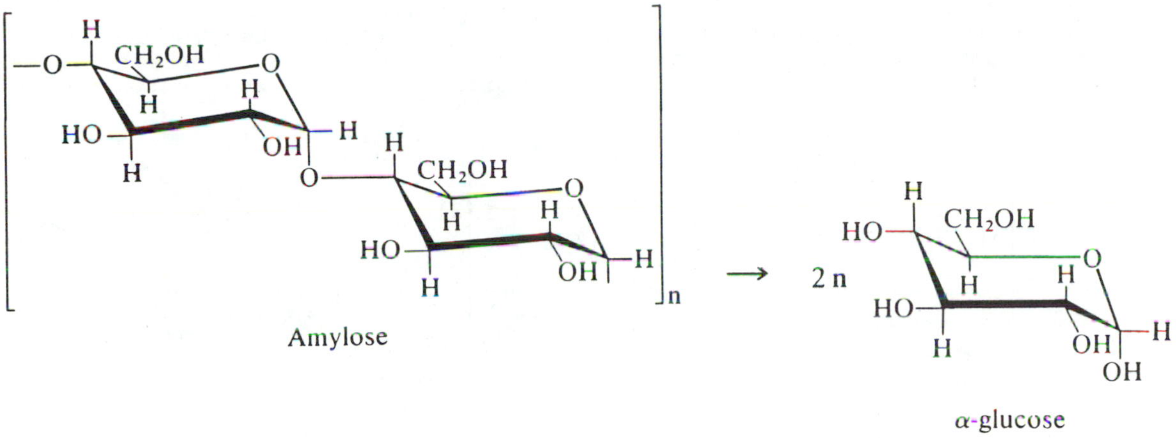

Amylose

α-glucose

You experience this depolymerization when you chew a piece of bread and hold it in your mouth for a few minutes until it begins to taste sweet.

In this experiment you will carry out the processes of addition polymerization, condensation polymerization, and depolymerization, and you will observe the changes in physical or chemical properties between monomers and polymers.

REFERENCES

(1) Kotz, J. C., and Purcell, K. F., <u>Chemistry and Chemical Reactivity</u>, Saunders College Publishing, Philadelphia, 1987, sections 22.2a, 23.3, 23.4, 24.4, and 24.6.

(2) Masterton, W. L., Slowinski, E. J., and Stanitski, C. L., <u>Chemical Principles</u>, 6th ed., Saunders College Publishing, Philadelphia, 1985, sections 13.1, 28.1, 28.3, and 28.4.

EXPERIMENTAL PROCEDURE

(Study this section and the PRE-LABORATORY QUESTIONS before coming to the laboratory. Wear safety goggles when performing this experiment. Most of the reagents in this experiment are flammable and/or toxic. Don't light Bunsen burners except when needed, and wash your hands thoroughly after each part.)

A. Addition Polymerization

1. **Lucite.** Lucite, or Plexiglas, is a high polymer resulting from the addition polymerization of methyl methacrylate in the presence of benzoyl peroxide initiator. The oxygen-oxygen bond in benzoyl peroxide is weak, and decomposition by heating or ultraviolet light yields a **free radical.**

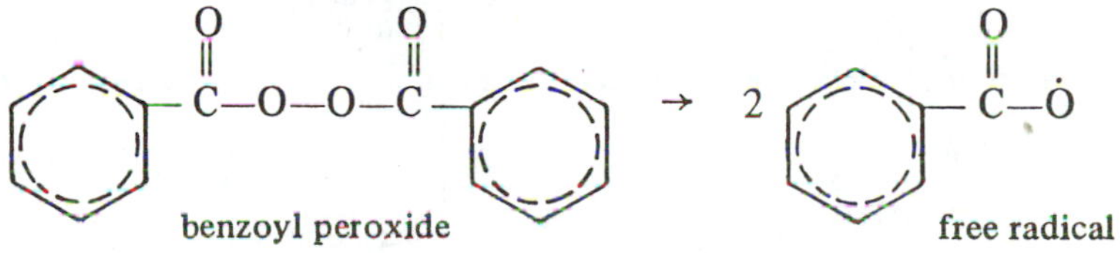

benzoyl peroxide

free radical

Such free radicals can add to a carbon-carbon double bond and form a new reactive free radical that is capable of adding onto another monomer unit. Thus, the polymerization is catalyzed by free radicals such as those produced by the decomposition of organic peroxides like benzoyl peroxide.

$$n \left[H_2C = \overset{\overset{\textstyle CH_3}{|}}{\underset{\underset{\textstyle CO_2CH_3}{|}}{C}} \right] \xrightarrow{\text{catalyst}} \cdots$$

methyl methacrylate monomer

Lucite polymer (Plexiglas)

Support a 100-mL beaker half-full of water on wire gauze on a ring stand, and heat the water to boiling (Laboratory Methods **D**). Place 1 mL of methyl methacrylate, $C_5H_8O_2$, in a clean 10 x 75-mm test tube, and have your instructor add about 5 mg of the catalyst, benzoyl peroxide, $C_{14}H_{10}O_4$. CAUTION: <u>**Methyl methacrylate is a moderate fire hazard when exposed to heat or flame and can react with oxidizing materials.**</u> CAUTION: <u>**Benzoyl peroxide is a moderate fire hazard, especially in contact with reducing agents, and may explode spontaneously, especially if heated in the absence of water near its melting point (104°C).**</u> Use a test tube holder, and hold the test tube in an upright position so that the tube is about half-immersed in the boiling water. Stir the mixture in the tube until the catalyst dissolves, and then continue to heat without stirring until the contents begin to become quite viscous. After placing the test tube on a slant in the beaker, cool the water bath to about 65°C by pouring some cold water into the beaker, and then allow the test tube to remain undisturbed until its contents become rigid.

Break the test tube carefully in a waste container, and describe in TABLE 32.1A1 how Lucite compares with the starting material. Test the solubility of Lucite in water, and drop a piece of it on your desktop to see if it shatters easily. Describe your results in TABLE 32.1A1.

Grasp a piece of Lucite with crucible tongs, and warm it gently above a Bunsen burner flame. Describe what happens in TABLE 32.1A1. Heat the Lucite more intensely, and describe what happens in TABLE 32.1A1.

2. **Polyacrylonitrile.** Acrylonitrile, C_3H_3N, can be addition polymerized in the presence of potassium persulfate, $K_2S_2O_8$, and sodium bisulfite, $NaHSO_3$. Potassium persulfate has a weak peroxide bond like benzoyl peroxide in part **A1**. The bond is easily broken to form free radicals that are capable of adding to the double bond in acrylonitrile to form a new reactive free radical, which can add to another reactive unit. The polymerization reaction is thus catalyzed by these free radicals. Sodium bisulfite is a reducing agent, which apparently reduces some of the $S_2O_8{}^{2-}$ to $SO_4{}^{2-}$ to help control the rate of the reaction once it has been initiated.

$$n \left[\underset{\underset{\textstyle CN}{|}}{CH_2 = CH} \right] \xrightarrow{\text{catalyst}} \cdots$$

acrylonitrile

polyacrylonitrile

Place 2 mL of acrylonitrile, C_3H_3N, in a clean 50-mL beaker. **CAUTION:** <u>**Acrylonitrile is highly poisonous; wash your hands thoroughly after using it. It should not be heated because it is a dangerous fire hazard when exposed to heat or flame.**</u> Add 2 mL of 5% potassium persulfate, $K_2S_2O_8$, solution and 2 mL of 5% sodium bisulfite, $NaHSO_3$, solution. **CAUTION:** <u>**Potassium persulfate evolves highly toxic fumes if decomposed by heating.**</u> Swirl the solution vigorously for several minutes until the polymer starts to form.

Check carefully for any temperature change, and record your results in TABLE 32.1A2.

Grasp the polymer with clean crucible tongs, and pull it upward out of the beaker. Describe what you observe in TABLE 32.1A2.

Grasp a piece of polyacrylonitrile with crucible tongs, and warm it gently with a Bunsen burner flame. Describe what happens in TABLE 32.1A2. Heat the polyacrylonitrile more intensely, and record what happens in TABLE 32.1A2.

Orlon is a highly polymeric fiber resulting from the catalyzed addition copolymerization of acrylonitrile with about 10% vinylpyridine, C_7H_7N. The vinylpyridine unit is incorporated into the polymer to enable dyes to be adsorbed more readily by the fiber. Normally the copolymer is dissolved in a solvent such as N,N-dimethylacetamide, from which it can be spun into threads. It can also be wet spun from concentrated salt solutions.

3. Sulfur. Orthorhombic sulfur, consisting of discrete S_8 molecules with the eight sulfur atoms in a nonplanar, crown-shaped ring, is the stable form of sulfur under ambient conditions. Orthorhombic sulfur is a yellow, brittle solid that is insoluble in water but readily soluble in carbon disulfide, CS_2.

If sulfur is heated to near its boiling point, the S_8 rings break open to form S_8 chains. These S_8 chains then react with each other to form much longer chains by addition polymerization. If the molten sulfur were cooled slowly, it would simply crystallize back to orthorhombic sulfur. However, by <u>rapid</u> quenching or cooling, the much longer chains can be preserved in a metastable state for further examination.

Fill a 10 x 75-mm test tube one-quarter full with powdered sulfur. Heat the test tube **gently** by passing it in and out of a cool, luminous Bunsen burner flame. Note any changes in phase, color, and viscosity, and record them in TABLE 32.1A3. Then continue to heat the sulfur, but **do not boil it. CAUTION:** <u>**Sulfur emits highly toxic fumes upon heating in excess oxygen; your very small-mouthed test tube has deficient oxygen.**</u> Stir the liquid sulfur with a glass rod during this stronger heating. Record any additional changes in TABLE 32.1A3.

Have ready a 100-mL or larger beaker at least half-filled with cold water. When the thick molten sulfur is near boiling, quench it by pouring it rapidly from the test tube into the beaker of cold water.

Remove the polymeric sulfur from the beaker of water. Examine the physical properties of the polymeric sulfur, and compare them with properties of a sample of orthorhombic crystals provided by your instructor. Use a magnifying glass, if available. Record the similarities and differences in TABLE 32.1A3.

Leave your polymeric sulfur in a beaker in your drawer until the next laboratory period. Then examine the properties of the sample again, noting carefully any changes over the week.

Experiment 32

B. Condensation Polymerization

   1. Glyptal resin. Glyptal resin is a high polymer resulting from the condensation polymerization of polyalcohols such as glycerol with anhydrides such as phthalic anhydride. Such resins are often used in the manufacture of paints and enamels.

glycerol   phthalic anhydride

water   Glyptal polymer

   Place 0.5 mL of glycerol, $C_3H_8O_3$, in a clean 10 x 75-mm test tube, and add 1.3 g of phthalic anhydride, $C_8H_4O_3$, weighed on a beam balance (Laboratory Methods **C**). Tilt the test tube at about a 30° angle, and heat the mixture **gently** over a cool, luminous Bunsen burner flame until it melts and bubbles slowly. **CAUTION:** <u>**Gentle heating is required because glycerol is a dangerous fire hazard when exposed to heat, flame, or powerful oxidizers**</u>. Describe what you observe in TABLE 32.1B1.

   Maintain the gentle heating until you notice a sudden change in the physical properties of the contents of the test tube. Then allow the product to cool. Examine the Glyptal polymer, and describe its properties in TABLE 32.1B1.

   2. Silicones. Silicones are high polymers which result from condensation polymerization of alkyl- and aryl-substituted silicon dihydroxides (silanediols). Since these polymers have a silicon-oxygen framework and have carbon atoms only in substituent groups, they have unique properties. The strong Si-O bonds make silicones thermally stable over a wide range of temperatures and resistant to oxidation. These properties make them useful as lubricants which must function at both low and high temperatures.   Organic substituents at the surface of the polymer make silicones highly water repellent.   This latter property makes them useful for waterproofing leather, various fabrics, and paper products, and as caulking materials.

   Dimethylsiliconedihydroxide, $(CH_3)_2Si(OH)_2$, can be obtained by the rapid hydrolysis in water, or even in moisture from the air, of dimethylsilicondichloride, $(CH_3)_2SiCl_2$.

   $(CH_3)_2SiCl_2(l)$ + 2 $H_2O(l)$ ---> $(CH_3)_2Si(OH)_2(l)$ + 2 $HCl(g)$

Rapid, spontaneous condensation polymerization then yields a silicone polymer.

$$n \begin{bmatrix} & CH_3 & \\ & | & \\ HO & -Si- & OH \\ & | & \\ & CH_3 & \end{bmatrix} \rightarrow nH_2O + \begin{array}{ccc} CH_3 & CH_3 & CH_3 \\ | & | & | \\ -Si-O- & Si-O- & Si-O- \\ | & | & | \\ CH_3 & CH_3 & CH_3 \end{array}$$

dimethylsilicondihydroxide   silicone polymer

Moisten a piece of filter paper **thoroughly** with a solution of dimethylsilicondichloride, $(CH_3)_2SiCl_2$, in n-hexane, C_6H_{14}. **CAUTION:** **n-Hexane is highly flammable.** Allow the filter paper to dry in the hood. By the time the paper is dry, the above two reactions should have occurred, and an invisible layer of silicone polymer should have formed on the filter paper.

Grasp the dry filter paper, and compare the feel of it with the feel of an untreated piece of filter paper. Record any differences in TABLE 32.1B2.

Drop several drops of water from a medicine dropper onto both treated and untreated filter paper. Describe in TABLE 32.1B2 any differences in behavior that you observe.

Place a drop of ink on both treated and untreated filter paper. Describe in TABLE 32.1B2 any differences in behavior that you observe.

Moisten a piece of previously folded filter paper **thoroughly** with a solution of dimethylsilicondichloride in hexane. Allow the filter paper to dry **in the hood.**

Filter a few drops of a mixture of an aqueous $CuSO_4$ solution and hexane through the treated filter paper using gravity filtration (Laboratory Methods I). Then filter the same mixture through an untreated piece of filter paper. Describe in TABLE 32.1B2 any differences in behavior that you observe.

3. **Nylon.** Nylon is a trade name for high-molecular weight polyamides that result from condensation polymerization of dibasic acids and diamines, or from ω-amino acids. Nylon can be extruded from a melt as monofilaments, or spun from a solution of formic acid, HCOOH, and phenol, C_6H_5OH. The resulting fibers have a low density, are elastic and lustrous, and mass for mass are stronger than steel. However, they are also low melting and difficult to dye.

$$n \begin{bmatrix} & O & & O & \\ & \| & & \| & \\ Cl-C- & (CH_2)_8 & -C-Cl \end{bmatrix} + n \begin{bmatrix} H_2N- & (CH_2)_6 & -NH_2 \end{bmatrix} \rightarrow nHCl$$

sebacoyl chloride   hexamethylenediamine

$$+ \begin{array}{c} O & O \\ \| & \| \\ -C-(CH_2)_8-C-NH-(CH_2)_6-NH- \end{array} \begin{array}{c} repeating\ unit \\ O & O \\ \| & \| \\ C-(CH_2)_8-C-NH-(CH_2)_6-NH- \end{array}$$

Nylon 6-10

Nylon 6-10, so named because one monomer has six carbon atoms and the other has ten carbon atoms, can be prepared by condensation polymerization of sebacoyl chloride and hexamethylenediamine. Formation of HCl increases the acidity enough to inhibit the reaction and hydrolyze the amide bonds. Thus, the reaction must be carried out in the presence of a base. The raw Nylon produced in this experiment is not highly enough polymerized to be strong enough for ordinary use. Even after appropriate polymerization the molecules must be drawn to four times their original length to gain strength by orienting the molecules along the axis of the fibers.

Add 1 mL of sebacoyl chloride, $C_{10}H_{16}Cl_2O_2$, and 50. mL of n-hexane, C_6H_{14}, to a 200-mL or larger beaker. **CAUTION: <u>Sebacoyl chloride is corrosive and is a lachrymator</u>. CAUTION: <u>n-Hexane is highly flammable</u>**. In a 100-mL beaker, dissolve in 25 mL of water 2 g of sodium carbonate, Na_2CO_3, and 2 g of hexamethylenediamine, $C_6H_{16}N_2$, both weighed on a beam balance. Without mixing any more than necessary, pour the water solution from the 100-mL beaker onto the n-hexane solution in the 200-mL beaker. Add two drops of phenolphthalein solution to make the organic-aqueous liquid junction more visible. Describe in TABLE 32.1B3 what you observe at the junction.

Grasp the film at the liquid junction with clean crucible tongs and slowly pull the film up through the water layer. Describe in TABLE 32.1B3 what happens at the organic-aqueous liquid junction.

Continue to lift the string of Nylon and wrap it around a glass rod. Wash the Nylon thoroughly with water. **CAUTION: <u>Avoid getting any of the solution on your hands</u>**. Examine the Nylon when it is dry, and describe its properties in TABLE 32.1B3.

Grasp a piece of the Nylon with your tongs, and warm it very gently above a cool, luminous Bunsen burner flame. Describe what you observe in TABLE 32.1B3.

Grasp a piece of Nylon stocking from the reagent bench, and warm it gently above a cool, luminous Bunsen burner flame. Describe what you observe in TABLE 32.1B3.

4. Rubber. Natural rubber is formed from the latex of certain trees or shrubs. It is a polymer made up of a large number of isoprene units.

repeating unit

$$+CH_2-\underset{CH_3}{\underset{|}{C}}=\underset{H}{\underset{|}{C}}-CH_2+CH_2-\underset{CH_3}{\underset{|}{C}}=\underset{H}{\underset{|}{C}}-CH_2+CH_2-\underset{CH_3}{\underset{|}{C}}=\underset{H}{\underset{|}{C}}-CH_2+$$

natural rubber

Unvulcanized rubber has an irregular noncrystalline structure, with the molecules being coiled and intertwined. It is sticky, and the molecules can be pulled apart from each other. A double bond is still available in each isoprene unit for addition polymerization. On heating with sulfur in the presence of appropriate catalysts, the three-dimensional, cross-linked structure of vulcanized rubber is formed.

$$-CH_2-CH=CH_2-CH_2-CH_2-\overset{\overset{\displaystyle CH_3}{|}}{\underset{\underset{\displaystyle CH_3}{|}}{C}}-\overset{\overset{\displaystyle S}{|}}{\underset{\underset{\displaystyle S}{|}}{CH}}-CH_2-$$

vulcanized rubber

Many attempts have been made to produce synthetic rubber-like polymers. One such polymer, **Thiokol**, which is very resistant to abrasion and to swelling by solvents, results from condensation polymerization of ethylene dichloride and sodium polysulfide.

$$n\text{Na}_2\text{S}_4 + n(\text{Cl}-\text{CH}_2\text{CH}_2-\text{Cl}) \rightarrow 2n\text{NaCl} + \left[CH_2-CH_2-S-\overset{\overset{S}{\|}}{\underset{\underset{S}{\|}}{S}}\right]$$

sodium polysulfide   ethylene dichloride

repeating unit

Thiokol

Dissolve 1 g of sodium hydroxide, NaOH, in 25 mL of water in a 100-mL or larger beaker. Heat this solution to boiling. Add 2 g of powdered sulfur, and stir until all of the sulfur has dissolved. Take the dark brown sodium polysulfide solution **to the hood**, and add 5 mL of ethylene dichloride, $C_2H_4Cl_2$, keeping the mixture hot but not boiling. **CAUTION: <u>Ethylene dichloride irritates the eyes, nose, and throat, and is highly flammable</u>.** Stir this mixture vigorously. Describe what you observe in TABLE 32.1B4.

After the reaction has been completed, decant the liquid (Laboratory Methods H), and wash the lump of spongy material several times with cold water. Carefully examine the product for stretch and bounce. Describe what you observe in TABLE 32.1B4.

C. Depolymerization

In order to study depolymerization, we must have an analytical method for showing when depolymerization occurs. When the depolymerization involves the breaking down of starch to glucose, two reagents - iodine, I_2, and Benedict's reagent - serve this function. You may have used starch-potassium iodide test paper in previous experiments and observed the intense purple-black complex that iodine forms in the presence of starch. Iodine forms no such colored complex with simple sugars such as glucose. Benedict's reagent oxidizes the free aldehyde groups of simple sugars such as glucose. There are no free aldehyde groups in starch because these groups are involved in the bonding to form the polymer chain. Therefore, starch does not react with Benedict's reagent. On the other hand, in the presence of glucose, the bright blue stabilized solution of Cu(II) in Benedict's reagent oxidizes the free aldehyde groups and forms an orange precipitate of copper(I) oxide, Cu_2O. Taken together, iodine indicates the presence of starch whereas Benedict's reagent indicates the presence of glucose.

Add 25 mL of distilled water to a 100-mL or larger beaker, and bring the water to a boil. Weigh 0.5 g of soluble starch on a beam balance, and stir it with 4 mL of distilled water to form a slurry. Pour the starch slurry into the boiling water, stir the mixture for 30. seconds, and then allow the mixture to cool to room temperature.

Transfer 1 mL of starch solution to each of two different 16 x 150-mm test tubes. To one of the test tubes, add a few drops of an iodine solution. **CAUTION:** **Iodine solution will stain your skin; wash your hands carefully if there are any spills**. Describe what you observe in TABLE 32.1C. To the second test tube, add 2 mL of Benedict's reagent, and heat the test tube in a boiling water bath for 5 minutes. **CAUTION:** **Benedict's reagent is toxic; wash your hands carefully after using it**. Describe what you observe in TABLE 32.1C.

Transfer 10. mL of starch solution to a 50-mL or larger beaker. Add 5 mL of 0.5 M hydrochloric acid, HCl, solution, and boil the solution gently for 5 minutes to hydrolyze the starch. Allow the solution to cool, and then repeat the tests of the previous paragraph on 1-mL portions. Describe what you observe in TABLE 32.1C.

Transfer 5 mL of starch solution to a 50-mL or larger beaker. Collect about 5 mL of your saliva, and transfer that to the same beaker. Heat the mixture **very gently**, maintaining the temperature of the mixture at about 40.°C for 10. minutes. Then repeat the tests of two paragraphs above on 1-mL portions. Describe what you observe in TABLE 32.1C.

Name _____

Student ID number _____

Section _____ Date _____

Instructor _____

TABLE 32.1. Experimental observations and related questions.

(Answer all QUESTIONS before handing in your report.)

A1 Observation Properties of Lucite versus methyl methacrylate

QUESTION Why do you suppose this product is called Lucite?

Observation Solubility of Lucite in water

Observation Ease of shattering Lucite

QUESTION Why do you suppose this product is called Plexiglas?

Observation Behavior of Lucite toward gentle heating

QUESTION What does this suggest about how Lucite may be formed into intricate shapes?

QUESTION Why is Lucite said to be a thermoplastic polymer?

Observation Behavior of Lucite toward intense heating

QUESTION What evidence do you observe for the breaking down of the Lucite at high temperatures?

TABLE 32.1 (continued)

| A2 | Observation | Evidence for temperature change while preparing polyacrylonitrile |

QUESTION   What does this indicate about the reaction?

Observation   Behavior of polyacrylonitrile upon pulling from beaker

Observation   Behavior of polyacrylonitrile toward gentle heating

Observation   Behavior of polyacrylonitrile toward intense heating

A3   Observation   Behavior of sulfur toward gentle heating

Observation   Behavior of sulfur toward stronger heating

Observation   Comparison of physical properties of orthorhombic sulfur and polymeric sulfur

Name _____

Student ID number _____

Section _____ Date _____

Instructor _____

TABLE 32.1 (continued)

B1 **Observation** Behavior of glycerol and phthalic anhydride toward gentle heating

QUESTION What volatile substance was formed in the tube? What is your evidence?

Observation Properties of the Glyptal polymer

B2 **Observation** Feel of filter paper treated with silicone polymer compared to untreated filter paper

Observation Differences in behavior toward water for treated versus un- treated filter paper

Observation Differences in behavior toward ink for treated versus un- treated filter paper

Observation Behavior of $CuSO_4$-hexane mixture upon filtering through treated versus untreated filter paper

B3 **Observation** Behavior at the organic-aqueous liquid junction upon com- bining the two solutions

Observation Behavior at the organic-aqueous liquid junction upon pull- ing up the film

Experiment 32

TABLE 32.1   (continued)

Observation   Properties of Nylon

Observation   Behavior of Nylon toward gentle heating

Observation   Behavior of Nylon stocking toward gentle heating

---

B4   Observation   Behavior during preparation of Thiokol rubber

Observation   Stretch and bounce properties of Thiokol rubber

---

C   Observation   Behavior of starch solution toward iodine solution

Observation   Behavior of starch solution toward Benedict's reagent

Observation   Behavior of HCl-hydrolyzed starch solution toward iodine solution

Observation   Behavior of HCl-hydrolyzed starch solution toward Benedict's reagent

Observation   Behavior of saliva-hydrolyzed starch solution toward iodine solution

Observation   Behavior of saliva-hydrolyzed starch solution toward Benedict's reagent

---

Instructor _____   PRE-LABORATORY QUESTIONS

1. Define clearly any of the following terms that your instructor desig-
    nates, illustrating each definition with a specific example.

    a. monomer   b. polymer   c. repeating unit   d. copolymer

    e. addition polymerization   f. condensation polymerization

    g. linear polymer   h. cross-linked polymer   i. depolymerization

    j. thermoplastic polymer   k. thermosetting polymer

2. Most substances become less viscous or more free-flowing when they are heated whereas sulfur exhibits exactly the opposite behavior when it is heated near its boiling point in part **A3**. Why should each of these behaviors be expected?

3. What functional group is present that makes Nylon 6-10 in part **B3** a polyamide?

4. A temperature of about 40.°C is required for the saliva-hydrolysis of starch in part **C**. Why should the temperature be neither higher nor lower?

5. Write a balanced net ionic equation for the oxidation of acetaldehyde to acetic acid by Benedict's solution.

$$Cu^{2+}(aq) \ + \ CH_3-\overset{\overset{O}{\|}}{C}-H(aq) \quad ---> \quad Cu_2O(s) \ + \ CH_3-\overset{\overset{O}{\|}}{C}-O-H(aq)$$

Name _____

Student ID number_____

Section _____ Date _____

Instructor _____

POST-LABORATORY QUESTIONS

1. Dynel is a copolymeric fiber resulting from addition polymerization of 50% acrylonitrile, CH_2＝$CHCN$, and 50% vinyl chloride, CH_2＝$CHCl$. It is used to make fabrics just like Nylon, Orlon, and Dacron fibers are, but it is very heat sensitive compared to the other three.

   a. Show, by analogy to the production of Orlon, how these monomers combine to form Dynel.

   b. Would you expect the product to consist of molecules all having the same molecular weight? Why, or why not?

2. Silicone II paintable sealant, commonly used as a caulking material, has the following statement under **Directions:** "Wipe hands and tools with a dry cloth or paper towel, before washing with soap and water." In fact, if you wash with soap and water first, you get a slimy mess. What is there about the structure of silicones that makes this direction important and the behavior expected?

3. In the synthesis of Glyptal resin, two different kinds of small molecules have reacted to form a unit that by repetition can build up an extended structure.

   a. What functional group of glycerol has taken part in this reaction?

   b. Are any of these functional groups remaining in the structure shown for the Glyptal polymer in part **B1**?

   c. What do you think might happen to the properties of the polymer if more phthalic anhydride were heated with the structure shown in part **B1**?

   d. What would be the effect upon natural rubber if a very large amount of sulfur were used in vulcanizing in part **B4** so that a very large number of cross-links were formed between chains?

   e. Account for the use of the term <u>thermosetting</u> to describe the polymers resulting in <u>c</u>. and <u>d</u>.

Stoichiometry of a Precipitation Reaction,

an Oxidation-Reduction Reaction,

and a Complexation Reaction

PURPOSE OF EXPERIMENT: Determine the mole ratio of reactants that produces the maximum yield of product in several types of reactions.

An important question for chemists is how the stoichiometry of a chemical reaction, or the amounts of materials consumed or produced as indicated by coefficients in a balanced equation, is determined for a new chemical reaction. The coefficients in a balanced equation indicate the limiting mole ratio of reactants that produces the maximum yield of product.

Consider a simple example that is analogous to amounts of chemical reactants. Suppose that you are in the construction business and need an answer to the following question. How many complete doors can be made if your warehouse contains 67 panels, 138 hinges, 83 knobs, and 366 panes of glass knowing that each door requires one panel, three hinges, one knob, and six panes of glass? The question can be answered by determining how many doors can be made assuming that each part is the limiting part and then choosing the smallest number from the set. Thus, 67 doors can be made if the 67 panels are limiting, 138/3 or 46 doors can be made if the 138 hinges are limiting, 83 doors can be made if the 83 knobs are limiting, and 366/6 or 61 doors can be made if the 366 panes of glass are limiting. The smallest number of doors, 46, can be made, and hinges are found to be the limiting part.

When a new chemical reaction is observed, the relative amounts of reactants consumed are uncertain. One of the reactants is almost always in excess because the reactants are not mixed in proportions that will react completely with each other. Some of this reactant remains unused. We cannot even know what the appropriate proportion is as long as we are uncertain about the chemical composition of the products. To find out which one of the reactants, and how much of it, remains unreacted, we could analyze the reaction mixture for the reactants. Often it is simpler and just as informative to analyze some physical property or chemical property that gives a measure of the quantity of a product formed and compare the product yield for many different mole ratios of reactants. This approach is particularly suitable for complex reaction mixtures. The general method, called the method of continuous variation, involves use of a constant total number of moles of reactants to determine the mole ratio of reactants that react completely with each other to produce the maximum yield of product.

In this experiment, you will carry out three simple aqueous chemical reactions: a precipitation reaction, an oxidation-reduction reaction, and a complexation reaction. For the precipitation reaction the product is easily separated as a solid, and the mass of the solid is determined after it is filtered and dried. For the exothermic oxidation-reduction reaction the change of temperature provides a measure of the heat evolved and thus of the quantity of product formed. For the complexation reaction the absorbance of visible light by the product provides a measure of the quantity of product formed.

The reactants for each of the three types of reaction will be mixed in a wide range of mole ratios. Each of you will be assigned the analysis of one of the mole ratios in triplicate for each pair of reactants. If the reactants are mixed in proportions far from that proportion where they react completely with each other, little product will form because one of the reactants will be present in a great excess and will be largely unused. If the reactants are mixed in proportions close to that where they react completely with each other, nearly the maximum yield of product will be formed. Since the individual results are meaningless alone, all members of the class will plot their results before the end of the period. The mole ratios of reactants at the maximum yields can be determined from the graphs.

REFERENCES

(1) Kotz, J. C., and Purcell, K. F., <u>Chemistry and Chemical Reactivity</u>, Saunders College Publishing, Philadelphia, 1987, sections 3.1-3.4, 4.1, 4.2, 4.4, 5.4, 15.6, 25.2, 25.3, and 25.6.

(2) Masterton, W. L., Slowinski, E. J., and Stanitski, C. L., <u>Chemical Principles</u>, 6th ed., Saunders College Publishing, Philadelphia, 1985, sections 2.6, 3.6-3.8, 5.1, 18.1, 18.2, 21.1, 21.2, and 23.2.

EXPERIMENTAL PROCEDURE

(Study this section and the PRE-LABORATORY QUESTIONS before coming to the laboratory. **Wear safety goggles when performing this experiment.**)

A. Stoichiometry of a Precipitation Reaction

In this experiment you will use reactants C_wA_y and $C'_xA'_z$, where C and C' are unknown cations; A and A' are unknown anions; and w, x, y, and z are subscripts in the chemical formulas. Various volumes of 0.50 $\underline{M}$ reactants will be mixed to give a total volume of 10.0 mL of reactants as shown in TABLE 33.1. You will be assigned one of these pairs of volumes by your instructor. Since the reactants are equimolar, the volume ratios also represent mole ratios. The quantity of product formed for each mole ratio will be determined by weighing the mass of dried precipitate produced.

TABLE 33.1. Volumes of reactants (mL).

0.50 $\underline{M}$ C_wA_y	1	2	3	4	5	6	7	8	9
0.50 $\underline{M}$ $C'_xA'_z$	9	8	7	6	5	4	3	2	1

Pipets that deliver up to 5.0 mL or up to 10.0 mL are available at the equipment bench. Obtain combinations of pipets that you require to measure the two volumes of reactants that you are assigned. Rinse the pipets carefully with distilled water from your wash bottle.

At the reagent bench obtain in a 10-mL or 100-mL graduated cylinder three times the required volume of 0.50 $\underline{M}$ C_wA_y solution plus a few milliliters in excess. Rinse the appropriate pipet twice with about 1-mL portions of 0.50 $\underline{M}$ C_wA_y solution, draining the solution into a beaker for waste liquid. **CAUTION: <u>Be certain to follow the procedure in Laboratory Methods B very carefully in order to draw liquids into pipets, preferably using a safety pipet filler</u>.** Then measure the required volume of 0.50 $\underline{M}$ C_wA_y solution in triplicate, and drain the portions into separate 50-mL or 100-mL beakers. Repeat the above procedure to add the appropriate volume of 0.50 $\underline{M}$ $C'_xA'_z$ solution to each beaker. <u>What do you observe?</u>

Heat the mixtures gently (below boiling) with stirring over a Bunsen burner flame for about 10. minutes (Laboratory Methods D). This heating helps to flocculate the precipitate or to convert very small particles to larger aggregates that can be filtered easily.

While the beakers and their contents are cooling to room temperature, label three pieces of filter paper that fit your long stem funnel or Buchner funnel 1, 2, and 3. Weigh each one on an analytical balance (Laboratory Methods C). Record the masses in TABLE 33.4.

Filter the mixtures in the three beakers through appropriately numbered and preweighed filter papers using either gravity filtration or vacuum filtration (Laboratory Methods I). In each instance use three successive 3-mL portions of cold water along with a rubber policeman to wash any solid remaining in each beaker onto the filter paper holding the crystalline residue (Laboratory Methods I).

After each filtration is complete, carefully remove the filter paper from the funnel with the help of a spatula, and place the filter paper and precipitate on the same watch glass labeled with your name. Dry the filter papers and precipitates by heating them either on a steam bath, with a heat lamp, or in an oven (Laboratory Methods M), as designated by your instructor. Continue to parts B and/or C while your samples are drying.

Let your filter papers and precipitates cool to room temperature, weigh them carefully on an analytical balance, and record the masses in TABLE 33.4.

Calculate the average mass of precipitate and average deviation, and plot the <u>average mass of precipitate</u> with error bars versus <u>mL of reactant</u> C_wA_y on the class graph provided by your instructor. After all students have plotted their data and linear portions of the graph have been extrapolated to form an intersection, determine from the graph the volume ratio (and thus mole ratio) of reactants at which maximum yield of product was obtained, and record that ratio in TABLE 33.4. Also record the maximum yield of product in TABLE 33.4.

B. Stoichiometry of an Oxidation-Reduction Reaction

In this experiment various volumes of 0.70 $\underline{M}$ hypochlorite, ClO^-, solution and a basic solution of 0.70 $\underline{M}$ thiosulfate, $S_2O_3^{2-}$, will be mixed to give a total volume of 50.0 mL of reactants as shown in TABLE 33.2. You will be assigned one of these pairs of volumes by your instructor. Since the reactants are equimolar, the volume ratios also represent mole ratios.

TABLE 33.2. Volumes of reactants (mL).

0.70 $\underline{M}$ ClO^-	2	5	10	15	20	25	30	35	40	45	48
0.70 $\underline{M}$ $S_2O_3^{2-}$	48	45	40	35	30	25	20	15	10	5	2

As a result of the graphical analysis that is used in this experiment, you need not determine the absolute quantity of product formed as you did in part A, but only some property that is directly proportional to that absolute quantity. For an exothermic reaction the amount of heat liberated is directly proportional to the number of moles (and thus the quantity) of product formed. Moreover, if the reaction is carried out in an insulated container that eliminates heat flow from the solution to the surroundings, all of the heat will be used to cause a rise in temperature of the solution. A styrofoam cup approximates such conditions. Therefore, the temperature rise is directly proportional to the quantity of product formed, and the maximum amount of heat is released producing a maximum temperature rise when the reactants are mixed in their correct stoichiometric ratio.

Labindustries repipet dispensers (Laboratory Methods B) located at the reagent bench may be used to measure and deliver the volumes of reactants that you are assigned. Otherwise use whatever method is designated by your instructor to deliver the required volume of 0.70 $\underline{M}$ ClO$^-$ solution in triplicate into three styrofoam cups. Similarly deliver the required volume of 0.70 $\underline{M}$ S$_2$O$_3^{2-}$ solution in triplicate into separate 50-mL or 100-mL beakers.

Measure the temperature of the 0.70 $\underline{M}$ ClO$^-$ solution in the first styrofoam cup, and record it in TABLE 33.5. Then measure and record the temperature of the 0.70 $\underline{M}$ S$_2$O$_3^{2-}$ solution in the first beaker. Be sure to rinse and dry your thermometer between measurements.

The two temperatures should be identical. However, if they are not, you can use a weighted average temperature (weighted for the volume of each solution used) to determine your initial temperature. Mix the reactants thoroughly with a swirling motion while adding S$_2$O$_3^{2-}$ solution in the first beaker to the first styrofoam cup. Measure and record the maximum temperature attained in TABLE 33.5. The mixture will warm rapidly at first and then gradually cool to room temperature so that you should have no difficulty observing the maximum temperature. Do you observe any other evidence for a chemical reaction?

Repeat the above procedure with the other two sets of solutions, and measure and record the maximum temperatures in TABLE 33.5.

Calculate the average temperature rise and average deviation, and plot the average temperature rise with error bars versus mL of ClO$^-$ solution on the class graph provided by your instructor. After all students have plotted their data and linear portions of the graph have been extrapolated to form an intersection, determine from the graph the volume ratio (and thus mole ratio) of reactants at which maximum yield of product was obtained, and record that ratio in TABLE 33.5. Also record the maximum temperature rise in TABLE 33.5.

C. Stoichiometry of a Complexation Reaction

In the absence of complexing agents, aqueous solutions of iron(II) contain the octahedral, pale blue-green hexaaquairon(II) cation, [Fe(OH$_2$)$_6$]$^{2+}$, shown in FIGURE 33.1a. Iron(II) also forms complexes with many other ligands. Although an iron(II) complex with ammonia, NH$_3$, as a ligand is unstable in water except in saturated aqueous ammonia, stable iron(II) complexes can be produced with chelating amine ligands that bind at more than one position simultaneously. One such ligand, 1,10-phenanthroline, shown in FIGURE 33.1b and often abbreviated phen, forms an intensely colored complex with iron(II) having an absorption maximum in the visible region of the spectrum at 490 nm. Thus, the formation of the complex can be followed spectrophotometrically.

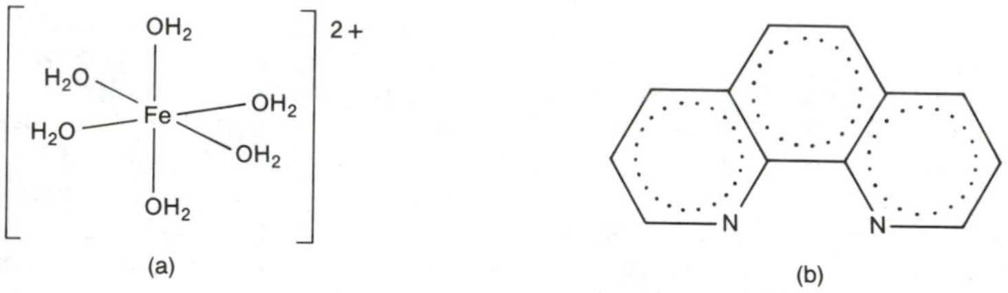

FIGURE 33.1. Structures of [Fe(OH$_2$)$_6$]$^{2+}$ and 1,10-phenanthroline.

As a result of the graphical analysis that is used in this experiment, you need not determine the absolute quantity of product formed as you did in part **A**, but only some property that is directly proportional to that absolute quantity. The intensity of color or the absorbance of the complex formed between Fe^{2+} and 1,10-phenanthroline is such a property because absorbance is directly proportional to molar concentration. Thus, for a constant volume, molar concentration is a measure of the number of moles of and the quantity of the complex. Therefore, the absorbance is directly proportional to the quantity of complex formed, and the maximum absorbance occurs when the reactants are mixed in their correct stoichiometric ratio.

In this experiment various volumes of 0.00070 $\underline{M}$ Fe^{2+} solution and 0.00070 $\underline{M}$ 1,10-phenanthroline solution will be mixed to give a total volume of 10.0 mL of reactants as shown in TABLE 33.3. You will be assigned one of these pairs of volumes by your instructor. Since the reactants are equimolar, the volume ratios also represent mole ratios. Because of the intense color of the resulting complex, the solutions will be diluted to 25.0 mL before determining absorbance spectrophotometrically.

TABLE 33.3. Volumes of reactants (mL).

0.00070 $\underline{M}$ Fe^{2+}	1	2	3	4	5	6	7	8	9
0.00070 $\underline{M}$ phen	9	8	7	6	5	4	3	2	1

Pipets that deliver up to 5.0 mL or up to 10.0 mL are available at the equipment bench. Obtain the combination of pipets that you require to measure the two volumes of reactants that you are assigned. Rinse the pipets carefully with distilled water from your wash bottle.

At the reagent bench obtain in a 10-mL or 100-mL graduated cylinder three times the required volume of 0.00070 $\underline{M}$ Fe^{2+} (from ferrous ammonium sulfate) plus a few milliliters in excess. Rinse the appropriate pipet twice with about 1-mL portions of 0.00070 $\underline{M}$ Fe^{2+} solution, draining the solution into a beaker for waste liquid. **CAUTION: Be certain to follow the procedure in Laboratory Methods B very carefully in order to draw liquids into pipets, preferably using a safety pipet filler**. Then measure the required volume of 0.00070 $\underline{M}$ Fe^{2+} solution in triplicate, and drain the portions into separate 25-mL volumetric flasks (Laboratory Methods **B**) obtained from the equipment bench. Repeat the above procedure to add first 5.0 mL of a pH 4.0 buffer solution, then 1.0 mL of 0.72 $\underline{M}$ hydroxylamine hydrochloride, $NH_2OH \cdot HCl$, solution, and finally the appropriate volume of 0.00070 $\underline{M}$ 1,10-phenanthroline solution to each volumetric flask. What do you observe?

Dilute each solution to the mark on the volumetric flask with distilled water, and mix each one thoroughly. Allow the solutions to stand for 10. minutes. Then using distilled water as a blank, measure the absorbance of each solution at 490 nm on a Bausch and Lomb Spectronic 20 spectrophotometer (Laboratory Methods **N**), and record absorbances in TABLE 33.6.

Calculate the average absorbance and average deviation, and plot the average absorbance with error bars versus mL of Fe^{2+} solution on the class graph provided by your instructor. After all students have plotted their data and linear portions of the graph have been extrapolated to form an intersection, determine from the graph the volume ratio (and thus mole ratio) at which maximum yield of product was obtained, and record that ratio in TABLE 33.6.

TABLE 33.4. Volumes and masses for a precipitation reaction.

	Run 1	Run 2	Run 3
Volume of 0.50 $\underline{M}$ C_wA_y (mL)			
Volume of 0.50 $\underline{M}$ $C'_xA'_z$ (mL)			
Mass of filter paper and precipitate (g)			
Mass of filter paper (g)			
Mass of precipitate (g)			
Average mass of precipitate (g)			
Average deviation (g)			
Mole(s) of C_wA_y per mole(s) of $C'_xA'_z$			
Maximum yield of precipitate (g)			

TABLE 33.5. Volumes and temperatures for an oxidation-reduction reaction.

	Run 1	Run 2	Run 3
Volume of 0.70 $\underline{M}$ ClO^- (mL)			
Volume of 0.70 $\underline{M}$ $S_2O_3^{2-}$ (mL)			
Temperature of 0.70 $\underline{M}$ ClO^- (^oC)			
Temperature of 0.70 $\underline{M}$ $S_2O_3^{2-}$ (^oC)			
Weighted average temperature (^oC)			
Maximum temperature (^oC)			
Temperature rise (^oC)			
Average temperature rise (^oC)			
Average deviation (^oC)			
Mole(s) of ClO^- per mole(s) of $S_2O_3^{2-}$			
Maximum temperature rise (^oC)			

TABLE 33.6. Volumes and absorbances for a complexation reaction.

	Run 1	Run 2	Run 3
Volume of 0.00070 $\underline{M}$ Fe^{2+} (mL)			
Volume of 0.00070 $\underline{M}$ phen (mL)			
Absorbance			
Average absorbance			
Average deviation			
Mole(s) of Fe^{2+} per mole(s) of phen			

Instructor's initials _____

434

Name _____

Student ID number _____

Section _____ Date _____

Instructor _____

PRE-LABORATORY QUESTIONS

1. What will be the effect on your mass of solid product recovered in part A if the pipet used for your reactant that is not in excess has a piece of inert solid of about 0.1-mL volume stuck to the wall just above the 3-mL mark? Indicate clearly your reasoning.

2. What will be the effect on your mass of solid product recovered in part A if you do not heat the mixture sufficiently to flocculate the precipitate? Indicate clearly your reasoning.

3. What will be the effect on your mass of solid product recovered in part A if you fail to dry your precipitate thoroughly? Indicate clearly your reasoning.

4. What will be the effect on your amount of heat evolved in part B if the scale on the Labindustries repipet dispenser used for your reactant that was in excess reads 0.1 mL too high above the 10-mL mark? Indicate clearly your reasoning.

5. Why is it important that the surface of your thermometer be rinsed carefully with distilled water and dried between room temperature measurements in part B?

6. When ClO^- acts as an oxidizing agent, two possible reduced products [$Cl_2(g)$ or Cl^-] can be formed. However, Cl^- is the favored product in both acidic and basic media. On the other hand, when a basic solution of $S_2O_3^{2-}$ acts as a reducing agent, at least three possible oxidized anions [tetrathionate ($S_4O_6^{2-}$), sulfite (SO_3^{2-}), and sulfate (SO_4^{2-})] can be formed depending on the oxidizing agent and concentrations. Write three possible balanced chemical equations for reactions in basic solution involving reduction of ClO^- to Cl^- in each and involving oxidation of $S_2O_3^{2-}$ to $S_4O_6^{2-}$, SO_3^{2-}, or SO_4^{2-}, respectively.

Name ______________________________ **Experiment 33**

Student ID number ________________

Section __________ Date __________

Instructor ____________________

POST-LABORATORY QUESTIONS

1. One student suggested plotting average mass of filter paper and precipitate versus mL of reactant C_wA_y on the class graph and claimed that the same mole ratio for maximum yield of product would be obtained. Do you agree with this claim? Why, or why not?

2. Considering any evidence available to you from this experiment or from the literature, is it possible that the precipitate that you formed in part **A** is an iodide salt? Why, or why not? Write a balanced chemical equation for an appropriate reaction if it is possible.

3. Answer POST-LABORATORY QUESTION 2 for a hydroxide compound.

4. Answer POST-LABORATORY QUESTION 2 for a Mn^{2+} salt.

5. Based on the experimental mole ratio of reactants, which of the reactions in PRE-LABORATORY QUESTION 6 must have dominated in part B? Indicate clearly your reasoning.

6. Given the following standard reduction potentials, is the reaction you chose in POST-LABORATORY QUESTION 5 thermodynamically the most favorable one under standard conditions? Indicate clearly your reasoning.

    ClO^-/Cl^- +0.89 V

    $S_4O_6^{2-}/S_2O_3^{2-}$ +0.08 V

    $SO_3^{2-}/S_2O_3^{2-}$ -0.58 V

    $SO_4^{2-}/S_2O_3^{2-}$ -0.76 V

7. For your chosen reaction in POST-LABORATORY QUESTION 5, calculate an approximate enthalpy change, ΔH_{rxn}, per mole of $S_2O_3^{2-}$, using the maximum temperature rise for the experimentally determined stoichiometric mole ratio. Assume that the density of your reaction mixture is 1.08 g/mL and that the specific heat of your reaction mixture is 3.9 J/g $^\circ$C.

8. Based on the experimental mole ratio of reactants in part C, write a balanced chemical equation describing the complexation reaction you observed assuming all reactants used (neglecting $NH_2OH \cdot HCl$ and buffer) actually formed complex.

9. Draw a three-dimensional structure for the complex formed in part C.

Temporarily and Permanently Hard Water

PURPOSE OF EXPERIMENT: Study some chemical properties of hard water and some ways for softening hard water.

Water containing Ca^{2+}, Mg^{2+}, and Fe^{2+} ions is known as <u>hard water</u> because, when soap is first added, a lather cannot be obtained. Common soap made from animal fat or vegetable oil is a mixtures of sodium and potassium salts of palmitic, stearic, and oleic acids. These salts are soluble and are dissociated in water. They readily form a lather with pure water and are widely used for cleansing purposes. However, Ca^{2+}, Mg^{2+}, and Fe^{2+} ions react with soap and form insoluble salts that separate as slimy, sticky precipitates. For example, Ca^{2+} ions react with stearate ions as follows.

$$Ca^{2+}(aq) \quad + \quad 2\ C_{17}H_{35}COO^-(aq) \quad ---> \quad Ca(C_{17}H_{35}COO)_2(s)$$

Before soap can form a lather or act as a cleansing agent, Ca^{2+}, Mg^{2+}, and Fe^{2+} ions must be removed by precipitation from the solution. Hence, the use of hard water for laundry purposes is very wasteful of soap, and the sticky residue left is undesirable.

When hard water is used in steam boilers where the water is heated and evaporated, any nonvolatile substances that are in solution in the water will precipitate on the walls of the boiler tubes when the water becomes saturated. In time, these tubes become thickly coated or even plugged with this deposit or scale. Only a 0.15-cm thickness of boiler scale lowers the efficiency of heat transfer by about 10.%.

The removal of ions that are responsible for the hardness of water is termed the <u>softening of water</u> and is an important application of chemistry. The hardness of water may be either temporary or permanent. If hydrogen carbonate, HCO_3^-, ions are present along with Ca^{2+}, Mg^{2+}, or Fe^{2+} ions, the water may be softened by boiling, and such water is called <u>temporarily hard water</u>. If no HCO_3^- ions are present, boiling will not soften the water, and it is said to be <u>permanently hard water</u>. Ground water that is found in limestone regions is particularly likely to be hard.

In this experiment you will prepare temporarily hard water; study some of the chemical properties of soft, temporarily hard, and permanently hard water; and study various processes available for softening hard water. The hardness of different water samples will be tested quantitatively by determining the volume of soap solution that must be added to a given volume of water in order to obtain a lather. Moreover, hard water will be treated by several methods designed to soften it, and the treated water will be titrated with soap solution to test the effectiveness of the methods. A study of the hardness of water, the action of soaps, and methods for softening water will illustrate characteristic chemical reactions and important differences in solubilities of some compounds of alkali metals and the alkaline earth metals. In addition, you will become familiar with a laboratory preparation for and properties of carbon dioxide gas.

REFERENCES

(1) Kotz, J. C., and Purcell, K. F., <u>Chemistry and Chemical Reactivity</u>, Saunders College Publishing, Philadelphia, 1987, sections 3.3, 15.1, 17.4, 17.5, 20.3, 20.4, and 21.2.

(2) Masterton, W. L., Slowinski, E. J., and Stanitski, C. L., <u>Chemical Principles</u>, 6th ed., Saunders College Publishing, Philadelphia, 1985, sections 12.3, 13.3, 18.1, 18.2, and 18.4.

EXPERIMENTAL PROCEDURE

(Study this section and the PRE-LABORATORY QUESTIONS before coming to the laboratory. **Wear safety goggles when performing this experiment.**)

You may work in pairs or use two laboratory periods for this experiment. Your instructor will make suggestions as to how to divide the work.

A. Preparation of Temporarily Hard Water

You will prepare temporarily hard water by bubbling carbon dioxide gas, CO_2, through an aqueous solution of calcium hydroxide, $Ca(OH)_2$, until the precipitate of calcium carbonate, $CaCO_3$, that first forms just dissolves as a result of formation of calcium hydrogen carbonate, $Ca(HCO_3)_2$.

The glass tubing in FIGURE 34.1 may have already been made in Experiment 1.

① 19-cm tubing with a 90° bend 7.5 cm from one end

② 24-cm tubing with a 90° bend 6 cm from one end

③ 12-cm tubing with a 90° bend 6 cm from one end

④ 23-cm straight tube

If so, obtain the three required 2-hole rubber stoppers (2 #6 and 1 #5) from the equipment bench or the stockroom. If not, the glass tubing will already be fitted in the appropriate 2-hole rubber stoppers when you get them.

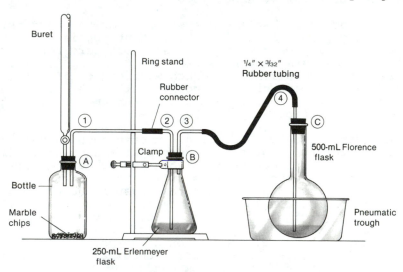

FIGURE 34.1. Apparatus for preparation of temporarily hard water.

Construct the apparatus shown in FIGURE 34.1, taking great care to avoid injury when assembling it. As an alternative your instructor may designate that a 250-mL Erlenmeyer flask be used as the bottle. **CAUTION: <u>If you must insert or remove glass tubing into or from rubber stoppers or rubber connectors, follow the instructions in Laboratory Methods F very carefully.</u>** Piece ① and the buret are inserted in that order into a #6 2-hole rubber stopper Ⓐ. Pieces ② and ③ are inserted in that order into a second #6 2-hole rubber stopper Ⓑ. Piece ④ is inserted into a #5 2-hole rubber stopper Ⓒ. The rubber connectors are attached as the apparatus is assembled. Some cautious sliding of rubber stoppers along glass tubing (**following the instructions in Laboratory Methods F very carefully**) may be required to make final adjustments to the apparatus in conformance with FIGURE 34.1.

Weigh 20. g of marble chips, $CaCO_3$; place them in the bottle, and replace the rubber stopper. Pour 40. mL of distilled water into the Erlenmeyer flask, and replace the rubber stopper. Be sure that the longer tube dips into the water. Pour 100. mL of 0.010 $\underline{M}$ calcium hydroxide, $Ca(OH)_2$, solution into the Florence flask, and replace the rubber stopper. Make certain that the tube dips into the solution. Place the Florence flask in an ice bath contained in a pneumatic trough. Finally, with the buret stopcock closed, add 30. mL of 3 $\underline{M}$ hydrochloric acid, HCl, solution to the buret. **Ask your instructor to inspect your apparatus.**

Add 5 mL of 3 $\underline{M}$ HCl solution from the buret into the bottle. Make certain that the level of the HCl solution never goes below the stopcock before it is closed. Describe in TABLE 34.1A what you observe around the marble chips in the bottle.

Regulate the stopcock of the buret so that the 3 $\underline{M}$ HCl solution is added dropwise in order to maintain a steady evolution of CO_2 in the bottle. The water in the Erlenmeyer flask washes any spattered HCl solution or vapor from the CO_2 gas stream that passes on into the Florence flask.

As CO_2 bubbles through the solution in the Florence flask, shake the flask vigorously in the ice bath so that the CO_2 gas is mixed thoroughly with the $Ca(OH)_2$ solution and is absorbed by it rapidly. The gas will also be absorbed faster if its partial pressure is increased by pressing your thumb lightly over the open hole in the stopper, **but be careful not to set up so much back pressure that one of the rubber stoppers pops loose**. Check the back pressure by watching the water level in tube 2 in the Erlenmeyer flask. Describe in TABLE 34.1A what you observe in the Florence flask and what is the product that forms.

Continue the CO_2 generation until the precipitate in the Florence flask redissolves and the solution becomes clear.

It might be necessary to swirl the bottle to generate more CO_2. If CO_2 generation ceases while there is still a significant amount of precipitate in the Florence flask, [remove the rubber stopper from the bottle; decant the reacted HCl solution from the bottle, discarding it into the sink; replace the rubber stopper; and add a fresh 30.-mL portion of 3 $\underline{M}$ HCl solution, first to the buret, and then in increments to the bottle].

After the solution in the Florence flask becomes clear, remove the rubber stopper from the Florence flask, and set the solution aside for later use as your temporarily hard water solution. Don't be concerned about a very slight cloudiness.

B. Carbon Dioxide Chemistry

To each of three different 13 x 100-mm test tubes, add 2 mL of distilled water and 1 drop of chlorophenol red indicator (color change from pH 5.2 to 6.8). Add 1 drop of 6 $\underline{M}$ aqueous ammonia, NH_3, solution to the first test tube and 1 drop of 3 $\underline{M}$ hydrochloric acid, HCl, solution to the third test tube so that the first and third test tubes will then serve as reference colors for the indicator. Bubble gas from your CO_2 generator through the solution in the second test tube for a few minutes. (If CO_2 generation is no longer strong, perform the bracketed procedure two paragraphs above.) Describe in TABLE 34.1B the colors of the three solutions.

Experiment 34

Dismantle the apparatus in FIGURE 34.1, discarding remaining solutions into the sink, but shaking and rinsing any remaining marble chips into a bottle on the reagent bench marked UNREACTED MARBLE CHIPS.

C. Characteristics of Soft Water

Obtain a 50-mL buret from the equipment bench or the stockroom, rinse the buret thoroughly with distilled water, and clamp it to your ring stand. Obtain about 40. mL of soap solution from the reagent bench, and pour it into your 50-mL buret (Laboratory Methods B). Record your initial buret reading in TABLE 34.2. Pour 25.0 mL of distilled water into a 250-mL Erlenmeyer flask, and add one drop of red food coloring. Then add the soap solution dropwise to the distilled water in the flask. Instead of swirling the solution, agitate the flask vigorously by swinging it like a pendulum after the addition of each drop. Titrate until a lather that persists for at least 30. seconds covers the _entire_ _surface_ of the solution. Record the final buret reading in TABLE 34.2. Only soap solution added beyond this volume that has already been used would be available for cleansing.

During this titration no precipitate formed to mask the lather endpoint. However, considerable precipitate forms during most titrations in this experiment, and it is easier to see the lather endpoint if the mixture is colored by the drop of food coloring.

D. Characteristics of Temporarily Hard Water

Compare the volume of soap solution required to obtain a lather in your temporarily hard water with the volume required for soft water by repeating the procedure in part C with 25.0 mL of temporarily hard water. Record your data and what you conclude in TABLE 34.2.

1. Boiling. Temporarily hard water containing HCO_3^- ions and Ca^{2+}, Mg^{2+}, or Fe^{2+} ions may be softened simply by boiling. Carbon dioxide is evolved, and an insoluble carbonate salt precipitates, thus removing the unwanted +2 cation.

Pour 25.0 mL of your temporarily hard water into a 100-mL or larger beaker, and boil the water gently for about 10. minutes (Laboratory Methods D). Describe what you observe in TABLE 34.1D.

Allow the mixture to cool, pour it into a 25-mL or larger graduated cylinder, add enough distilled water to give a total volume of 25.0 mL, and then pour the mixture into a 250-mL Erlenmeyer flask. Don't be concerned about slight losses of solid. Compare the volume of soap solution required to obtain a lather in the boiled mixture with the volumes required for soft water and for temporarily hard water by repeating the procedure in part C with the boiled mixture. Record your data and what you conclude in TABLE 34.2.

2. Chemical softening. Temporarily hard water may also be softened by reaction with a basic substance such as calcium hydroxide, $Ca(OH)_2$, sodium carbonate, Na_2CO_3, or aqueous ammonia, NH_3, solution. The HCO_3^- ions are converted to carbonate, CO_3^{2-}, ions, and insoluble carbonate salts precipitate, thus removing the unwanted +2 cations.

Pour 25.0 mL of your temporarily hard water into a 250-mL Erlenmeyer flask, and add 2 mL of 6 $\underline{M}$ aqueous ammonia, NH_3, solution. Describe what you observe in TABLE 34.1D.

Stir the mixture, and allow it to stand for 10. minutes. Compare the volume of soap solution required to obtain a lather in the NH_3 mixture with the volume required for soft water and for temporarily hard water by repeating the procedure in part C with the NH_3 mixture. Record your data and what you conclude in TABLE 34.2.

442

E. Characteristics of Permanently Hard Water

Compare the volume of soap solution required to obtain a lather in permanently hard water with the volumes required for soft water and for temporarily hard water by repeating the procedure in part C with 25.0 mL of permanently hard water from the reagent bench. Record your data, what you conclude, and what precipitate forms in TABLE 34.2.

1. Boiling. Pour 25.0 mL of permanently hard water into a 100-mL or larger beaker, and boil the water gently for about 10. minutes. Describe what you observe in TABLE 34.1E.

Allow the mixture to cool, pour it into your 25-mL or larger graduated cylinder, add enough distilled water to give a total volume of 25.0 mL, and then pour the solution into a 250-mL Erlenmeyer flask. Compare the volume of soap solution required to obtain a lather in the boiled solution with the volumes required for soft water and for temporarily hard water softened by boiling by repeating the procedure in part C with the boiled solution. Record your data and what you conclude in TABLE 34.2.

2. Chemical softening. Permanently hard water containing Ca^{2+}, Mg^{2+}, or Fe^{2+}, but no HCO_3^- ions, may be softened by the addition of various chemical water softeners, three of which may be tried below. Each of them precipitates an insoluble salt, thus removing the unwanted +2 cations.

Pour 25.0 mL of permanently hard water into a 250-mL Erlenmeyer flask, add 0.1 g of sodium carbonate, Na_2CO_3, and stir. Describe what you observe in TABLE 34.1E.

Stir the mixture, and allow it to stand for 10. minutes. Compare the volume of soap solution required to obtain a lather in the Na_2CO_3 mixture with the volumes required for soft water and for permanently hard water by repeating the procedure in part C with the Na_2CO_3 mixture. Record your data and what you conclude in TABLE 34.2.

Repeat the procedure of the above two paragraphs using first 0.4 g of borax, $Na_2B_4O_7 \cdot 10H_2O$, and then 0.2 g of sodium metaphosphate, $NaPO_3$, instead of 0.1 g of Na_2CO_3. Describe what you observe in each case in TABLE 34.1E, and record your data and what you conclude in TABLE 34.2.

3. Ion exchange. Permanently hard water may also be softened by the use of a cation exchanger that removes the unwanted +2 cations. Ion exchange involves a reversible exchange of ions between a porous, stationary solid, the exchanger, and a percolating solution, with no change occurring in the framework of the solid. The ion exchanger is a complex, insoluble, three-dimensional network into which water and ions in solution can easily penetrate. A large number of ionizable groups is attached to the ion exchanger. The nature of these ionizable groups determines the chemical behavior of the ion exchanger. If the ionizable group is an anion, the ion exchanger is called an anion exchanger. If the ionizable group is a cation, the ion exchanger is called a cation exchanger. You will use a cation exchanger in which Na^+ ions comprise the ionizable groups. Thus, when permanently hard water flows through the cation exchanger, the Na^+ ions are easily displaced by the polyvalent cations in the aqueous solution that percolates through the pores of the cation exchanger.

$$2 \text{ Na(exchanger)} + Mg^{2+}(aq) \longrightarrow Mg(\text{exchanger})_2 + 2 Na^+(aq)$$

A higher concentration of polyvalent cations forces the above equilibrium to the right and causes a greater displacement of Na^+ ions on the cation exchanger. Moreover, it is generally found that cations with higher charge and/or larger radius displace Na^+ ions more readily. The reaction can be reversed by increasing the Na^+ concentration in solution. Hence, when the exchanger capacity has been exhausted, the exchanger may be regenerated or changed back to the Na^+ form by passing a concentrated sodium chloride, NaCl, solution through it. A similar reaction occurs in a home water softener where complex aluminum silicates, called zeolites, are used. Hard water is allowed to flow through a bed of the zeolite, and Ca^{2+}, Mg^{2+}, and Fe^{2+} ions are exchanged for Na^+ ions from the zeolite. Regeneration of the zeolite can be accomplished by treatment with a 10.% by mass NaCl solution. A typical household of four people uses over 5,000 gallons of softened water per month requiring up to 100. lb of sodium chloride for regeneration of the zeolite. As a result many nonfarm households purchase more sodium chloride than any other chemical except for water and fuels. This is the most efficient practical method for softening large quantities of natural water.

Set up a cation exchange column using a 50-mL buret that has been rinsed thoroughly with distilled water. Clamp the buret to your ring stand. With the aid of a wooden dowel, place a small wad of cotton about the size of a pea just above the stopcock at the bottom of the buret. Place a teaspoon of cation exchange resin in a 250-mL beaker. Add distilled water, and stir to form a slurry. Pour the slurry into the buret, and allow the cation exchanger to settle. The wad of cotton should retain the cation exchanger, but also allow free passage of water through the stopcock of the buret. Continue adding slurry until the buret is filled with settled resin up to the 45-mL mark (unless a different mark is designated by your instructor). **Never let the liquid level dip below the top level of the resin**. Activate the resin (unless your instructor says that this procedure is unnecessary) by washing it with 100. mL of 1 M sodium chloride, NaCl, solution, using a flow rate of about 15 mL/min. Then wash the resin with 50. mL of distilled water using the same flow rate.

Pour 25.0 mL of permanently hard water through the cation exchange resin using the same flow rate. Collect the effluent in a 250-mL Erlenmeyer flask. Then wash the cation exchanger with 25.0 mL of distilled water using the same flow rate. Compare the volume of soap solution required to obtain a lather in the effluent with the volumes required for soft water, permanently hard water, and permanently hard water treated with chemical water softeners by repeating the procedure in part C with the effluent. Don't forget to correct your volume of soap solution required for the additional 25.0 mL of distilled water that you added to wash the resin. This can be done by simply subtracting the volume of soap solution required in part C to cause a lather in distilled water. Record your data and what you conclude about the effectiveness of softening permanently hard water by a cation exchanger in TABLE 34.2.

Empty the slurry of your cation exchange resin from your buret into a bottle on the reagent bench marked USED RESIN.

F. Characteristics of City Water

Laboratory tap water should be representative of your city water. Compare the volume of soap solution required to obtain a lather in laboratory tap water with the volume required for soft water, temporarily hard water, and permanently hard water by repeating the procedure in part C with laboratory tap water. Record your data and what you conclude in TABLE 34.2.

Name _____

Student ID number _____

Section _____ Date _____

Instructor _____

TABLE 34.1.   Experimental observations and related questions.

(Answer all **QUESTIONS** before handing in your report.)

A   Observation   Behavior of reaction of marble chips in the bottle

QUESTION   Write a balanced complete equation describing the reaction that occurs in the bottle.

Observation   Behavior observed in the Florence flask

QUESTION   Write a balanced complete equation describing the reaction that occurs in the Florence flask.

QUESTION   Write a balanced complete equation describing the reaction involving the dissolving of the white precipitate in the Florence flask.

TABLE 34.1 (continued)

___

B Observation   Colors of the three indicator solutions

QUESTION   Explain the behavior in the second test tube with reference to the first and third test tubes using one or more balanced net ionic equations.

___

D1 Observation   Behavior of temporarily hard water softened by boiling

QUESTION   Write a balanced complete equation describing the reaction that occurred during boiling of temporarily hard water.

QUESTION   How does boiling favor this reaction?

2 Observation   Behavior of temporarily hard water softened by 6 $\underline{M}$ NH_3

QUESTION   Write a balanced net ionic equation describing the reaction that occurred assuming that Fe^{2+} is present in the temporarily hard water.

___

TABLE 34.1 (continued)

E1 Observation   Behavior of permanently hard water during boiling

2 Observation   Behavior of permanently hard water softened by Na_2CO_3

QUESTION   Write a balanced complete equation describing the reaction that occurred assuming that Ca^{2+} is present in the permanently hard water.

Observation   Behavior of permanently hard water softened by $Na_2B_4O_7 \cdot 10H_2O$ and by $NaPO_3$

QUESTION   Write a balanced complete equation for each of the reactions that occurred assuming that Mg^{2+} is present in the permanently hard water.

TABLE 34.2. Volumes of soap solution for 25.0-mL
water samples, and conclusions.

Water samples		Initial volume (mL)	Final volume (mL)	Volume used (mL)
			CONCLUSIONS	
C	Soft water (distilled)			
D	Temporarily hard water			
1	Temporarily hard water softened by boiling			
2	Temporarily hard water softened by 6 $\underline{M}$ NH$_3$			
E	Permanently hard water			
1	Permanently hard water softened by boiling			
2	Permanently hard water softened by Na$_2$CO$_3$			
	Permanently hard water softened by Na$_2$B$_4$O$_7$·10H$_2$O			
	Permanently hard water softened by NaPO$_3$			
3	Permanently hard water softened by ion exchange			
F	Laboratory tap water			

Instructor's initials _____

Name ______________________________ **Experiment 34**

Student ID number _____________

Section __________ Date __________

Instructor ________________________

PRE-LABORATORY QUESTIONS

1. Why is it important that the level of 3 <u>M</u> HCl solution never goes below the stopcock in your buret once the generation of CO_2 starts in part **A**?

2. Why would you expect hydrogen chloride, HCl(g), and carbon dioxide, $CO_2(g)$, to differ markedly in solubility in water?

3. What would happen if hydrogen chloride, HCl(g), were carried over into the Florence flask in part **A**? Write balanced complete equations to describe the several possible reactions that could occur.

4. Why is a reduced temperature for the Florence flask helpful in the preparation of your temporarily hard water solution in part **A**? Cite literature data to support your reasoning. What relationship does this have to a bottle of soda pop?

5. Ethanol, C_2H_5OH, one of the alcohols added to raise the octane level in gasoline, and phenol, C_6H_5OH, an aromatic alcohol which is a weak acid, both form CO_2 on complete combustion. Write balanced complete equations describing the combustion of ethanol and phenol.

6. People working in the area of water quality consider water with high levels of magnesium and calcium (usually greater than 200. mg total Mg^{2+} and Ca^{2+} per mL) to be hard and water with low levels of magnesium and calcium (usually less than 100. mg total Mg^{2+} and Ca^{2+} per mL) to be soft. Assuming that a given water sample contains equal masses of Mg^{2+} and Ca^{2+}, calculate the minimum molarity of Ca^{2+} for hard water and the maximum molarity of Ca^{2+} for soft water.

Name _____

Student ID number _____

Section _____ Date _____

Instructor _____

POST-LABORATORY QUESTIONS

1. Find in a current chemical catalog the prices for equivalent purities and quantities of sodium carbonate, Na_2CO_3, borax, $Na_2B_4O_7 \cdot 10H_2O$, and sodium metaphosphate, $NaPO_3$. Based on this data and your titration data, arrange the three chemical water softeners in order of decreasing cost effectiveness assuming all other factors are equal. Show clearly your reasoning.

2. Write a balanced complete equation for any of the following where reaction occurs.

    a. $Na_2CO_3(aq)$ + $BaCl_2(aq)$ --->

    b. $CaCO_3(s)$ + $HBr(aq)$ --->

    c. $KCl(aq)$ + $Na_2CO_3(aq)$ --->

    d. $KHCO_3(aq)$ + $HCl(aq)$ --->

    e. $Ca(OH)_2(aq)$ + $NH_3(aq)$ --->

3. An acid form of a cation exchange resin, H(exchanger), can be used in place of the sodium form described in part **E**. Hydrogen ions, H^+, are then displaced by the +2 cations. Suppose that 25.00 mL of permanently hard water are eluted through an acid cation exchange resin followed by washing of the resin. The eluent then requires 11.26 mL of 0.1038 $\underline{M}$ sodium hydroxide, NaOH, solution for titration to a phenolphthalein endpoint.

    a. What molarity of +2 cation would be required to produce this result? Show clearly your calculations and reasoning.

    b. If the only +2 cation present is Ca^{2+}, what mass of $CaCl_2 \cdot 2H_2O$ must be dissolved per liter of hard water to produce the molarity in <u>a</u>.? Show clearly your calculations and reasoning.

4. People who heat their homes by means of a woodstove frequently keep a container of water on the stove throughout the heating season in order to put water vapor into the air to raise the humidity to a comfortable level. However, the container eventually develops on its inner surface a crust that is both unsightly and an insulator which inhibits vaporization. What are likely to be some of the constituents of this crust, and how would you remove them without damage to the container?

Boron and Its Compounds

PURPOSE OF EXPERIMENT: Study some properties and reactions of boron and some boron compounds, and analyze a mixture containing borax.

Group III elements manifest as clearly as any group important changes in chemical characteristics as one moves down the group. For example, group I elements are all metals whereas group VII elements are all nonmetals; however, group III elements shift from nonmetallic to metallic behavior going down the group. As another example, group I elements form basic oxides and hydroxides exhibiting primarily ionic bonding whereas group VII elements form acidic oxides and hydroxy compounds exhibiting primarily covalent bonding; however, the oxides and hydroxides of the group III elements shift from acidic to amphoteric to basic behavior and from covalent to ionic character going down the group. Even studying just the chemistry of boron illustrates the beginnings of some of the trends.

A primary source of boron and one of the most important compounds of boron is borax, sodium tetraborate decahydrate, $Na_2B_4O_7 \cdot 10H_2O$. This compound occurs in nature in desert regions, but it must be purified before it can be used commercially. Purification usually entails reacting borax with water to form orthoboric acid, H_3BO_3, then reconverting H_3BO_3 to borax by reaction with sodium carbonate, Na_2CO_3. You may have already tried borax as a water softener in Experiment 34. Borax is also valuable as a flux for soldering and welding, for making artificial gems and pigments, and in the manufacture of glass, enamels, soap, and preservatives for wood and meat. Elemental boron can be prepared by dehydrating H_3BO_3 to boric oxide, B_2O_3, and then reducing the oxide with magnesium. The purposes of this experiment are to examine some of the more important properties and reactions of orthoboric acid, borax, and boron, and to analyze an unknown impure sample of borax by titrating the hydroxide ion and the boric acid formed from its hydrolysis.

REFERENCES

(1) Kotz, J. C., and Purcell, K. F., _Chemistry and Chemical Reactivity_, Saunders College Publishing, Philadelphia, 1987, sections 2.5, 3.4, 4.4, 12.1, 16.4, 16.5a, 17.1a, 17.3, 17.4, 21.1, and 22.2b.

(2) Masterton, W. L., Slowinski, E. J., and Stanitski, C. L., _Chemical Principles_, 6th ed., Saunders College Publishing, Philadelphia, 1985, sections 2.5, 2.6, 4.2, 8.4, 12.2, 19.3, 19.4, 19.6, 19.7, 23.2, and 26.1.

EXPERIMENTAL PROCEDURE

(Study this section and the PRE-LABORATORY QUESTIONS before coming to the laboratory. **Wear safety goggles when performing this experiment.)**

A. Properties and Reactions of Orthoboric Acid

Orthoboric acid forms white, needle-like crystals in which $B(OH)_3$ units are linked together by hydrogen bonds to form infinite sheets. The ortho prefix indicates that this particular oxyacid of B(III) is formally the most fully hydrated one. This nomenclature also carries over to oxyacids of other nonmetals such as silicon, phosphorus, and sulfur.

To each of three different 13 x 100-mm test tubes, add 2 mL of distilled water and 2 drops of bromocresol green indicator (color change from pH 3.8 to 5.4). Add 1 drop of 6 $\underline{M}$ aqueous ammonia, NH_3, solution to the first test tube and 1 drop of 3 $\underline{M}$ hydrochloric acid, HCl, solution to the third test tube. The first and third test tubes will then serve as reference colors for the indicator.

Add a couple dozen crystals of orthoboric acid, H_3BO_3, to the second test tube, and dissolve the crystals with stirring. Describe what you observe in TABLE 35.1A.

Solutions of H_3BO_3 are acidic because the concentration of OH^- has been reduced, not because the concentration of H^+ has been increased by dissociation from H_3BO_3. Thus, H_3BO_3 acts not as a Brønsted-Lowry acid (proton donor), but as a Lewis acid (lone pair acceptor, from OH^-), as illustrated by the following equation.

$$H_3BO_3(aq) + H_2O(l) \rightleftharpoons H^+(aq) + B(OH)_4^-(aq) \qquad K_a = 7.3 \times 10^{-10}$$

Add a sample of mannitol about the size of a pea to the above solution, and stir carefully. Describe the color change for the solution in TABLE 35.1A.

Mannitol is a polyfunctional alcohol. It reacts with $B(OH)_4^-$ and stabilizes that anion according to the following schematic equation.

$$B(OH)_4^-(aq) + \text{mannitol}(aq) \rightleftharpoons \left[\begin{array}{c} -C \\ | \\ -C \end{array} \begin{array}{c} O \\ \\ O \end{array} B \begin{array}{c} O \\ \\ O \end{array} \begin{array}{c} C- \\ | \\ C- \end{array} \right]^- (aq) + 2\,H_2O(l)$$

B. Properties and Reactions of Borax, $Na_2B_4O_7 \cdot 10H_2O$

The oxyacids and oxyanions of boron exhibit a multitude of formulas and structures. Trigonal BO_3 units are most common, but tetrahedral BO_4 units also occur.

Several of the most common oxyacids or oxyanions can be thought of as being derived formally by various degrees of dehydration of H_3BO_3. The pyro acid or anion is formed by removing one water molecule per pair of ortho molecules. The meta acid or anion is formed by removing one water molecule per molecule of ortho acid. Chain polymers or rings form at the meta stage. Further dehydration of H_3BO_3 yields tetraboric acid or the tetraborate anion. Complete dehydration of H_3BO_3 yields the normal oxide, boric oxide in this case. You will study two of these oxyanions in this part.

Place 0.2 gram of borax, $Na_2B_4O_7 \cdot 10H_2O$, in a 16 x 150-mm test tube. Heat the borax gently over a Bunsen burner flame (Laboratory Methods D), making certain to hold the test tube at about a 30°-angle from the horizontal. Describe any evidence showing that the substance is a hydrate in TABLE 35.1B.

Allow the borax sample to cool until the test tube is warm to the touch. Then dissolve the sample by adding 3 mL of distilled water and stirring vigorously. Test the solution with red and blue litmus paper (Laboratory Methods P), and record your observations in TABLE 35.1B. Borax hydrolyzes to form H_3BO_3 when it dissolves in water.

Borax is valuable as a flux - that is, a substance that will react with difficultly fusible substances to form compounds that fuse readily. The action of borax as a flux in soldering and welding depends upon the reaction of sodium tetraborate with metallic oxides on the surface of the metal. Metaborates that are formed are easily melted, and the metallic oxides are thus removed from the surface, leaving clean metallic surfaces that will adhere firmly to each other or to solder.

Make a loop about 2 mm in diameter at one end of a piece of copper wire that has been polished with emery cloth. Grasp the other end of the wire with crucible tongs, and heat the wire loop in the tip of a cool, luminous Bunsen burner flame (Laboratory Methods D). Remove the wire from the flame. Describe what you observe in TABLE 35.1B.

Heat the loop again, and while it is still hot, dip it into borax, $Na_2B_4O_7 \cdot 10H_2O$, contained on a watch glass so that a considerable quantity of borax adheres to the loop. Place it back in the flame, and heat it vigorously until the borax forms a molten bead in the loop. Repeat the dipping into borax several times if necessary. Allow the loop and borax bead to cool, and bend the wire away from the borax bead. Describe the change in color of the borax and the change in appearance of the wire in TABLE 35.1B.

The color of the borax is due to copper metaborate, $Cu(BO_2)_2$, that has been formed. The color of many metallic borates is characteristic of the particular metal that is present. Thus, the presence of certain metals is sometimes determined by what are called borax bead tests. Sodium metaborate, $NaBO_2$, was the other product in the borax bead reaction. A different borate of sodium will be prepared below.

C. Preparation and Reactions of Sodium Perborate Tetrahydrate, $NaBO_3 \cdot 4H_2O$

A number of salts contain an O_2^{2-} anion in place of an oxide anion, O^{2-}, in the normal compound and also possess the oxidizing and reducing properties of peroxides. One such compound is sodium perborate, $NaBO_3$, which is related to $NaBO_2$ in the same way that hydrogen peroxide, H_2O_2, is related to water.

Sparingly soluble $NaBO_3 \cdot 4H_2O$ crystallizes from a dilute basic $NaBO_2$ solution to which H_2O_2 has been added and can be prepared by the following procedure (**Blanchard, A. A., Phelan, J. W., and Davis, A. R., Synthetic Inorganic Chemistry, 5th ed., Wiley, New York, 1936, pages 210-212**). To 0.5 gram of borax, $Na_2B_4O_7 \cdot 10H_2O$, and two sodium hydroxide, NaOH, pellets contained in a 16 x 150-mm test tube, add 3 mL of distilled water. Stir, and heat the mixture gently, to dissolve the components. Cool the solution to near room temperature under a water tap, and then add slowly with stirring 25 drops of 30% hydrogen peroxide, H_2O_2, solution. **CAUTION: Hydrogen peroxide of this concentration is corrosive and bleaches and burns your skin; wear rubber gloves when handling it. If any spills or spatters occur onto your skin, rinse the affected area thoroughly with water.** Cool the solution further by immersing the test tube in an ice bath contained in a 250-mL beaker. If no crystals have formed in that time span, add a small chip of ice, and stir. Be patient, and occasionally scratch the inside wall of the test tube with your stirring rod until crystals form. If necessary, add a few drops of chilled 95% ethanol, C_2H_5OH, to aid crystallization. Prepare for the following tests if you have to wait for crystals to form.

Filter the crystals using either gravity filtration or suction filtration (Laboratory Methods I). Wash additional crystals from the test tube with three successive 1-mL portions of 95% ethanol and use that ethanol to wash the crystals on the filter paper to remove excess H_2O_2 (Laboratory Methods I).

Several simple tests illustrate the oxidizing and reducing ability of $NaBO_3$. Place two strips of moistened starch-potassium iodide test paper opposite each other in the concave dip of a watch glass. Place some crystals of your sodium perborate tetrahydrate, $NaBO_3 \cdot 4H_2O$, on one of the strips, and then add a drop of 6 $\underline{M}$ acetic acid, $HC_2H_3O_2$, solution over the crystals. Describe what you observe in TABLE 35.1C. Place a drop of 30%

hydrogen peroxide, H_2O_2, solution on the other strip. Describe what you observe in TABLE 35.1C.

Dissolve crystals of sodium perborate tetrahydrate, $NaBO_3 \cdot 4H_2O$, about the size of a pea in 1 mL of distilled water in a 16 x 150-mm test tube, stirring and heating under the warm water tap if necessary. Place 10. drops of 30% hydrogen peroxide, H_2O_2, solution in a second 16 x 150-mm test tube. Add 4 drops of 3 $\underline{M}$ sulfuric acid, H_2SO_4, solution to each test tube, and then add several drops of 0.01 $\underline{M}$ potassium permanganate, $KMnO_4$, solution to each test tube. Describe what you observe in each case in TABLE 35.1C.

D. Properties and Reactions of Amorphous Boron

Place a pinch of amorphous boron in six different 10 x 75-mm test tubes. To the first tube, add 1 mL of distilled water. To the second test tube, add 1 mL of 3 $\underline{M}$ hydrochloric acid, HCl, solution. To the third test tube, add 1 mL of 3 $\underline{M}$ nitric acid, HNO_3, solution. To the fourth test tube, add 1 mL of 3 $\underline{M}$ sulfuric acid, H_2SO_4, solution. To the fifth test tube, add 1 mL of 6 $\underline{M}$ sodium hydroxide, NaOH, solution. **CAUTION: <u>These acids and base cause burns on your skin and holes in your clothing. If there are any spills or spatters onto your skin or clothing, rinse the affected area thoroughly with water</u>.** Stir the contents of each test tube to see if any changes occur. Warm those test tubes in which no apparent reaction occurs first <u>gently</u>, and, if necessary, in a boiling water bath (Laboratory Methods D), stirring the contents occasionally. Describe what you observe in each case in TABLE 35.1D.

Heat the sixth sample of amorphous boron first gently and then vigorously over a Bunsen burner flame. Describe what you observe in TABLE 35.1D.

E. Analysis of a Mixture Containing Borax

Using an analytical balance (Laboratory Methods C), weigh onto weighing paper four 0.35-g samples of your unknown mixture. Record your unknown number and your masses in TABLE 35.2. Put samples 1 and 2 into 250-mL Erlenmeyer flasks and samples 3 and 4 into 250-mL or larger beakers. Dissolve each sample in 100. mL of distilled water. Add to **only sample 1** 8 drops of Methyl Red indicator. In another 250-mL Erlenmeyer flask, prepare a solution for color comparison by adding 100. mL of a 0.1 $\underline{M}$ boric acid, H_3BO_3-0.05 $\underline{M}$ sodium chloride, NaCl, stock solution and 8 drops of Methyl Red indicator.

Set up a ring stand and buret clamp, clean a buret carefully with soap solution, and rinse it thoroughly with distilled water. Practice reading the buret and manipulating the stopcock or pinch clamp (Laboratory Methods B) before starting the following titration.

Rinse your buret with several 1-mL portions of an approximately 0.1 $\underline{M}$ hydrochloric acid, HCl, solution provided on the reagent bench. Then titrate sample 1 until the color of the solution matches that of your prepared stock solution of 0.1 $\underline{M}$ H_3BO_3-0.05 $\underline{M}$ NaCl. Record in TABLE 35.2 the molarity and the volume of HCl solution that you have used.

On the basis of the masses of samples 1 and 2 and the volume of 0.1 $\underline{M}$ HCl solution added to sample 1, calculate the volume of 0.1 $\underline{M}$ HCl solution required by sample 2 to just reach the Methyl Red endpoint. Add this precise volume of 0.1 $\underline{M}$ HCl to sample 2, then add 10. g of mannitol with stirring, and finally 4 drops of phenolphthalein. Titrate the solution with approximately 0.1 $\underline{M}$ sodium hydroxide, NaOH, solution, and record in TABLE 35.2 the molarity and volume of NaOH solution that you have used. Repeat the procedure of this paragraph on samples 3 and 4.

Perform the calculations in TABLE 35.3 including sample calculations for sample 2 at the bottom of TABLE 35.3.

Name ____________________

Student ID number ____________

Section ________ Date ________

Instructor ____________________

Experiment 35

DATA

TABLE 35.1. Experimental observations and related questions.

(Answer all **QUESTIONS** before handing in your report.)

A Observation Behavior upon dissolving H_3BO_3 in water with indicator

QUESTION Is the solution acidic or basic? Estimate the pH of the solution, indicating clearly your reasoning.

Observation Color change upon adding mannitol to H_3BO_3 solution

QUESTION Estimate the pH of the solution. Indicate clearly your reasoning.

QUESTION Should the reaction of mannitol with $B(OH)_4^-$ make H_3BO_3 a weaker acid or a stronger acid? Why?

QUESTION Does your conclusion agree with your experimental results? Why, or why not?

TABLE 35.1 (continued)

B   **Observation**   Evidence that borax is a hydrate

Observation   Behavior upon testing a solution of borax with litmus paper

QUESTION   Is the solution acidic or basic?

QUESTION   Write balanced chemical equations to illustrate first the dissolving of borax in water and then the hydrolysis of the tetraborate anion.

Observation   Behavior of copper wire heated in a luminous Bunsen burner flame

QUESTION   What has happened chemically to the copper wire? Explain the observed behavior by means of a balanced chemical equation.

Observation   Change in color of borax and change of appearance of copper wire upon heating copper wire in a molten bead of borax and then cooling

TABLE 35.1   (continued)

C   **Observation**   Behavior of $NaBO_3$ toward moist starch-KI test paper

Observation   Behavior of 30% H_2O_2 toward moist starch-KI test paper

QUESTION   Write balanced net ionic equations to account for the behavior of H_2O_2 and $NaBO_3$ toward starch-KI test paper.

QUESTION   Are H_2O_2 and $NaBO_3$ acting as oxidizing agents or reducing agents?

Observation   Behavior of 0.01 $\underline{M}$ $KMnO_4$ solution toward acidified $NaBO_3$ solution

Observation   Behavior of 0.01 $\underline{M}$ $KMnO_4$ solution toward acidified 30% H_2O_2 solution

QUESTION   Write balanced net ionic equations to account for the behavior of $KMnO_4$ toward H_2O_2 solution and $NaBO_3$ solution.

QUESTION   Are H_2O_2 and $NaBO_3$ acting as oxidizing agents or reducing agents?

TABLE 35.1 (continued)

| D | Observation | Behavior of amorphous boron toward |

Distilled water

3 <u>M</u> HCl

3 <u>M</u> HNO₃

3 <u>M</u> H₂SO₄

6 <u>M</u> NaOH

QUESTION Write balanced complete equations describing any chemical reactions that occur with any of the reagents above.

Observation Behavior of amorphous boron toward heat

QUESTION Write a balanced complete equation describing any chemical reaction that occurs.

460

TABLE 35.2.   Masses of borax samples and volumes of acid and base.

Unknown number __	Sample 1	Sample 2	Sample 3	Sample 4
Mass of weighing paper (g)				
Mass of weighing paper and sample (g)				
Molarity of HCl solution (M)				
Initial buret reading (HCl) (mL)				
Final buret reading (HCl) (mL)				
Calculated volume of HCl solution added (mL)				
Molarity of NaOH solution (M)				
Initial buret reading (NaOH) (mL)				
Final buret reading (NaOH) (mL)				

Instructor's initials _____

TABLE 35.3. Percent borax and boron in the borax samples.

	Sample 1	Sample 2	Sample 3	Sample 4
Volume of HCl solution added (mL)				
Moles of HCl added (mol)				
Moles of OH^- neutralized (mol)				
Moles of borax present (mol)				
Mass of borax present (g)				
Percent borax by mass (%)				
Volume of NaOH solution added (mL)				
Moles of NaOH added (mol)				
Moles of H_3BO_3 present (mol)				
Mass of boron present (g)				
Percent boron by mass (%)				
Average percent boron (%)				

Sample calculations for sample 2

Moles of NaOH added

Moles of H_3BO_3 present

Mass of boron present

Name ______________________________ **Experiment 35**

Student ID number ______________

Section __________ Date __________

Instructor ______________________

PRE-LABORATORY QUESTIONS

1. Draw and describe in words the three-dimensional structure of a molecule of orthoboric acid, H_3BO_3, in the solid state.

2. Draw and describe in words the three-dimensional structure of $B(OH)_4^-$.

3. Balance the following equations to gain a better understanding of the various stages of dehydration of orthoboric acid, H_3BO_3.

$H_3BO_3(s)$ ---> $H_2O(l)$ + $H_4B_2O_5(s)$
pyroboric acid

$H_3BO_3(s)$ ---> $H_2O(l)$ + $HBO_2(s)$
metaboric acid

$H_3BO_3(s)$ ---> $H_2O(l)$ + $H_2B_4O_7(s)$
tetraboric acid

$H_3BO_3(s)$ ---> $H_2O(l)$ + $B_2O_3(s)$
boric oxide

4. Borax reacts with copper(II) oxide on strong heating to form sodium metaborate and copper(II) metaborate. Write a balanced chemical equation describing this reaction.

5. The preparation of sodium perborate tetrahydrate, $NaBO_3 \cdot 4H_2O$, can be considered formally in two steps. Balance the following equations to help understand these steps. Also write the overall reaction neglecting the intermediate formation of sodium metaborate, $NaBO_2$.

$$Na_2B_4O_7 \cdot 10H_2O(aq) \quad + \quad NaOH(aq) \quad ---> \quad NaBO_2(aq) \quad + \quad H_2O(l)$$

$$NaBO_2(aq) \quad + \quad H_2O_2(aq) \quad + \quad H_2O(l) \quad ---> \quad NaBO_3 \cdot 4H_2O(s)$$

6. Why is it important to wash the $NaBO_3 \cdot 4H_2O$ crystals with ethanol, C_2H_5OH, to remove excess hydrogen peroxide, H_2O_2, before performing the tests for oxidizing or reducing ability?

Name_____

Student ID number _____

Section _____ Date _____

Instructor _____

POST-LABORATORY QUESTIONS

1. How does your estimated pH for your H_3BO_3 solution in part **A** compare with the pH value you would calculate for a 0.10 $\underline{M}$ H_3BO_3 solution? Indicate clearly your calculations and reasoning.

2. Show clearly the splitting out of water in the dehydration of ortho-phosphoric acid to yield the pyro and meta forms.

3. Write a balanced net ionic equation describing the reaction in acid solution of sodium perborate, $NaBO_3$, with each of the following, and indicate any color changes that you would expect.

   a. $K_2Cr_2O_7(aq)$.

   b. $Cr^{2+}(aq)$.

4. Consider each of the following pairs of elements or compounds, and write a brief outline describing clearly both how they are similar and how they are different in terms of physical and chemical properties. Write balanced chemical equations to illustrate your points wherever appropriate.

    a. B and Al.

    b. $B(OH)_3$ and $Al(OH)_3$.

Aluminum and Aluminum Hydroxide

PURPOSE OF EXPERIMENT: Study some properties and reactions of aluminum and aluminum hydroxide.

Group IIIA elements manifest as clearly as any group important changes in chemical characteristics as one moves down the group. For example, group IA elements are all metals whereas group VIIA elements are all nonmetals; however, group IIIA elements shift from nonmetallic to metallic behavior going down the group. As another example, group IA elements form basic oxides and hydroxides exhibiting primarily ionic bonding whereas group VIIA elements form acidic oxides and hydroxy compounds exhibiting primarily covalent bonding; however, the oxides and hydroxides of group IIIA elements shift from acidic to amphoteric to basic behavior and from covalent to ionic character going down the group. Aluminum and its compounds are frequently at the crossover point in many of these trends and have chemical and physical characteristics that are intermediate between those of the top and bottom members of group IIIA and their compounds.

Aluminum is the third most abundant element in the earth's crust following silicon and oxygen. The primary source of aluminum is bauxite, $Al_2O_3 \cdot xH_2O$. The major impurities in bauxite are iron(III) oxide, Fe_2O_3, and silica, SiO_2. They are removed by taking advantage of the amphoteric nature of the hydrated oxide of aluminum and converting it first to $Al(OH)_4^-$ and then back to aluminum hydroxide, $Al(OH)_3$.

$$
\begin{array}{llllll}
Fe_2O_3(s) & concentrated & Fe_2O_3(s) & filter \to & Fe_2O_3(s) \\
SiO_2(s) & \text{------------} \to & SiO_3^{2-}(aq) & \text{-------}\Big[& \\
Al_2O_3(s) & NaOH & Al(OH)_4^-(aq) & \to & SiO_3^{2-}(aq) \\
& & & & Al(OH)_4^-(aq) \\
& & & & \quad | \\
& & & & \quad | CO_2 \\
& & & & \quad \downarrow \\
& heat & SiO_3^{2-}(aq) & \gets \text{filter} & SiO_3^{2-}(aq) \\
Al_2O_3(s) & \gets\text{----} & Al(OH)_3(s) & \gets \text{-------} & Al(OH)_3(s) \\
& & & \gets
\end{array}
$$

Elemental aluminum is prepared by electrolysis of molten Al_2O_3 in the Hall-Heroult process.

$$2\ Al_2O_3(l) \xrightarrow{\text{electrolysis}} 4\ Al(l) + 3\ O_2(g)$$

In this experiment you will examine some of the more important properties and reactions of aluminum and aluminum hydroxide.

REFERENCES

(1) Kotz, J. C., and Purcell, K. F., <u>Chemistry and Chemical Reactivity</u>, Saunders College Publishing, Philadelphia, 1987, section 21.1.

(2) Masterton, W. L., Slowinski, E. J., and Stanitski, C. L., <u>Chemical Principles</u>, 6th ed., Saunders College Publishing, Philadelphia, 1985, sections 4.3, 22.1, and 22.3.

EXPERIMENTAL PROCEDURE

(Study this section and the PRE-LABORATORY QUESTIONS before coming to the laboratory. **Wear safety goggles when performing this experiment.**)

A. Properties and Reactions of Aluminum

Hold a 0.5-cm strip of aluminum with crucible tongs, and heat the aluminum first gently and then more vigorously in a Bunsen burner flame (Laboratory Methods **D**). Describe what you observe in TABLE 36.1A.

Polish a piece of aluminum with emery cloth, and then cut the aluminum into six smaller pieces. Add one of the pieces to a 16 x 150-mm test tube and the other five pieces to separate 13 x 100-mm test tubes.

Add 5 mL of water to the larger test tube. Then heat the water gradually to boiling over a Bunsen burner flame. Describe in TABLE 36.1A any evidence for a chemical reaction either before or during heating.

Add separately to three smaller test tubes containing aluminum 2-mL portions of 3 M hydrochloric acid, HCl, solution; 3 M nitric acid, HNO_3, solution; and 3 M sulfuric acid, H_2SO_4, solution. **CAUTION: These strong acids at these concentrations are mildly corrosive and cause minor burns on your skin and holes in your clothing. If any spills or spatters occur onto your skin or clothing, immediately rinse the affected area thoroughly with water**. Describe any evidence for reaction in any of the acids in TABLE 36.1A. Warm the test tubes in a boiling water bath. Record any additional observations in TABLE 36.1A.

Add 2 mL of 3 M aqueous ammonia, NH_3, solution to the fourth small test tube containing aluminum. Describe what you observe in TABLE 36.1A. Warm the test tube in a boiling water bath. Record any additional observations in TABLE 36.1A.

Add 2 mL of 6 M sodium hydroxide, NaOH, solution to the fifth small test tube containing aluminum. **CAUTION: This concentration of NaOH solution is caustic and corrosive toward your skin and clothing. If any spills or spatters occur onto your skin or clothing, immediately rinse the affected area thoroughly with water**. Describe what you observe in TABLE 36.1A. Potassium hydroxide reacts with aluminum in a similar way. $Al(OH)_4^-$ is the aluminum-containing product. This simple tetrahedral anion is favored at pH > 13, but more complex aluminate anions involving polymers with octahedral Al and with OH bridges are formed in less basic pH ranges.

B. Properties and Reactions of Aluminum Hydroxide

Place 10. mL of 0.1 M aluminum sulfate, $Al_2(SO_4)_3$, solution in a 50-mL or larger beaker. Test a drop of this solution with blue litmus paper. Describe what you observe in TABLE 36.1B.

Add to the beaker 10. mL of 3 M aqueous ammonia, NH_3, solution. Describe what you observe in TABLE 36.1B.

Filter the precipitate using either gravity filtration or suction filtration (Laboratory Methods **I**). Wash additional precipitate from the beaker with four successive 1-mL portions of hot water, and use that water to wash the precipitate on the filter paper to remove excess NH_3 (Laboratory Methods **I**). Describe the appearance of the precipitate in TABLE 36.1B.

Use a spatula or wooden splint to transfer four small portions of the $Al(OH)_3$ precipitate to separate 10 x 75-mm test tubes. To one of the test tubes, add a few drops of 3 M hydrochloric acid, HCl, and stir. Describe what you observe in TABLE 36.1B.

To the second test tube containing Al(OH)$_3$ precipitate, add a few drops of 6 <u>M</u> sodium hydroxide, NaOH, solution, and stir. **CAUTION:** <u>**This concentration of NaOH solution is caustic and corrosive toward your skin and clothing. If any spills or spatters occur onto your skin or clothing, immediately rinse the affected area thoroughly with water**</u>. Describe what you observe in TABLE 36.1B.

A substance that both reacts with acids, thus behaving as a base, and that reacts with bases, thus behaving as an acid, is termed <u>amphoteric</u>. The relative acid and base strength of aluminum hydroxide, Al(OH)$_3$, can be illustrated in several simple tests.

Place strips of moistened red and blue litmus paper opposite each other in the concave dip of a watch glass. Use a spatula or wooden splint to transfer a very small amount of Al(OH)$_3$ precipitate from your filter paper to each litmus paper. Describe what you observe in TABLE 36.1B. This experiment illustrates both the insolubility of Al(OH)$_3$ and its weakness as an acid and a base, but still does not tell whether it is stronger as an acid or as a base.

To the third 10 x 75-mm test tube containing Al(OH)$_3$ precipitate, add 1 mL of 3 <u>M</u> NH$_3$ solution, heat gently, and stir. Describe what you observe in TABLE 36.1B. To the fourth test tube containing Al(OH)$_3$ precipitate, add 1 mL of 3 <u>M</u> acetic acid, HC$_2$H$_3$O$_2$, solution, heat gently, and stir. Describe what you observe in TABLE 36.1B.

Place 1 mL of 0.1 <u>M</u> Al$_2$(SO$_4$)$_3$ solution in a 16 x 150-mm test tube. Add dropwise up to 10. drops of 6 <u>M</u> NaOH solution, stirring carefully between each drop. Describe in TABLE 36.1B what you observe as more and more NaOH is added. To the same test tube, add <u>**carefully dropwise**</u> up to 12 drops of 6 <u>M</u> nitric acid, HNO$_3$, solution, stirring between each drop. **CAUTION:** <u>**Nitric acid of this concentration is corrosive and causes burns on your skin and holes in your clothing. If any spills or spatters occur onto your skin or clothing, immediately rinse the affected area thoroughly with water**</u>. Describe in TABLE 36.1B what you observe as more and more HNO$_3$ is added.

A number of compounds form <u>lakes</u> which are colored solids produced when a dye combines with or is adsorbed by the compounds. Aluminum hydroxide is one such substance. Add 2 mL of 3 <u>M</u> NH$_3$ solution to a 20 x 150-mm test tube. Then add 1 mL of aluminon reagent, a dye used for the detection and colorimetric determination of aluminum in water, foods, and tissues. Finally, add 2 mL of 0.1 <u>M</u> Al$_2$(SO$_4$)$_3$ solution, and stir. Describe what you observe in TABLE 36.1B. Filter the precipitate using either gravity filtration or suction filtration. Note the color of the filtrate in TABLE 36.1B. Wash the precipitate carefully with two 1-mL portions of water. Can you wash all the color out? Describe the color of the <u>lake</u> in TABLE 36.1B. Clean the test tube thoroughly at once.

Many dyes, such as aluminon, cannot be used directly to dye cotton cloth because the dyes are not held firmly by the fibers and are removed by washing. When aluminum hydroxide is precipitated on cotton, it is strongly adsorbed. If aluminon is then applied, a colored lake is formed with the aluminum hydroxide on the fiber, and this color is <u>fast</u>. The aluminum hydroxide is called a <u>mordant</u>.

Place a piece of cotton cloth in a 20 x 150-mm test tube. Add 2 mL of 0.1 <u>M</u> Al$_2$(SO$_4$)$_3$ solution and 5 mL of water. Heat the solution gently to boiling, and then add 5 mL of 3 <u>M</u> NH$_3$. Describe in TABLE 36.1B what you observe on cooling.

Stopper the test tube with a cork stopper, and shake the contents vigorously. Then decant the liquid. The cloth is now said to be <u>mordanted</u>. Add 1 mL of aluminon reagent, and swirl it around the piece of cloth. Then decant the liquid, and rinse the cloth in the test tube three times with

10-mL portions of water. Can you wash all the color out of the cloth? Compare its color with that of the lake formed previously, and record your observations in TABLE 36.1B. Clean the test tube thoroughly at once.

Textbooks indicate that aluminum carbonate has not been prepared and that it certainly cannot exist in aqueous solution because it would hydrolyze completely. Test this information as follows. Add 2 mL of 0.1 $\underline{M}$ $Al_2(SO_4)_3$ solution to a 20 x 150-mm test tube. Then add 3 mL of 1 $\underline{M}$ sodium carbonate, Na_2CO_3, solution. Describe what you observe in TABLE 36.1B.

To show whether the precipitate is a hydroxide, or perhaps a carbonate or basic carbonate, filter it by either gravity filtration or suction filtration, and wash it <u>very thoroughly</u> with water to remove excess Na_2CO_3. Then use a wooden splint to transfer two small portions of the precipitate to separate 10 x 75-mm test tubes. To one test tube, add a few drops of 3 $\underline{M}$ HCl. Is there any effervescence? Describe what you observe in TABLE 36.1B. To the other test tube, add a few drops of aluminon reagent, stir, and centrifuge. Describe what you observe in TABLE 36.1B.

Name _____

Student ID number _____

Section _____ Date _____

Instructor _____

DATA

TABLE 36.1.   Experimental observations and related questions.

(Answer all QUESTIONS before handing in your report.)

A   Observation   Behavior of Al in a Bunsen burner flame

QUESTION   Does massive Al burn easily?

QUESTION   Write a balanced chemical equation describing the oxidation
of Al in air.

Observation   Behavior of Al in cold water   In hot water

QUESTION   Is hydrogen gas evolved in either cold or hot water?

Observation   Behavior of Al toward

Cold 3 $\underline{M}$ HCl   Hot 3 $\underline{M}$ HCl

Cold 3 $\underline{M}$ HNO_3 Hot 3 $\underline{M}$ HNO_3

Cold 3 $\underline{M}$ H_2SO_4 Hot 3 $\underline{M}$ H_2SO_4

QUESTION   What is the relative reactivity of Al toward the three
acids?

QUESTION   Write balanced complete equations describing any reactions
that occur.

Experiment 36

TABLE 36.1 (continued)

Observation   Behavior of Al toward cold 3 $\underline{M}$ NH$_3$ Hot 3 $\underline{M}$ NH$_3$

QUESTION   Write a balanced net ionic equation describing any reaction that occurs.

Observation   Behavior of Al toward 6 $\underline{M}$ NaOH

QUESTION   Write a balanced net ionic equation describing the reaction that occurs.

B   Observation   Behavior of 0.1 $\underline{M}$ Al$_2$(SO$_4$)$_3$ toward litmus paper

QUESTION   Explain the behavior of Al$_2$(SO$_4$)$_3$ toward litmus by means of a balanced net ionic equation.

Observation   Behavior of 0.1 $\underline{M}$ Al$_2$(SO$_4$)$_3$ toward 3 $\underline{M}$ NH$_3$

QUESTION   Write a balanced net ionic equation describing the reaction that occurs.

Observation   Appearance of the precipitate, Al(OH)$_3$(s)

Observation   Behavior of Al(OH)$_3$(s) toward 3 $\underline{M}$ HCl

QUESTION   Write a balanced net ionic equation describing the reaction that occurs.

Name _____

Student ID number _____

Section _____ Date _____

Instructor _____

DATA

TABLE 36.1 (continued)

Observation   Behavior of $Al(OH)_3(s)$ toward 6 $\underline{M}$ NaOH

QUESTION   Write a balanced net ionic equation describing the reaction that occurs.

Observation   Behavior of $Al(OH)_3(s)$ toward red and blue litmus

Observation   Behavior of $Al(OH)_3(s)$ toward 3 $\underline{M}$ NH_3

QUESTION   Write a balanced net ionic equation describing any reaction that occurs.

Observation   Behavior of $Al(OH)_3(s)$ toward 3 $\underline{M}$ $HC_2H_3O_2$

QUESTION   Write a balanced net ionic equation describing any reaction that occurs.

Observation   Behavior of 0.1 $\underline{M}$ $Al_2(SO_4)_3$ toward 6 $\underline{M}$ NaOH

QUESTION   Write balanced net ionic equations describing two consecutively occurring reactions to explain this behavior.

Experiment 36

TABLE 36.1 (continued)

Observation Behavior of product from above toward 6 $\underline{M}$ HNO_3

QUESTION Write balanced net ionic equations describing two consecutively occurring reactions to explain this behavior.

Observation Behavior from mixing 3 $\underline{M}$ NH_3, aluminon, and 0.1 $\underline{M}$ $Al_2(SO_4)_3$

Observation Color of filtrate   Color of "lake"

Observation Behavior of cotton with 0.1 $\underline{M}$ $Al_2(SO_4)_3$ and 3 $\underline{M}$ NH_3

QUESTION Write a balanced net ionic equation describing the reaction that occurs.

Observation Color of mordanted cloth versus the "lake" formed previously

Observation Behavior of 0.1 $\underline{M}$ $Al_2(SO_4)_3$ toward 1 $\underline{M}$ Na_2CO_3

QUESTION What must be the gas evolved?

Observation Behavior of the precipitate toward 3 $\underline{M}$ HCl

Observation Behavior of the precipitate toward aluminon

QUESTION What do you conclude as to the identity of the precipitate? Why?

Instructor's initials _____

Student ID number _____

Section _____ Date _____

Instructor _____   **PRE-LABORATORY QUESTIONS**

1. Look up the standard reduction potentials for sodium, magnesium, and aluminum. What do these potentials indicate in terms of their relative strengths as reducing agents? Should aluminum liberate hydrogen readily from water? Indicate clearly your reasoning.

2. Why is it important to polish aluminum with emery cloth before reacting it with acids or base in part **A**?

3. Aluminum wire has only 64% of the conductivity of copper wire of the same diameter. Aluminum has a lower density (2.70 g/cm^3) than copper (8.69 g/cm^3). Aluminum is more expensive ($49.35/lb) than copper ($35.80/lb). Use this data and any other information you consider important to decide whether you would use aluminum or copper wiring in a new house. Would your answer change if you wanted long distance transmission lines rather than house wiring? Why, or why not?

4. Describe in detail the specific procedures that you would use to test for the behavior of 0.1 $\underline{M}$ $Al_2(SO_4)_3$ solution toward litmus paper. Why would you use this particular procedure?

5. In several parts of this experiment aluminum hydroxide precipitates even in acidic solution. To gain some perspective on this, calculate each of the following.

   a. The pH of a saturated solution of $Al(OH)_3(s)$ ($K_{sp} = 5.0 \times 10^{-33}$).

   b. The pH at which $Al(OH)_3(s)$ will first precipitate from a 0.1 $\underline{M}$ $Al_2(SO_4)_3$ solution.

POST-LABORATORY QUESTIONS

1. A mixture of $Al_2(SO_4)_3$, $NaHCO_3$, detergent, and water produce a liquid foam composed of a liquid and a gas. Assuming that the mixture is not stirred, account for the formation of the liquid foam, using a net ionic equation wherever appropriate.

2. Contrasting your conclusion in PRE-LABORATORY QUESTION 1 and your observations of aluminum in cold and hot water, explain why aluminum can be used for cooking utensils even though it has a highly negative reduction potential.

3. Sodium carbonate hydrolyzes to form a basic solution. Write a balanced net ionic equation illustrating that hydrolysis. What will happen then if a strong aqueous solution of Na_2CO_3 is boiled in an aluminum pan? Indicate clearly your reasoning.

4. Write a balanced net ionic equation describing the reaction that occurs when aqueous solutions of $Al_2(SO_4)_3$ and Na_2CO_3 are mixed. Label the Lewis acid and Lewis base on each side of the equation.

5. Compare and contrast the behavior of boron and aluminum toward 3 $\underline{M}$ HCl, 3 $\underline{M}$ HNO$_3$, 3 $\underline{M}$ H$_2$SO$_4$, and 6 $\underline{M}$ NaOH. Write balanced chemical equations to illustrate your points wherever appropriate.

6. Compare and contrast the behavior of H$_3$BO$_3$ and Al(OH)$_3$ toward 3 $\underline{M}$ HC$_2$H$_3$O$_2$, 3 $\underline{M}$ HCl, 3 $\underline{M}$ NH$_3$, and 6 $\underline{M}$ NaOH. Write balanced complete equations to illustrate your points wherever appropriate.

Tin Compounds and Covalent Character

PURPOSE OF EXPERIMENT: Prepare tin(II) iodide, SnI_2, tin(IV) iodide, SnI_4, and study some physical and chemical properties of these compounds and of potassium iodide, KI.

Many metals exhibit more than one oxidation state in compounds that they form with a given nonmetal. Tin is one such metal, forming compounds in both the +2 state and the +4 state with a number of nonmetals. Studies of these compounds show that the lower valent ones exhibit greater ionic character compared to the greater covalent character of the higher valent ones. This behavior can be understood starting from either ionic bonding or covalent bonding as an extreme. Suppose that two tin compounds involving the same anion are assumed to be ionic. The tin(IV) compound, having the cation with higher charge density by virtue of both higher charge and smaller size, will exhibit stronger polarization of the anion by the cation. This will result in sharing of electron density between the cation and anion and greater covalent character for the tin(IV) compound. Suppose that the two tin compounds with the same anion are assumed to be covalent. The tin(II) compound, having the cation of lower charge density by virtue of both lower charge and larger size, will exhibit a lower effective electronegativity. This will result in a greater electronegativity difference between tin and the nonmetal and greater ionic character for the tin(II) compound.

Ionic and covalent compounds often have relatively large differences in the strengths of their intermolecular forces and frequently can be distinguished easily by their physical properties. Ionic compounds have strong electrostatic attraction in three dimensions between oppositely charged cations and anions. Therefore, they tend to have high melting points and boiling points, to be nonvolatile, and to be hard but easily shattered. They also tend to be more soluble in water or strongly polar solvents, but are less likely to be soluble in nonpolar solvents. Many covalent compounds have strong covalent bonds within molecules, but only weak London forces between nonpolar molecules or fairly weak dipole-dipole attractions between weakly polar molecules. Since the intermolecular forces determine physical properties, many covalent compounds tend to have low or moderate melting points and boiling points, to be fairly volatile, and to be relatively soft. They also tend to be more soluble in nonpolar or weakly polar solvents, but are less likely to be soluble in water unless they react with water.

In this experiment you will synthesize tin(II) iodide, SnI_2, and/or tin(IV) iodide, SnI_4, gain experience with aqueous and/or nonaqueous preparations, examine the properties of these compounds along with potassium iodide, KI, and attempt to understand the impact of ionic and covalent character on properties.

REFERENCES

(1) Kotz, J. C., and Purcell, K. F., <u>Chemistry and Chemical Reactivity</u>, Saunders College Publishing, Philadelphia, 1987, sections 3.3, 3.4, 4.1-4.3, 9.2, 9.4, 9.6, 11.1, 11.2, 12.2, and 21.2.

(2) Masterton, W. L., Slowinski, E. J., and Stanitski, C. L., <u>Chemical Principles</u>, 6th ed., Saunders College Publishing, Philadelphia, 1985, sections 9.1, 9.2, 10.2, 11.3, 11.4, and 12.3.

Experiment 37

EXPERIMENTAL PROCEDURE

(Study this section and the PRE-LABORATORY QUESTIONS before coming to the laboratory. **Wear safety glasses when performing this experiment.**)

A. Preparation of Tin(IV) Iodide, SnI_4

Dichloromethane, CH_2Cl_2, is used as a solvent in the following preparation. This solvent presents no fire hazard and is only a moderate local irritant to skin or mucous membranes. However, the high volatility of CH_2Cl_2 (bp, 39.8°C; vapor pressure, 380 mmHg at 22°C) is both an advantage and a disadvantage. High volatility is an advantage because it allows for easy removal of the solvent by evaporation after the preparation is completed. However, it is also a disadvantage. CAUTION: CH_2Cl_2 is very dangerous to the eyes, and its vapors are highly toxic upon inhalation by virtue of their strong narcotic powers. Therefore, every possible precaution should be taken during the preparation at your desk to prevent the vaporization of CH_2Cl_2. CAUTION: CH_2Cl_2 is also a suspected carcinogen; avoid skin contact, and wash your hands thoroughly before leaving the laboratory.

Weigh 0.5 g of granular tin on a beam balance (Laboratory Methods C), and add the tin to a clean, dry 50-mL or larger Erlenmeyer flask. Take your flask and an appropriately fitting cork stopper to the hood. Working in the hood, measure 15. mL of dichloromethane, CH_2Cl_2, into a graduated cylinder, keeping the mouth of the reagent bottle as close to the graduated cylinder as you can to minimize vaporization of CH_2Cl_2. Then immediately pour the CH_2Cl_2 into your Erlenmeyer flask, stopper the flask loosely, and return it to your desk. Keep a partially filled 400-mL or larger beaker of cold water nearby so that if the CH_2Cl_2 solvent starts to boil during the reaction below, you can immediately cool the flask in the water and stop the bubbling.

Weigh 1.5 g of iodine, I_2, directly into a 50-mL or larger beaker using a beam balance. Do not let iodine touch the weighing pan or an aluminum weighing dish because iodine may react with aluminum on contact. Cover the beaker with a watch glass so that iodine vapor does not diffuse readily into the air. CAUTION: Iodine can also be an irritant and cause systemic difficulties; however, this is seldom a problem because of its low volatility (vapor pressure, 1 mmHg at 38.7°C).

Add a few crystals of iodine to the mixture of tin and CH_2Cl_2 by tapping the beaker containing iodine, and immediately restopper the flask. Record what you observe in TABLE 37.1A.

Swirl the flask gently for 5 to 10. minutes until the CH_2Cl_2 solution takes on a red-brown color. Based on the results from your procedure in PRE-LABORATORY QUESTION 4, what causes the red-brown color?

Start the reaction in part **B** while continuing the reaction in part **A**.

Repeat the cycle of adding a few crystals of I_2, swirling the flask gently, and waiting for the color change, until all the I_2 has been added. CAUTION: Keep the watch glass over the beaker, and keep the flask stoppered except when you are actually transferring I_2 from the beaker to the flask. Continue swirling the contents of the flask for at least 10. minutes after the last color change indicating that all the I_2 apparently has reacted. The total reaction time should be at least 1 hour from the first addition of I_2.

Working in the hood, carefully decant (Laboratory Methods H) the CH_2Cl_2 solution of SnI_4 into a watch glass labeled with your name, leaving the unreacted tin in the flask. Add a couple mL of CH_2Cl_2 to the flask, and wash the last traces of SnI_4 solution into the watch glass. Then empty any unreacted tin into a bottle marked UNREACTED TIN.

To obtain solid SnI_4, evaporate CH_2Cl_2 from the solution in the following manner. Place your watch glass just behind the lip of the hood designated by your instructor. Assuming that nobody else is working in the hood at the time, lower the hood window to about 1 cm above the lip of the hood. As air is drawn into the hood, it will pass over the solution and cause slow evaporation of CH_2Cl_2. While this evaporation is taking place, continue part **B**, and complete the preparation of SnI_2.

When the SnI_4 is dry, use a spatula to scrape it out of the beaker onto a previously weighed piece of filter paper. Break up large pieces of the solid to allow CH_2Cl_2 trapped inside to evaporate. Do <u>not</u> weigh the SnI_4 or examine its properties in part **C** until you are certain that your SnI_4 sample is completely free of CH_2Cl_2. A good check is to sniff the SnI_4 sample **cautiously**; if you can still detect the odor of CH_2Cl_2, the sample is not yet free of CH_2Cl_2.

Weigh your filter paper and SnI_4 crystals, and determine the mass of SnI_4 formed. Record the mass in TABLE 37.1A.

B. Preparation of Tin(II) Iodide, SnI_2

Weigh 1.0 g of granular tin on a beam balance, and add the tin to a 50-mL Erlenmeyer flask. Take your flask **to the hood**, and add 5 mL of 12 **M** hydrochloric acid, HCl, solution. **CAUTION: <u>Concentrated HCl causes burns on your skin and holes in your clothing. Wear rubber gloves when handling it. If there are any spills or spatters onto your skin or clothing, rinse the affected area thoroughly with water</u>.** Heat the flask on a hotplate <u>in the hood</u>, maintaining gas evolution but keeping the solution well below boiling so that the HCl does not boil away. Placing a small watch glass, if available, over the Erlenmeyer flask also inhibits loss of HCl solution. Record what you observe in TABLE 37.1B.

If at any time either the reaction ceases before all the tin is dissolved or the solution boils away leaving white crystals, add several mL of 12 **M** HCl to the flask. Try, however, to keep added excess HCl to a minimum.

Weigh the amount of sodium iodide, NaI, calculated in PRE-LABORATORY QUESTION 5 into a 100-mL or larger beaker using a beam balance. Dissolve the NaI with stirring in 10. mL of water.

After the tin has dissolved completely in the 12 **M** HCl solution, pour the clear, hot Sn^{2+} solution into the beaker containing the NaI solution. Place the beaker in an ice-water bath, and stir for about 5 minutes until no additional crystallization takes place. Then filter the crystals using a preweighed piece of filter paper either by gravity filtration or by suction filtration (Laboratory Methods I). When as much water as possible has been removed, use a spatula to help remove the filter paper to a watch glass. Dry the crystals by heating on a steam bath, with a heat lamp, or in an oven (Laboratory Methods M) for about 20. minutes. Complete the drying and weighing of SnI_4 in part **A** while you are waiting for the SnI_2 to dry.

Weigh your filter paper and dried SnI_2 crystals, and determine the mass of SnI_2 formed. Record your masses in TABLE 37.1B.

Experiment 37

C. Properties of KI, SnI$_2$, and SnI$_4$

Water, ethanol (CH$_3$CH$_2$OH), and xylene (C$_8$H$_{10}$), or actually a mixture of three isomeric xylenes, will be used as solvents in the following solubility tests. CAUTION: <u>Ethanol and xylene are both dangerous when exposed to heat or an open flame, and ethanol also reacts vigorously with oxidizing materials</u>. As a result <u>your water bath should be heated on a hotplate rather than over an open flame</u>.

Test the solubilities of potassium iodide, KI, from the reagent bench and of your freshly prepared SnI$_2$ and SnI$_4$ in water, ethanol, and xylene using the following procedures. Obtain a small amount of KI on the tip of a spatula, and place it in a 10 x 75-mm or larger test tube; place the same amount of SnI$_2$ in a second test tube; and place the same amount of SnI$_4$ in a third test tube. Add 1 mL of distilled water to each tube. Heat each tube in a water bath contained in a 50-mL or larger beaker and held at about 75°C on a <u>hotplate</u>. Stir the contents of each tube vigorously. Test the pH of each aqueous solution with wide range pH paper (Laboratory Methods P). Record all observations in TABLE 37.1C. Save the aqueous solutions for later experiments.

Repeat the same procedures except for the pH paper test using ethanol, CH$_3$CH$_2$OH, and xylene, C$_8$H$_{10}$, as solvents, and record your observations in TABLE 37.1C.

Centrifuge the aqueous solutions of SnI$_2$ and SnI$_4$ (Laboratory Methods J), and note carefully the color of any solid present in each solution. Record your observations in TABLE 37.1C.

Transfer 7 drops of the <u>aqueous</u> KI, SnI$_2$, and SnI$_4$ solutions from the solubility tests to three separate, clean 10 x 75-mm or larger test tubes. To each tube add 0.1 $\underline{M}$ mercury(II) chloride, HgCl$_2$, solution dropwise until a precipitate just forms and remains. CAUTION: <u>HgCl$_2$ is poisonous. Wash your hands thoroughly after using it</u>. <u>Do not add excess HgCl$_2$ solution</u>. Record all observations in TABLE 37.1C.

Again transfer 7 drops of the <u>aqueous</u> KI, SnI$_2$, and SnI$_4$ solutions from the solubility tests to three separate, clean 10 x 75-mm or larger test tubes. To each tube add 5 drops of 1 $\underline{M}$ sodium sulfide, Na$_2$S, solution, stir for about 2 minutes, and allow any precipitate to settle. Record what you observe in TABLE 37.1C.

Empty any remaining SnI$_2$ and SnI$_4$ crystals into appropriately marked bottles on the reagent bench.

Name ______________________________

Student ID number ______________

Section __________ Date __________

Instructor ________________________

Experiment 37

DATA

TABLE 37.1. Experimental observations and data, and related questions.

(Answer all **QUESTIONS** before handing in your report.)

A Observation Behavior upon adding I_2 to a mixture of Sn in CH_2Cl_2

QUESTION Is I_2 or SnI_4 responsible for the color in CH_2Cl_2?

Observation Check your conclusion using your procedure in PRE-LABORATORY QUESTION 4, and describe your results.

Mass of filter paper (g) ____________

Mass of filter paper and SnI_4 (g) ____________

QUESTION Write a balanced complete equation describing the reaction between tin and I_2 to produce SnI_4.

QUESTION Calculate your theoretical and percent yield of SnI_4 based on this equation.

TABLE 37.1   (continued)

B   Observation   Behavior upon heating Sn in 12 $\underline{M}$ HCl

QUESTION   Write a balanced complete and a balanced net ionic equation describing the   reaction that   occurs   between   Sn and   HCl solution.

Mass of filter paper (g)   _____

Mass of filter paper and SnI$_2$ (g) _____

QUESTION   Write   a   balanced complete equation describing the precipitation of SnI$_2$.

QUESTION   Calculate the theoretical and percent yield of SnI$_2$ based on the two previous equations.

QUESTION   What are some probable reasons for your low yield?

Name ______________________________ **Experiment 37**

Student ID number ________________

Section __________ Date __________

Instructor ________________________ **DATA**

TABLE 37.1 (continued)

C Observation Solubilities of KI, SnI_2, and SnI_4

QUESTION Write balanced complete equations to describe reactions that formed the new solid products in the aqueous solubility tests.

TABLE 37.1 (continued)

Observation Behavior upon reacting aqueous solutions with $HgCl_2$ solution

QUESTION Write balanced complete equations to describe any reactions that occurred with $HgCl_2$ solution.

Observation Behavior upon reacting aqueous solutions with Na_2S solution

QUESTION Write balanced net ionic equations to describe any reactions that occurred with Na_2S solution.

Instructor's initials _____

Name _____

Student ID number _____

Section _____ Date _____

Instructor _____

PRE-LABORATORY QUESTIONS

1. Describe the structure of and the kind of intermolecular forces in dichloromethane, CH_2Cl_2. Why should it have a low boiling point and a high volatility?

2. Write a balanced complete equation describing the heterogeneous reaction between aluminum and iodine solids.

3. Describe the kind of intermolecular forces in iodine, I_2. Why should I_2 have a lower vapor pressure than dichloromethane, CH_2Cl_2?

4. What simple experimental test(s) could you perform to determine whether iodine, I_2, or tin(IV) iodide, SnI_4, is responsible for the initial color in dichloromethane, CH_2Cl_2?

5. Calculate the volume of 12 $\underline{M}$ hydrochloric acid, HCl, solution required to react stoichiometrically with tin and also the mass of sodium iodide, NaI, required to react stoichiometrically with the tin dissolved in 12 $\underline{M}$ HCl solution in order to form tin(II) iodide, SnI_2. Show clearly your calculations.

6. a. Describe the kind(s) of intermolecular forces in each of the following.

   (1) Water, H_2O

   (2) Ethanol, CH_3CH_2OH.

   (3) $\underline{p}$-Xylene, C_8H_{10}.

   b. Look up the boiling points of ethanol and $\underline{p}$-xylene in the Handbook of Chemistry and Physics (Chemical Rubber Company), and explain clearly why the boiling points of water, ethanol, and $\underline{p}$-xylene should increase in the order that they do.

Name ______________________________

Student ID number ______________

Section __________ Date __________

Instructor ______________________

Experiment 37

POST-LABORATORY QUESTIONS

1. Considering the polarities of the three solvents used (water, ethanol, and p-xylene), write a brief essay describing what your solubility observations indicate about the strength of and the nature of intramolecular and intermolecular bonding forces in potassium iodide, KI, tin(II) iodide, SnI_2, and tin(IV) iodide, SnI_4. Indicate clearly your reasoning.

2. Look up the melting points of potassium iodide, KI, tin(II) iodide, SnI_2, and tin(IV) iodide, SnI_4, in the <u>Handbook of Chemistry and Physics</u> (Chemical Rubber Company), and write a brief essay describing what the melting points indicate about the strength of and the nature of intramolecular and intermolecular bonding forces in KI, SnI_2, and SnI_4. Indicate clearly your reasoning.

Compounds of Nitrogen

PURPOSE OF EXPERIMENT: Prepare and/or study some properties and reactions of oxyacids, oxyanions, and oxides of nitrogen, and determine the unknown metal in a metal oxide that you prepare.

Much of the common chemistry of nitrogen involves oxyacids of nitrogen, oxyanions of nitrogen, and oxides of nitrogen; however, these species are seldom produced directly from nitrogen itself. The reason for this is that elemental nitrogen in the form of N_2 diatomic molecules is very unreactive because the atoms are held together by a very strong triple bond. One manifestation of this inertness is the fact that nitrogen and oxygen, O_2, a more reactive gas, coexist in the earth's atmosphere.

Nitric acid, HNO_3, is the most common substance that acts as both a strong acid and as a good oxidizing agent. The salts of HNO_3, called nitrates, are all quite soluble in water and are convenient sources for preparing aqueous solutions of many cations. Nitrates are also major water pollutants because the NO_3^- anion is an effective nutrient for aiding the growth of algae. Nitrous acid, HNO_2, is less stable than HNO_3 and is a weak acid. The salts of HNO_2 are called nitrites.

Common oxides of nitrogen are obtained readily by a variety of methods. For example, nitrous oxide, N_2O, can be produced by thermal decomposition of ammonium nitrate, NH_4NO_3, at about 250°C.

$$NH_4NO_3(s) \quad ---> \quad N_2O(g) \quad + \quad 2 H_2O(g)$$

Nitric oxide, NO, can be prepared by the reduction of <u>dilute</u> HNO_3 by a relatively inactive metal such as tin. Tin is oxidized to the +2 oxidation state in the form of $Sn(NO_3)_2(aq)$.

$$3 Sn(s) \quad + \quad 8 HNO_3(aq) \quad ---> \quad 3 Sn(NO_3)_2(aq) \quad + \quad 2 NO(g) \quad + \quad 4 H_2O(l)$$

Nitrogen dioxide, NO_2, can be prepared by the reduction of <u>concentrated</u> HNO_3 by a relatively inactive metal such as tin. Concentrated HNO_3 oxidizes tin to the +4 oxide, SnO_2, and no nitrate is formed. This is the only case of a common metal forming a precipitate rather than a solution of the nitrate upon oxidation by HNO_3.

$$Sn(s) \quad + \quad 4 HNO_3(aq) \quad ---> \quad SnO_2(s) \quad + \quad 4 NO_2(g) \quad + \quad 2 H_2O(l)$$

Both bismuth (the lowest member of group VA and the only group VA element that has dominantly metallic characteristics) and copper can also be used as the relatively inactive metal in these preparations. Nitrogen dioxide spontaneously dimerizes to set up an equilibrium with dinitrogen tetroxide, N_2O_4. Dinitrogen trioxide, N_2O_3, can be prepared by combining NO_2 and NO below about 0°C. Oxides of nitrogen from automobile exhaust, along with hydrocarbons, are primary precursors of photochemical smog which is a serious problem in many of our larger cities.

The purposes of this experiment are to study some of the properties of HNO_3, NO_3^-, and NO_2^-, and to prepare and study some properties and reactions of several oxides of nitrogen. In addition, you will prepare a metal oxide by oxidation of an unknown metal with HNO_3, and you will prepare the same metal oxide by thermal decomposition of a nitrate salt. From your mass data you will determine the empirical formula of the metal oxide, the atomic mass of the metal, and thus the unknown metal that you have used.

Experiment 38

REFERENCES

(1) Kotz, J. C., and Purcell, K. F., <u>Chemistry and Chemical Reactivity</u>, Saunders College Publishing, Philadelphia, 1987, sections 2.5, 4.1, 4.2, 14.5a, 16.4, 16.5, and 22.1c.

(2) Masterton, W. L., Slowinski, E. J., and Stanitski, C. L., <u>Chemical Principles</u>, 6th ed., Saunders College Publishing, Philadelphia, 1985, sections 2.5, 3.1-3.3, 13.3, 15.1, 19.3-19.6, 26.2, and 26.4.

EXPERIMENTAL PROCEDURE

(Study this section and the PRE-LABORATORY QUESTIONS before coming to the laboratory. **Wear safety goggles when performing this experiment.**)

A. Properties of Nitric Acid, HNO_3

Among the common strong acids, HNO_3 with $K_a = 24$ is the weakest ($HClO_4$, $K_a \sim 10^9$; HI, $K_a \sim 10^7$; HBr, $K_a \sim 10^6$; HCl, $K_a \sim 10^5$; and H_2SO_4 and $HClO_3$, $K_a \sim 10^3$).

The oxidizing properties of HNO_3 can be shown by a very simple experiment. Place a drop of 3 $\underline{M}$ nitric acid, HNO_3, solution on a strip of starch-potassium iodide test paper that has been placed in the concave dip of a watch glass. Describe what you observe in TABLE 38.1A.

The change of oxidation state that occurs when HNO_3 is reduced depends on a number of factors, but particularly the concentration of the acid and the strength of the reducing agent. This will be illustrated in parts **D** and **E** during the preparations of NO_2 and NO.

B. Properties of Some Nitrate Salts

Place 0.2 g of solid ammonium nitrate, NH_4NO_3, weighed on a beam balance (Laboratory Methods **C**), in a 10 x 75-mm test tube. Hold the test tube in the palm of your hand, and add about 5 drops of distilled water with stirring to dissolve the NH_4NO_3 (Laboratory Methods **K**). Describe what you feel and what happens in TABLE 38.1B. A chemical cold pack takes advantage of this behavior of NH_4NO_3 in water.

Test the pH of the NH_4NO_3 solution with wide range pH paper (Laboratory Methods **P**), and record your result in TABLE 38.1B.

Place 0.2 g of solid sodium nitrate, $NaNO_3$, weighed on a beam balance, in a 10 x 75-mm test tube. Add about 10. drops of distilled water with stirring to dissolve the $NaNO_3$. Moisten your stirring rod several times with the $NaNO_3$ solution, and trace the letter **T** on a piece of filter paper with the solution. Save the $NaNO_3$ solution for additional tests in part **C**. Dry the paper by letting it sit in the air or by holding it some distance from a Bunsen burner flame. When the filter paper is thoroughly dry, touch a glowing (<u>but not flaming</u>) wooden splint to one end of the letter **T**, and blow gently on the splint and paper. Describe what you observe in TABLE 38.1B.

This observation is evidence of the strong oxidizing ability of the nitrate ion when associated with combustible materials. Advantage is taken of this property in the manufacture of matches, fireworks, and explosives.

C. Properties of Nitrate, NO_3^-, versus Nitrite, NO_2^-

Place 0.2 g of solid sodium nitrite, $NaNO_2$, weighed on a beam balance, in a 10 x 75-mm test tube. Add about 10. drops of water with stirring to

492

dissolve the $NaNO_2$. Compare the properties of NO_3^- and NO_2^- by performing the following tests on both this freshly prepared $NaNO_2$ solution and your remaining $NaNO_3$ solution from part **B**.

Test the pH of both $NaNO_3$ and $NaNO_2$ solutions with wide range pH paper, and record your results in TABLE 38.1C.

Divide each of the remaining solutions into two 10 x 75-mm test tubes. To one portion of both $NaNO_3$ and $NaNO_2$ solutions, add 3 drops of 6 **M** sulfuric acid, H_2SO_4, solution. **CAUTION:** <u>**Sulfuric acid of this concentration causes burns on your skin and holes in your clothing. If there are any spills or spatters onto your skin or clothing, rinse the affected area thoroughly with water**</u>. Describe what you observe in each case in TABLE 38.1C.

To the second portion of both $NaNO_3$ and $NaNO_2$ solutions, add 3 drops of 6 **M** acetic acid, $HC_2H_3O_2$, solution. Then place pieces of starch-potassium iodide test paper in the concave dip of a watch glass, and add a drop of the acidified $NaNO_3$ and $NaNO_2$ solutions onto separate pieces of test paper. Describe what you observe in each case in TABLE 38.1C. Finally, add a couple drops of 0.1 **M** potassium permanganate, $KMnO_4$, solution to the acidified $NaNO_3$ and $NaNO_2$ solutions, and shake the test tubes. Describe what you observe in TABLE 38.1C.

D. Preparation of Nitrogen Dioxide, NO_2

Nitrogen dioxide can be prepared using the apparatus shown in FIGURE 38.1 which will be constructed sequentially as described below. One piece of the apparatus in FIGURE 38.1 may have already been made in Experiment 1. If not, you will either make it now, or it will be available for you at the equipment bench.

① 43-cm tube with 90°-bends 8.5 cm from one end and 18.5 cm from the other end

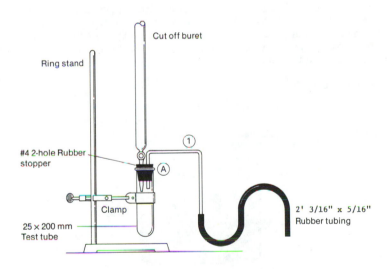

FIGURE 38.1. Apparatus for the preparation of nitrogen dioxide, NO_2.

Insert the short end of the 43-cm glass tubing having two 90°-bends ①
into a #4 2-hole rubber stopper Ⓐ. **CAUTION: <u>Be careful to follow rigorously the procedures for inserting glass tubing into rubber stoppers that
are given in Laboratory Methods F</u>.** Attach a 2-foot length of 3/16" x 5/16"
rubber tubing to the long end of the 43-cm glass tubing. Then weigh either
0.5 g of copper metal or 1.5 g of bismuth metal (as designated by your
instructor), place it in a dry 25 x 200-mm test tube, and complete the
construction of the apparatus in FIGURE 38.1. Finally, put about 5 mL of
16 $\underline{M}$ nitric acid, HNO₃, solution into your buret. **CAUTION: <u>Concentrated
HNO₃ is very corrosive and causes burns on your skin and holes in your
clothing. Wear rubber gloves when handling it. If any spills or spatters
occur onto your skin or clothing, rinse the affected area thoroughly with
water</u>.**

Place two <u>dry</u> 20 x 150-mm test tubes and one 20 x 150-mm test tube
containing 1 mL of water in a test tube rack near the above apparatus. Make
certain that three #2 solid rubber stoppers are close at hand. Also place a
1-L beaker full of water next to the test tube rack. Put the loose end of
the rubber tubing into bottom of the first test tube, and begin NO₂ generation by adding 16 $\underline{M}$ HNO₃ dropwise from the buret into the large test tube.
**CAUTION: <u>NO₂ is a poisonous gas; be careful not to generate it into the
laboratory</u>.** As the generation proceeds, quickly fill the two dry test tubes
in the rack with the dense brown gas, stoppering each test tube as soon as
it is filled. The test tubes are not filled until the gas has a brown
color, not just a pale yellow-brown. When both test tubes are filled, put
the loose end of the rubber tubing into the third test tube containing water
to saturate the water with NO₂ by dissolving the NO₂ and preventing it from
escaping into the laboratory. If the solution becomes saturated before NO₂
generation ceases (as evidenced by brown NO₂ escaping from the solution),
put the loose end of the rubber tubing into the beaker of water. If the gas
generation is insufficient for all three test tubes, add another mL or more
of the 16 $\underline{M}$ HNO₃ to your buret and then into the large test tube. Describe
in TABLE 38.1D what you observe in the large test tube as NO₂ is generated.

E. Preparation of Nitric Oxide, NO

Nitric oxide can be prepared using the apparatus shown in FIGURE 38.1
along with the modification at the exit tube shown in FIGURE 38.2. One
piece of the apparatus in FIGURE 38.2 may have already been made in
Experiment 1. If not, you will either make it now, or it will be available
for you at the equipment bench.

① 17-cm U-tube

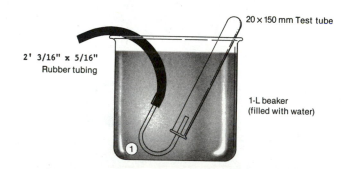

2' 3/16" x 5/16"
Rubber tubing

20 x 150 mm Test tube

1-L beaker
(filled with water)

FIGURE 38.2. Exit tube modification for the preparation of nitric oxide.

Clean the apparatus used in the preparation of NO₂, emptying any HNO₃
solution into a bottle in the hood marked USED HNO₃. Weigh either 1.0 g of
copper metal or 3.0 g of bismuth metal (as designated by your instructor),
place it in the 25 x 200-mm test tube, and reassemble the apparatus shown in

FIGURE 38.1. Then pour about 10. mL of 8 $\underline{M}$ nitric acid, HNO_3, solution into the buret. CAUTION: $\underline{HNO_3}$ of this concentration is corrosive and causes burns on your skin or holes in your clothing. Wear rubber gloves when handling it. If any spills or spatters occur onto your skin or clothing, rinse the affected area thoroughly with water.

Attach the glass U-tube to the loose end of the rubber tubing. Fill the 1-L beaker with water. Fill three 20 x 150-mm test tubes with water, and place them in the test tube rack. Make certain that three #2 solid rubber stoppers are close at hand. Place your thumb over the first test tube, and invert its mouth under water in the beaker without allowing any air to enter. Then insert the glass U-tube into the test tube as shown in FIGURE 38.2, and allow the assembly to rest on the bottom of the beaker.

Begin NO gas generation by allowing a few mL of 8 $\underline{M}$ HNO_3 from the buret into the large test tube. Gas will eventually enter the collection test tube via the glass U-tube and will begin displacing the water already present. You must immediately hold the collection tube with your hand to prevent it from popping out of the beaker. As the water is displaced from the test tube, do not allow the water level in the test tube to drop $\underline{below}$ the exit of the U-tube. This is easily accomplished by slowly lifting the test tube off the U-tube, and finally, still holding the mouth of the test tube under water, catching the last few bubbles of NO gas required to fill the test tube as these bubbles rise upward through the water. When all the water in the test tube has been displaced, take a rubber stopper in your other hand, and reach underwater and stopper the test tube without allowing the NO gas in the test tube to be exposed to the air. Then repeat the appropriate procedure of the last two paragraphs in order to fill the two additional 20 x 150-mm test tubes with NO gas. If the gas generation is insufficient to fill three test tubes, allow an additional mL or more of 8 $\underline{M}$ HNO_3 from the buret into the large test tube. Describe in TABLE 38.1E what you observe in the large test tube as NO is generated. When you are finished, take the apparatus apart carefully and clean it, emptying any HNO_3 solution into the bottle in the hood marked USED HNO_3.

F. Properties of Nitrogen Dioxide, NO_2

Place a piece of moistened starch-potassium iodide test paper in the concave dip of a watch glass. Remove the stopper of the first 20 x 150-mm test tube containing NO_2, and place the watch glass snugly over the mouth of the test tube so that the starch-potassium iodide test paper is exposed to the brown NO_2 gas. Describe what you observe in TABLE 38.1F.

Then quickly invert the test tube into a 50-mL beaker partially filled with distilled water so that the mouth of the test tube is just below the surface of the water. Agitate the test tube slightly. Describe what you observe in TABLE 38.1F.

Using the procedure described in Laboratory Methods O, prepare care-fully in a hood a slush bath composed of 200. mL of 1-hexanol, $CH_3(CH_2)_5OH$, and sufficient liquid nitrogen, contained in a 250-mL Dewar flask. This slush bath is a solid-liquid equilibrium mixture of 1-hexanol and maintains a temperature of about $-52^{\circ}C$ as long as both phases are present. If you make a thick slush, the equilibrium will be maintained long enough for use in this experiment. CAUTION: Liquid nitrogen is very cold ($-196^{\circ}C$) and can cause severe burns if spilled on your hands; wear the gloves that are provided while preparing the slush bath. CAUTION: Liquid 1-hexanol is handled in a hood because its vapor is a moderate local irritant to mucous membranes of the eyes and the upper respiratory tract. At its melting point ($-52^{\circ}C$) the vapor pressure of 1-hexanol is so low that once the slush bath is made, it can be utilized at your laboratory desk.

Below about $150^{\circ}C$, brown, paramagnetic NO_2 gas exists in equilibrium with its colorless, diamagnetic, dimerized form, dinitrogen tetroxide, N_2O_4.

Place the second 20 x 150-mm test tube containing NO_2 into the 1-hexanol slush bath contained in the 250-mL Dewar flask. Examine the tube periodically while allowing it to cool for a few minutes. Describe what you observe in TABLE 38.1F.

Remove the test tube containing NO_2 from the slush bath, and allow it to warm to room temperature. Describe what you observe in TABLE 38.1F.

Test the pH of the aqueous solution in the third test tube through which NO_2 has been bubbled using wide range pH paper, and record your result in TABLE 38.1F.

Test the same solution for both nitrate and nitrite ions by the following methods, performing the tests first with standard NO_3^- and NO_2^- solutions so that you become familiar with the positive tests for these ions.

1. Nitrate ion, NO_3^-. Place 3 drops of test solution (either 0.1 $\underline{M}$ sodium nitrate, $NaNO_3$, solution, or your solution from shaking NO_2 with water) in a 10 x 75-mm test tube. Add 3 drops of barium diphenylamine sulfonate solution. Then add **dropwise** 10. drops of 18 $\underline{M}$ sulfuric acid, H_2SO_4, solution **carefully** down the inside of the sloping test tube. CAUTION: **Concentrated H_2SO_4 is very corrosive and causes burns on your skin and holes in your clothing. Handle the dropper with care so as not to spill any drops on you or the desktop. If any spills or spatters occur onto your skin or clothing, rinse the affected area thoroughly with water.** Stir the solution gently. After mixing, a deep blue color slowly changing to violet, and possibly to wine red, indicates the presence of NO_3^-. Record your observations for both test solutions in TABLE 38.1F. Save these solutions for the nitrite test below.

2. Nitrite ion, NO_2^-. **Without** mixing, add 5 drops of test solution (either 0.1 $\underline{M}$ sodium nitrite, $NaNO_2$, solution, or your solution from shaking NO_2 with water) **carefully** down the side of the test tube already having the blue-violet solution from using 0.1 $\underline{M}$ $NaNO_3$ solution or your NO_2 solution in the nitrate test above. A yellow layer forming on the top of the solution indicates the presence of NO_2^-. Record your observations for both test solutions in TABLE 38.1F.

G. Properties of Nitric Oxide, NO

Dissolve a pinch of solid iron(II) sulfate, $FeSO_4$, in 3 mL of distilled water contained in a 13 x 100-mm test tube. Add this $FeSO_4$ solution to the first 20 x 150-mm test tube containing NO, and restopper the tube quickly. Shake the test tube gently, and watch it carefully. Describe what you observe in TABLE 38.1G. The color is due to the pentaaquanitrosoiron(II) cation, $[Fe(H_2O)_5NO]^{2+}$, in which an NO molecule has displaced a single, octahedrally oriented water molecule of the original hexaaquairon(II) cation, $[Fe(H_2O)_6]^{2+}$.

Place a piece of moistened starch-potassium iodide test paper in the concave dip of a watch glass. Remove the stopper of the second 20 x 150-mm test tube containing NO, and place the watch glass snugly over the mouth of the test tube so that the starch-potassium iodide test paper is exposed to the colorless NO gas. Describe changes, if any, in TABLE 38.1G.

Remove the watch glass, swing the test tube (mouth forward) through the air to force air into the tube, and replace the stopper. Describe what you observe in TABLE 38.1G. Repeat the swing through the air, if necessary.

Remove the stopper of this same test tube, add 1 mL of water, replace the stopper, and shake the test tube. Describe what you observe in TABLE 38.1G.

H. Preparation and Properties of Dinitrogen Trioxide, N_2O_3

Regenerate the 1-hexanol slush bath, if necessary.

Remove the stopper of the third 20 x 150-mm test tube containing NO, swing the test tube (mouth forward) through the air, and replace the stopper. Place the test tube in the 1-hexanol slush bath contained in the Dewar flask, and allow the test tube to cool for a few minutes. Remove the test tube, wipe it free of frost and of solvent with a tissue, and note carefully its contents. Describe in TABLE 38.1H what you observe on the sides of the test tube. Any blue liquid is N_2O_3.

Allow the test tube to warm to room temperature, and then note its contents carefully. Describe what you observe in TABLE 38.1H. The dissociation of N_2O_3 is significant above -30°C and is almost complete at its boiling point (3.5°C).

I. Preparation of an Oxide of an Unknown Metal

Support a clean porcelain crucible without a cover on a wire triangle, and heat it for a couple minutes with a Bunsen burner flame until the crucible is thoroughly dry (Laboratory Methods D). Allow the crucible to cool, then weigh it on an analytical balance, and record the mass in TABLE 38.2I. Transfer to the crucible just over 0.1 g of an unknown metal obtained from your instructor, and record your unknown number in TABLE 38.2I. Weigh the crucible and contents, and record the mass in TABLE 38.2I.

Place the crucible on the wire triangle again, and place the entire apparatus in the hood. Add slowly to the metal 6 drops of 16 M nitric acid, HNO_3, solution. CAUTION: <u>**Concentrated nitric acid is very corrosive and causes burns on your skin and holes in your clothing. If there are any spills or spatters onto your skin or clothing, immediately rinse the affected area thoroughly with water**</u>. When the reaction has ceased, heat the crucible **gently** to drive off the excess nitric acid and water. Apply heat gently to the crucible just above the water layer. **Too rapid heating will cause spattering and loss of material.** When the solid that is formed appears dry, heat the crucible in the full flame of the Bunsen burner for about 10. minutes. Then use your crucible tongs to place the hot crucible on a ceramic board to cool. When the crucible and oxide have cooled to room temperature, weigh them, and record your mass in TABLE 38.2I.

J. Thermal Decomposition of a Nitrate Salt

Nitrates, except those of the very reactive metals, readily undergo decomposition when heated. Such reactions are excellent for preparing small amounts of oxides of nitrogen, and also may yield metal oxides. However, some nitrates, especially ammonium nitrate, may decompose explosively if heated too rapidly. Obviously, none of these are used in this experiment.

Support a clean porcelain crucible without a cover on a wire triangle, and heat for a couple minutes with a Bunsen burner flame until the crucible is thoroughly dry. Allow the crucible to cool, then weigh it on an analytical balance, and record its mass in TABLE 38.2J. Transfer to the crucible about 0.7 g of an unknown metal nitrate hexahydrate, $M(NO_3)_x \cdot 6H_2O$, obtained from your instructor. (The metal present is the same as that used in part I.) Weigh the crucible and contents, and record the mass in TABLE 38.2J.

Place the crucible on the wire triangle again, and place the entire apparatus in the hood. Heat the crucible **gently**, and record what you observe in TABLE 38.2J. Continue to heat the crucible gently for about 5 minutes, and then heat it in the full flame of the Bunsen burner for about 10. minutes. Use your crucible tongs to place the hot crucible on a ceramic board to cool. When the crucible and oxide have cooled to room temperature, weigh them, and record your mass in TABLE 38.2J. [The metal oxide formed by oxidation of your unknown metal is also the only nonvolatile product formed by the thermal decomposition of your metal nitrate hexahydrate. It is useful to think about the thermal decomposition as $M(NO_3)_x \cdot 6H_2O$ being converted to $MO_{x/2}$.]

Perform the calculations in TABLE 38.3 including the sample calculations on the back of TABLE 38.3.

Student ID number _____

Section _____ Date _____

Instructor _____   D A T A

TABLE 38.1.   Experimental observations and related questions.

(Answer all **QUESTIONS** before handing in your report.)

A **Observation**   Behavior of 3 $\underline{M}$ HNO_3 toward starch-KI test paper

QUESTION   Write a balanced net ionic equation describing the reaction that occurs, and point out the oxidizing agent and the reducing agent as well as the initial and final oxidation states for nitrogen. Why did you choose your particular nitrogen-containing product?

B **Observation**   Behavior of NH_4NO_3(s) in water

QUESTION   What is responsible for this behavior?

Observation   pH of the NH_4NO_3 solution

QUESTION   Write a balanced net ionic equation to explain your pH.

Observation   Behavior of filter paper impregnated with $NaNO_3$

C **Observation**   pH of the $NaNO_3$ solution

Observation   pH of the $NaNO_2$ solution

QUESTION   Write a balanced net ionic equation to explain the difference in pH between $NaNO_3$ and $NaNO_2$ solutions.

TABLE 38.1 (continued)

Observation Behavior of the $NaNO_3$ solution toward 6 $\underline{M}$ H_2SO_4

Observation Behavior of the $NaNO_2$ solution toward 6 $\underline{M}$ H_2SO_4

QUESTION Write a balanced net ionic equation describing the redox reaction that occurred.

Observation Behavior of the $NaNO_3$ solution toward starch-KI test paper

QUESTION Write a balanced net ionic equation describing the redox reaction that occurred.

Observation Behavior of the $NaNO_2$ solution toward starch-KI test paper

QUESTION Write a balanced net ionic equation describing the redox reaction that occurred.

Observation Behavior of the $NaNO_3$ solution toward 0.1 $\underline{M}$ $KMnO_4$

Observation Behavior of the $NaNO_2$ solution toward 0.1 $\underline{M}$ $KMnO_4$

QUESTION Write a balanced net ionic equation describing the redox reaction that occurred.

TABLE 38.1 (continued)

D Observation Behavior of either Cu or Bi in 16 $\underline{M}$ HNO_3

        QUESTION Write balanced complete and net ionic equations describing the primary reaction that occurs during the preparation of NO_2 gas.

E Observation Behavior of either Cu or Bi in 8 $\underline{M}$ HNO_3

        QUESTION Write balanced complete and net ionic equations describing the primary reaction that occurs during the preparation of NO gas.

        QUESTION The presence of a brown gas in the exit tube indicated that a small amount of $NO_2(g)$ was also generated during the preparation of NO(g); however, no $NO_2(g)$ was collected in the NO(g) test tube. Why?

F Observation Behavior of $NO_2(g)$ toward moist starch-KI test paper

        QUESTION Write a balanced net ionic equation to explain this behavior.

TABLE 38.1 (continued)

Observation   Behavior of the $NO_2(g)$ test tube inverted in water

QUESTION   What does this behavior indicate about the solubility of $NO_2(g)$ in water?

Observation   Behavior of $NO_2(g)$ in the 1-hexanol slush bath

QUESTION   Explain your observation in terms of the $NO_2-N_2O_4$ equilibrium. What does this behavior tell you about the endothermic or exothermic nature of this equilibrium? Why?

Observation   Behavior of N_2O_4 on warming

QUESTION   Does this behavior agree with your previous conclusion about the endothermic or exothermic nature of the $NO_2-N_2O_4$ equilibrium? Why, or why not?

Observation   pH of the NO_2 solution

QUESTION   Write a balanced complete equation to account for this pH.

Observation   Behavior of 0.1 $\underline{M}$ $NaNO_3$ in the NO_3^- test

Observation   Behavior of your NO_2 solution in the NO_3^- test

Name ______________________________

Student ID number ______________

Section __________ Date __________

Instructor ________________________

Experiment 38

DATA

TABLE 38.1 (continued)

Observation Behavior of 0.1 M $NaNO_2$ in the NO_2^- test

Observation Behavior of your NO_2 solution in the NO_2^- test

QUESTION Based on your results in the NO_3^- and NO_2^- tests, write a balanced net ionic equation describing the reaction of $NO_2(g)$ on being shaken with water.

G **Observation** Behavior of $FeSO_4(aq)$ toward $NO(g)$

Observation Behavior of $NO(g)$ toward moist starch-KI test paper

QUESTION Explain any differences between $NO_2(g)$ and $NO(g)$ as to their behavior toward moist starch-KI test paper.

Observation Behavior of $NO(g)$ toward air

QUESTION Write a balanced complete equation to explain this behavior.

Observation Behavior of the gas mixture toward water

QUESTION Write a balanced complete equation to explain this behavior.

TABLE 38.1 (continued)

H **Observation** Behavior of the gas mixture in the 1-hexanol slush bath

   QUESTION Write a balanced complete equation describing the formation of N_2O_3.

   Observation Behavior of N_2O_3 on warming

   QUESTION Write a balanced complete equation describing the dissociation of $N_2O_3(g)$.

TABLE 38.2. Masses for preparation of an oxide and thermal decomposition of an unknown metal nitrate hexahydrate.

I	Mass of crucible (g)	
	Mass of crucible and metal (g)	
	Mass of crucible and metal oxide (g)	
J	Mass of crucible (g)	
	Mass of crucible and metal nitrate hexahydrate (g)	
	Mass of crucible and metal oxide (g)	

Observation Behavior of metal nitrate hexahydrate upon gentle heating

Instructor's initials _____

Name _____

Experiment 38

Student ID number _____

Section _____ Date _____

CALCULATIONS

Instructor _____

TABLE 38.3. Percent metal by mass and determination of the metal.

I	Mass of metal oxide formed (g)	
	Mass of metal in metal oxide (g)	
	Percent metal in metal oxide (%)	
J	Mass of metal nitrate hexahydrate reacted (g)	
	Mass of metal oxide formed (g)	
	Mass of metal in metal oxide (g)	
	Mass of metal in metal nitrate hexahydrate (g)	
	Percent metal in metal nitrate hexahydrate (%)	
	Atomic mass of metal (g/mol)	
	Metal used	
	Empirical formula of metal oxide	
	Empirical formula of metal nitrate hexahydrate	

Experiment 38

Sample calculations

I Percent metal in metal oxide

J Mass of metal in metal oxide formed from metal nitrate hexahydrate

   Mass of metal in metal nitrate hexahydrate

   Percent metal in metal nitrate hexahydrate

   Atomic mass of metal

   Empirical formula of metal oxide

1. Write a balanced complete equation describing the thermal decomposition of nitric acid, $HNO_3(l)$. Show clearly that oxidation and reduction occur.

2. Dehydration of nitric acid, $HNO_3(l)$, with phosphorus(V) oxide, $P_4O_{10}(s)$, produces nitrogen(V) oxide, $N_2O_5(s)$. N_2O_5 is the anhydride of HNO_3 and is an oxide of nitrogen that you will not prepare in this experiment. Write a balanced complete equation describing this dehydration. Show clearly that this is not a redox reaction.

3. Hydration of nitrogen(V) oxide, $N_2O_5(s)$, produces nitric acid, $HNO_3(aq)$. Write a balanced complete equation describing this reaction.

4. The nitrate anion, NO_3^-, which is a major water pollutant, is very difficult to remove from a water supply. Why should this be the case? By what method would you try to remove NO_3^- from a water supply if that were your responsibility?

5. Write Lewis electron dot structures to show the bonding and to explain the magnetic properties of nitrogen dioxide, NO_2, and dinitrogen tetroxide, N_2O_4. Then compare the magnetic properties of nitric oxide, NO, using molecular orbital theory for your prediction. Indicate clearly your reasoning.

Name ______________________________ **Experiment 38**

Student ID number ________________

Section __________ Date __________

Instructor _______________________

POST-LABORATORY QUESTIONS

1. Calculate the pH expected for a 5.0 M ammonium nitrate, NH_4NO_3, solution, and compare the answer with your experimental result in part **B**. $K_b(NH_3) = 1.8 \times 10^{-5}$. Indicate clearly your calculations and reasoning.

2. At 100°C the equilibrium constant in terms of molar concentrations, K_c, is 1.99 for the following equilibrium.

$$2\ NO_2(g) \rightleftharpoons N_2O_4(g)$$

Calculate the ratio of $[NO_2]/[N_2O_4]$ under these conditions. Does this ratio increase or decrease as temperature is decreased? Indicate clearly your calculations and reasoning.

3. Why was it important that no air be present in the test tubes used to collect nitric oxide, $NO(g)$, in part **E**?

4. When nitrogen dioxide, $NO_2(g)$, is shaken with water and heated, an additional reaction beyond that observed in part **F** occurs. The additional reaction involves the decomposition or disproportionation of nitrous acid, $HNO_2(aq)$. Write balanced molecular and net ionic equations for that reaction.

5. Of what oxyacid is dinitrogen trioxide, $N_2O_3(g)$, an anhydride? Write a balanced complete equation to indicate clearly your reasoning.

6. What would be the effect on the calculated atomic mass of your metal if the thermal decomposition of your metal nitrate hexahydrate had not gone to completion? Indicate clearly your reasoning.

Halogens and Hydrogen Halides

PURPOSE OF EXPERIMENT: Study some preparations, properties, and reactions of some halogens and hydrogen halides.

The halogens - fluorine, chlorine, bromine, iodine, and astatine - are the family of elements comprising group VIIA and preceding the noble gases in the periodic table. All of these halogens are very reactive and volatile, fluorine and chlorine being gases at room temperature, bromine being a liquid, and iodine being a solid. Atoms of all of these elements have seven electrons in their outermost energy levels. The octet can be completed by gaining one electron completely to form a -1 ion, the state of common occurrence in nature, or by sharing one additional electron forming one covalent bond to give a diatomic molecule, the state of the elemental halogens. The chemistry of the halogens, except fluorine, is also characterized by the positive oxidation states +1, +3, +5, and +7.

The hydrogen halides make up one of many sets of compounds in which the halogen is formally in the -1 oxidation state. The hydrogen halides are all gases at room temperature. The hydrogen halides are acids comprising only hydrogen and a halogen. They are thus part of a rather small group of hydrogen-containing acids that have no oxygen. Such acids can be prepared by the reaction of one of their salts with a second acid of lower volatility than the one desired. The second acid furnishes hydrogen ions, which associate with the anions of the salt. By using the proper temperature, the volatile product acid distills and can be collected. When they are dry, the hydrogen halides are rather inert. They are quite soluble in water and undergo dissociation to H^+ and X^- ions, producing acidic solutions. Hydrochloric, hydrobromic, and hydriodic acids are almost completely dissociated, but hydrofluoric acid is only weakly dissociated, and in general is quite different from the other hydrogen halides.

In this experiment you will prepare chlorine, bromine, and iodine and will investigate their physical and chemical properties. The preparative reaction involves the oxidation of the -1 halide anion by potassium permanganate in an acidic solution. In addition, you will prepare the hydrogen halides and study their physical and chemical properties.

REFERENCES

(1) Kotz, J. C., and Purcell, K. F., <u>Chemistry and Chemical Reactivity</u>, Saunders College Publishing, Philadelphia, 1987, sections 16.4-16.6, 19.7c, and 22.3.

(2) Masterton, W. L., Slowinski, E. J., and Stanitski, C. L., <u>Chemical Principles</u>, 6th ed., Saunders College Publishing, Philadelphia, 1985, sections 4.2, 13.2, 15.4, 19.3, 23.2, 26.1, and 26.3.

EXPERIMENTAL PROCEDURE

(Study this section and the PRE-LABORATORY QUESTIONS before coming to the laboratory. **Wear safety goggles when performing this experiment.**)

A. Preparation, Properties, and Reactions of Halogens

1. **Chlorine, Cl_2.** Place moistened, single strips of blue litmus paper, starch-potassium iodide paper, and dyed cotton cloth into the concave dip of a watch glass, and have the watch glass ready when you begin generating Cl_2. Place about 5 small crystals of potassium permanganate, $KMnO_4$, and the same volume of sodium chloride, $NaCl$, in a dry 16 x 150-mm test tube. Clamp the test tube, mouth up, in a vertical position. Add 0.5 mL of 6 $\underline{M}$ sulfuric acid, H_2SO_4, solution, and invert the watch glass over the test tube so that the test papers are toward any evolved gases. **CAUTION: <u>Never add concentrated</u>** (18 $\underline{M}$) **<u>sulfuric acid to $KMnO_4$.</u> CAUTION: <u>H_2SO_4 of this concentration is corrosive and causes burns on your skin or holes in your clothing. If any spills or spatters occur onto your skin or clothing, immediately rinse the affected area thoroughly with water</u>.**

Working in the hood, gently warm the contents of the tube until gas is evolved; then stop heating. **CAUTION: <u>Keep the test tube covered with the watch glass, and avoid inhaling the gas because Cl_2 is poisonous</u>**; however, you can essentially slide the watch glass carefully over the mouth of the test tube without removing it. Record in TABLE 39.1A1 what evidence you notice for the consumption of MnO_4^-. Also record in TABLE 39.1A1 what happens to the blue litmus paper, the starch-potassium iodide paper, and the dyed cotton cloth. Note the color of Cl_2 gas by holding white paper behind the test tube, and record the color in TABLE 39.1A1.

Grasp a piece of copper foil with a pair of crucible tongs, and heat one end of the foil to redness. Immediately remove the watch glass, and dip the hot Cu foil into the test tube of Cl_2 gas, being careful not to drop the foil into the tube. Then remove the foil, and cover the test tube with the watch glass. Describe what happened to the Cu foil in TABLE 39.1A1.

Place your test tube of Cl_2 gas **in the hood** in a beaker labeled with your name, remove the watch glass, add 2 mL of water, and swirl the test tube. Describe in TABLE 39.1A1 what you observe regarding the solubility of Cl_2 gas in water. The resulting solution is called **chlorine water.** Since such a solution affords a convenient supply of chlorine for certain reactions, it is a common laboratory reagent. When chlorine dissolves in water, some of the chlorine reacts chemically with the water and **disproportionates** to form species in which the chlorine atoms have higher and lower oxidation states than they have in Cl_2. Thus Cl_2 has been both oxidized and reduced.

Obtain 1 mL of chlorine water from the reagent bench. Test its effect on blue litmus paper, starch-potassium iodide paper, and dyed cotton cloth (Laboratory Methods **P**), and describe your results in TABLE 39.1A1. Add 10. drops of cyclohexane to the chlorine water contained in a test tube, stopper the test tube with a cork, shake the contents, and then let the contents stand a few moments. Describe what you observe in TABLE 39.1A1.

2. **Bromine, Br_2.** Repeat all the tests in part **A1**, substituting sodium bromide, $NaBr$, for $NaCl$ in the initial preparation of Br_2 and substituting bromine water for chlorine water in the final test. Record your observations in the appropriate parts of TABLE 39.1A2. Note carefully the differences in results in parts **A1** and **A2**.

3. **Iodine, I_2.** Repeat all the tests in part **A1**, substituting sodium iodide, NaI, for NaCl in the initial preparation of I_2 and substituting a saturated aqueous solution of I_2 for chlorine water in the final test. Record your observations in the appropriate parts of TABLE 39.1A3. Note carefully the differences in results in parts **A1**, **A2**, and **A3**.

4. **Relative reactivity of Cl_2, Br_2, and I_2.** Place a crystal of sodium bromide, NaBr, and a crystal of sodium iodide, NaI, in separate 10 x 75-mm test tubes. Add 0.5 mL of distilled water and 5 drops of cyclohexane to each tube, shake the tubes, and record what you observe in TABLE 39.1A4. Add a few drops of chlorine water to each tube, shake the tubes again, and record what you observe in TABLE 39.1A4.

Repeat the procedure of the previous paragraph, substituting sodium chloride, NaCl, for NaBr and substituting bromine water for chlorine water. Record your observations in TABLE 39.1A4.

B. Preparation, Properties, and Reactions of Hydrogen Halides

1. **Hydrogen fluoride, HF.** HF is prepared by the action of concentrated sulfuric acid, H_2SO_4, solution (a substance of low volatility), upon fluorspar, which is a naturally occurring form of calcium fluoride, CaF_2. The HF dissolves in water to form a solution, which is commercial hydrofluoric acid.

Place 0.5 g of powdered fluorspar, CaF_2, weighed on a beam balance (Laboratory Methods C), in the bottom of a 10 x 75-mm test tube, and moisten the powder with a few drops of 18 M sulfuric acid, H_2SO_4, solution. **CAUTION: Concentrated H_2SO_4 is very corrosive and causes burns on your skin or holes in your clothing. If any spills or spatters occur onto your skin or clothing, immediately rinse the affected area thoroughly with water.** Carefully place the test tube in a 50-mL or larger beaker labeled with your name, and put them **in the hood. CAUTION: Don't let any of the paste from the test tube touch your hands, and do not inhale any of the gas that is evolved. Hydrofluoric acid in the paste causes severe burns to your skin that are very slow to heal, and any traces of it should be rinsed immediately with copious amounts of water. HF gas is poisonous and also easily forms hydrofluoric acid in the presence of moisture in the nasal passages.** At the end of the laboratory period, add 1 mL of water dropwise to the test tube, empty the solution into the sink, and rinse the test tube thoroughly with considerable quantities of water. Dry the test tube, and examine the inner surface of the glass at the bottom of the tube. Describe what you observe in TABLE 39.1B1. This action of hydrofluoric acid on glass is called "etching." Hydrofluoric acid should not be stored in glass bottles because of etching. However, polymeric organic materials such as polypropylene and Teflon are resistant to attack by hydrofluoric acid and are suitable for containing it.

2. **Hydrogen chloride, HCl.** HCl gas resembles HF gas but is neither as corrosive toward glass or skin nor as toxic, and therefore it is safer and easier to handle. HCl gas may be prepared conveniently by adding concentrated sulfuric acid, H_2SO_4, to sodium chloride, NaCl, and warming the mixture.

The glass tubing in FIGURE 39.1 may have already been made in Experiment 1.

(1)   2 12-cm glass tubes with a 90°-bend 6 cm from one end

(2)   46-cm glass tube with a 90°-bend 6 cm from one end

(3)   20-cm glass tube with a 90°-bend 6 cm from one end

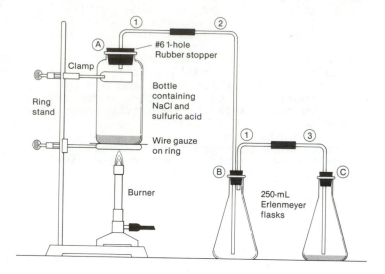

FIGURE 39.1. Apparatus for preparation of HCl.

If so, obtain the required rubber stoppers (one #6 1-hole and two #6 2-hole) from the equipment bench or the stockroom, and skip to the next paragraph. If not, the glass tubing will be already fitted in the appropriate 2-hole stoppers, and you can skip the next paragraph.

    Following rigorously the procedures for inserting glass tubing into rubber stoppers that are given in Laboratory Methods F, carefully insert one end of one 12-cm glass tube with a 90°-bend ① into the #6 1-hole rubber stopper Ⓐ until the glass tube just projects below the rubber stopper. Insert one end of the second 12-cm glass tube with a 90°-bend ① into one of the #6 2-hole rubber stoppers Ⓑ in an identical fashion. Into that same #6 2-hole rubber stopper Ⓑ, insert the long end of the 46-cm glass tube with a 90°-bend 6 cm from one end ② until the glass tube projects about 11 cm below the rubber stopper and nearly reaches the bottom of a 250-mL Erlenmeyer flask when the stopper is inserted snugly into the mouth of the flask. Finally, insert the long end of the 20-cm glass tube with a 90°-bend 6 cm from one end ③ into the remaining #6 2-hole rubber stopper Ⓒ until it projects 10 cm below the rubber stopper.

    Place 25 mL of distilled water and a small piece of blue litmus paper in what will be the right-hand 250-mL Erlenmeyer flask. Insert rubber stopper Ⓒ, and make certain that the glass tube extending into the flask is just above but does not touch the water. Clamp the bottle in place, and add to it 5 g of sodium chloride, NaCl, weighed on a beam balance. As an alternative, your instructor may designate that you use a 500-mL Florence flask as the bottle. Then complete the construction of the apparatus shown in FIGURE 39.1, using rubber connectors where appropriate. Finally, ask your instructor to check your apparatus.

    Add 12 mL of 18 $\underline{M}$ sulfuric acid, H_2SO_4, solution, **slowly and cautiously**, to 9 mL of distilled water. **CAUTION: Concentrated H_2SO_4 is very corrosive and causes burns on your skin or holes in your clothing. If any spills or spatters occur onto your skin or clothing, rinse the affected area thoroughly with water. CAUTION: Concentrated H_2SO_4 is also a very good dehydrating agent and reacts violently and exothermically with water.**

Therefore, concentrated H_2SO_4 should always be diluted by adding it to water, never the reverse. Remove stopper (A) from the bottle in FIGURE 39.1, pour the diluted H_2SO_4 into the bottle, and immediately replace the stopper.

Gently warm the contents of the bottle and note any changes that occur to the blue litmus paper in TABLE 39.1**B2**. Occasionally agitate the water in the right-hand 250-mL Erlenmeyer flask by swirling it. Continue to heat the generator **gently** for five minutes after the litmus paper changes color, and then allow the bottle to cool. Disconnect the two Erlenmeyer flasks, and cover the left-hand one with a watch glass. Wash out the bottle with water.

The following tests should be performed on the HCl gas collected in the left-hand 250-mL Erlenmeyer flask in FIGURE 39.1. Remove the watch glass only long enough to perform the tests, and then replace it. Observe the color of the gas by holding white paper behind the flask, and record the color in TABLE 39.1**B2**.

Blow your breath across the open mouth of the flask, and describe in TABLE 39.1**B2** what happens. This behavior in which the gas combines with water vapor from your breath to form tiny droplets of solution is characteristic of the hydrogen halides.

Dip a stirring rod into a 6 <u>M</u> aqueous ammonia, NH_3, solution, and bring it to the open mouth of the flask. Describe in TABLE 39.1**B2** what happens.

Invert the same 250-mL Erlenmeyer flask in water contained in a large beaker or in a pneumatic trough. Remove the watch glass, and describe in TABLE 39.2**B2** what happens.

The following tests should be performed on the HCl solution that formed in the right-hand Erlenmeyer flask in FIGURE 39.1.

Place a few pieces of granulated zinc, Zn, in a 10 x 75-mm test tube. Add dropwise 1 mL of the HCl solution. Describe in TABLE 39.1**B2** what you observe.

Place about 0.5 g of manganese dioxide, MnO_2, weighed on a beam balance, in a 16 x 150-mm test tube, and add 5 mL of the HCl solution. Warm the test tube gently in a water bath. Describe the color of the evolved gas in TABLE 39.1**B2**. Test the effect of the evolved gas on moist starch-potassium iodide paper and moist dyed cotton cloth placed in the concave dip of a watch glass and inverted over the mouth of the test tube. Describe in TABLE 39.1**B2** what you observe.

Place 5 drops of 0.1 <u>M</u> silver nitrate, $AgNO_3$, solution in a 10 x 75-mm test tube. **CAUTION:** <u>**$AgNO_3$ stains your skin by reacting with protein and is toxic. Wash your hands thoroughly after use or if there is any spill.**</u> Add a few drops of the HCl solution. Describe in TABLE 39.1**B2** what you observe.

3. Hydrogen bromide, HBr. Hydrogen bromide is a gas that also may be formed by the action of concentrated sulfuric acid on a salt. However, bromide ion is more easily displaced from its salts than chloride ion, and molecular bromine is more easily liberated by oxidizing agents. Thus, while concentrated sulfuric acid cannot oxidize hydrogen chloride, it is a strong enough oxidizing agent to oxidize hydrogen bromide.

Place 0.2 g of solid sodium bromide, NaBr, weighed on a beam balance, in a 16 x 150-mm test tube. Have at hand the materials and reagents to carry out all of the gas tests in the next paragraph, and then move to the hood. <u>Working at the hood</u>, add 1 mL of concentrated (18 <u>M</u>) sulfuric acid, H_2SO_4, solution. **CAUTION:** <u>**Concentrated (18 <u>M</u>) H_2SO_4 is very corrosive and**</u>

causes burns on your skin or holes in your clothing. If any spills or spatters occur onto your skin or clothing, rinse the affected area thoroughly with water. Describe in TABLE 39.1B3 what happens.

Observe the color of the gas evolved, and record the color in TABLE 39.1B3. Blow your breath across the mouth of the test tube, and describe in TABLE 39.1B3 what you observe. Dip a stirring rod into a 6 $\underline{M}$ aqueous ammonia, NH_3, solution, and bring it to the mouth of the test tube. Describe in TABLE 39.1B3 what happens. Finally, test the effect of the evolved gas on moistened starch-potassium iodide paper placed in the concave dip of a watch glass and inverted over the mouth of the test tube. Describe in TABLE 39.1B3 what you observe.

4. Hydrogen iodide, HI. Iodide ion is even more easily displaced from its compounds than bromide ion, and hydrogen iodide is more easily oxidized than is hydrogen bromide.

Repeat the preparation in part **B3** substituting solid sodium iodide, NaI, for NaBr and warming the test tube gently in warm tap water contained in a beaker after the 18 $\underline{M}$ H_2SO_4 has been added. Describe what happens in TABLE 39.1B4.

Also repeat the gas tests in part **B3**, and record your observations in TABLE 39.1B4.

Even though you could not obtain very much HI in the preparation above, reasonably pure HI can be obtained when NaI is heated with concentrated (15 $\underline{M}$) phosphoric acid, H_3PO_4, solution.

Repeat the preparation in part **B3**, substituting NaI for NaBr and 15 $\underline{M}$ H_3PO_4 for 18 $\underline{M}$ H_2SO_4, and heating the test tube in hot tap water contained in a beaker after the 15 $\underline{M}$ H_3PO_4 has been added. Describe what happens in TABLE 39.1B4.

Also repeat the gas tests in part **B3**, and record your observations in TABLE 39.1B4.

Name _____

Student ID number _____

Section _____ Date _____

Instructor _____

DATA

TABLE 39.1. Experimental observations and related questions.

(Answer all QUESTIONS before handing in your report.)

A1 **Observation** Evidence for consumption of MnO_4^-

QUESTION Write a balanced net ionic equation describing the redox reaction that occurred.

Observation Behavior of Cl_2 toward blue litmus paper

QUESTION Write a balanced net ionic equation describing the reaction that accounts for this behavior.

Observation Behavior of Cl_2 toward starch-KI paper

QUESTION Write a balanced net ionic equation describing the redox reaction that occurred.

Observation Behavior of Cl_2 toward dyed cotton cloth

Observation Color of Cl_2 gas

Observation Behavior of Cl_2 toward Cu foil

TABLE 39.1 (continued)

QUESTION   Write a balanced complete equation describing the redox reaction that occurred.

Observation   Solubility of Cl_2 gas in water

QUESTION   Write a balanced net ionic equation describing the redox reaction that occurred when Cl_2 gas dissolved.

Observation   Behavior of chlorine water toward blue litmus paper

QUESTION   Write a balanced net ionic equation describing the reaction that accounts for this behavior.

Observation   Behavior of chlorine water toward starch-KI paper

QUESTION   Write a balanced net ionic equation describing the redox reaction that occurred.

Observation   Behavior of chlorine water toward dyed cotton cloth

Observation   Behavior of chlorine water shaken with cyclohexane

QUESTION   What do you conclude about the relative solubility of Cl_2 in water versus cyclohexane?

Name _____

Student ID number _____

Section _____ Date _____

Instructor _____

DATA

TABLE 39.1 (continued)

A2 **Observation** Evidence for consumption of MnO_4^-

QUESTION Write a balanced net ionic equation describing the redox reaction that occurred.

Observation Behavior of Br_2 toward blue litmus paper

QUESTION Write a balanced net ionic equation describing the reaction that accounts for this behavior.

Observation Behavior of Br_2 toward starch-KI paper

QUESTION Write a balanced net ionic equation describing the redox reaction that occurred.

Observation Behavior of Br_2 toward dyed cotton cloth

Observation Color of Br_2 gas

Observation Behavior of Br_2 toward Cu foil

TABLE 39.1 (continued)

QUESTION   Write a balanced complete equation describing the redox reaction that occurred.

Observation   Solubility of Br_2 gas in water

QUESTION   Write a balanced net ionic equation describing the redox reaction that occurred when Br_2 gas dissolved.

Observation   Behavior of bromine water toward blue litmus paper

QUESTION   Write a balanced net ionic equation describing the reaction that accounts for this behavior.

Observation   Behavior of bromine water toward starch-KI paper

QUESTION   Write a balanced net ionic equation describing the redox reaction that occurred.

Observation   Behavior of bromine water toward dyed cotton cloth

Observation   Behavior of bromine water shaken with cyclohexane

QUESTION   What do you conclude about the relative solubility of Br_2 in water versus cyclohexane?

Name _____

Student ID number _____

Section _____ Date _____

Instructor _____

DATA

TABLE 39.1 (continued)

A3 **Observation** Evidence for consumption of MnO_4^-

QUESTION Write a balanced net ionic equation describing the redox reaction that occurred.

Observation Behavior of I_2 toward blue litmus paper

QUESTION Write a balanced net ionic equation describing the reaction that accounts for this behavior.

Observation Behavior of I_2 toward starch-KI paper

QUESTION How do you account for this behavior?

Observation Behavior of I_2 toward dyed cotton cloth

Observation Color of I_2 gas

Observation Behavior of I_2 toward Cu foil

TABLE 39.1   (continued)

QUESTION   Write a balanced complete equation describing the redox reaction that occurred.

Observation   Solubility of I_2 gas in water

QUESTION   Write a balanced net ionic equation describing the redox reaction that occurred when I_2 gas dissolved.

Observation   Behavior of aqueous iodine toward blue litmus paper

QUESTION   Write a balanced net ionic equation describing the reaction that accounts for this behavior.

Observation   Behavior of aqueous iodine toward starch-KI paper

QUESTION   How do you account for this behavior?

Observation   Behavior of aqueous iodine toward dyed cotton cloth

Observation   Behavior of aqueous iodine shaken with cyclohexane

QUESTION   What do you conclude about the relative solubility of I_2 in water versus cyclohexane?

Name _____

Student ID number _____

Section _____ Date _____

Instructor _____

TABLE 39.1 (continued)

A4 **Observation** Behavior of NaBr with water and cyclohexane

Observation Behavior of NaI with water and cyclohexane

Observation Behavior of NaBr, water, cyclohexane, and chlorine water

QUESTION Write a balanced net ionic equation for the redox reaction that occurred.

Observation Behavior of NaI, water, cyclohexane, and chlorine water

QUESTION Write a balanced net ionic equation for the redox reaction that occurred.

Observation Behavior of NaCl with water and cyclohexane

Observation Behavior of NaCl, water, cyclohexane, and bromine water

Observation Behavior of NaI, water, cyclohexane, and bromine water

QUESTION Write a balanced net ionic equation describing the redox reaction, if any, that occurred.

QUESTION Arrange Br_2, Cl_2, and I_2 in order of increasing oxidizing strength.

TABLE 39.1 (continued)

B1 Observation Behavior of hydrofluoric acid toward glass

QUESTION Write a balanced complete equation describing the reaction that occurred, treating glass as SiO_2.

B2 Observation Behavior of blue litmus paper in right-hand 250-mL Erlenmeyer flask during preparation of HCl

QUESTION How do you account for this behavior?

Observation Color of HCl gas

Observation Behavior of your breath mixing with HCl gas

Observation Behavior of HCl gas toward 6 $\underline{M}$ NH_3

QUESTION Write a balanced complete equation describing the reaction that occurred, and name the product.

Observation Behavior when Erlenmeyer flask containing HCl gas is inverted in water

QUESTION Should HCl gas be termed slightly soluble or very soluble in water?

Name _____

Student ID number _____

Section _____ Date _____

Instructor _____

Experiment 39

DATA

TABLE 39.1 (continued)

QUESTION   Why were you told, in the preparation of HCl, to be certain that the tube extending into the right-hand Erlenmeyer flask in FIGURE 39.1 did not dip into the water?

Observation   Behavior of HCl solution toward granulated Zn

QUESTION   Write a balanced net ionic equation describing the redox reaction that occurred.

Observation   Color of gas from reacting MnO_2 and the HCl solution

QUESTION   Write a balanced net ionic equation describing the redox reaction that occurred.

Observation   Behavior of the gas toward moist starch-KI paper

QUESTION   Write a balanced net ionic equation describing the redox reaction that occurred.

Observation   Behavior of the gas toward dyed cotton cloth

Observation   Behavior of HCl solution toward 0.1 $\underline{M}$ $AgNO_3$

QUESTION   Write a balanced net ionic equation describing the reaction that occurred.

Experiment 39

TABLE 39.1 (continued)

B3 **Observation** Behavior of NaBr and 18 $\underline{M}$ H_2SO_4

    Observation Color of gas evolved

    Observation Behavior of your breath mixing with the gas evolved

    Observation Behavior of the gas toward 6 $\underline{M}$ NH_3

        QUESTION Write a balanced complete equation describing the reaction
                      that occurred, and name the product.

    Observation Behavior of the gas toward moist starch-KI paper

        QUESTION Write a balanced net ionic equation describing the redox
                      reaction that occurred.

        QUESTION Write two balanced net ionic equations describing the
                      possible production of different gases in this preparation.

Name ____________________ **Experiment 39**

Student ID number ____________________

Section __________ Date __________

Instructor ____________________ **DATA**

TABLE 39.1 (continued)

B4 Observation Behavior of NaI and 18 M H_2SO_4

Observation Color of gas evolved

Observation Behavior of your breath mixing with the gas evolved

Observation Behavior of the gas toward 6 M NH_3

QUESTION Write a balanced complete equation describing the reaction that occurred, and name the product.

Observation Behavior of the gas toward moist starch-KI paper

QUESTION How do you account for this behavior?

QUESTION Write two balanced net ionic equations describing the possible production of different gases in this preparation.

TABLE 39.1   (continued)

Observation   Behavior of NaI and 15 $\underline{M}$ H_3PO_4

Observation   Color of gas evolved

Observation   Behavior of your breath mixing with the gas evolved

Observation   Behavior of the gas toward 6 $\underline{M}$ NH_3

QUESTION   Write a balanced complete equation describing the reaction
that occurred, and name the product.

Observation   Behavior of the gas toward moist starch-KI paper

QUESTION   Write   a   balanced   net   ionic   equation   describing   the
production of gas when NaI reacted with 15 $\underline{M}$ H_3PO_4.

Instructor's initials _____

Name _____

Student ID number _____

Section _____ Date _____

Instructor _____

PRE-LABORATORY QUESTIONS

1. Several types of precautions for working with H_2SO_4 are given in this experiment. What are these precautions, and why are they necessary?

2. Tests on evolved gases are often described as inherently insensitive. Describe clearly several reasons why this should be the case.

3. Why is it important that the hot Cu foil not be dropped into the test tube containing Cl_2 gas in part **A1**? Write a balanced complete equation describing an additional chemical reaction that would occur.

4. Starch-potassium iodide test paper is used a number of times in this experiment. Describe clearly the principles underlying the use of starch-KI test paper including the nature of the species for which you are testing, what behavior you expect for positive and negative tests, and what chemical reactions or processes are responsible for this behavior.

5. a. Starting with the equation of state for an ideal gas, $PV = nRT$, derive an expression for the density of a gas, ρ, in terms of the molecular weight of the gas, M.

   b. The apparatus in FIGURE 39.1 allows you to collect HCl gas in the left-hand Erlenmeyer flask by displacement of air. Based on your answer in $\underline{a}$., and assuming that the average molecular weight of the air mixture is 29 g/mol, why is it important to have the tubes in the left-hand Erlenmeyer flask inserted to the levels shown in FIGURE 39.1?

6. a. Write a balanced complete equation describing the reaction of fluorspar, CaF_2, with concentrated sulfuric acid, H_2SO_4. Name the products.

   b. How do you rationalize the occurrence of the reaction in $\underline{a}$. in terms of the driving force due to sparingly soluble solids, sparingly soluble gases, and weak electrolytes?

   c. Should an analogous equilibrium substituting calcium chloride, $CaCl_2$, for CaF_2 lie farther toward reactants or products than the reaction in $\underline{a}$.? Write the analogous reaction, and indicate clearly your reasoning.

Name _____

Student ID number _____

Section _____ Date _____

Instructor _____

POST-LABORATORY QUESTIONS

1. Fluorine, F_2, is not prepared in this experiment. Describe clearly how F_2 is prepared commercially, and write a balanced chemical equation for the reaction involved.

2. a. If chlorine had been put into an aqueous solution of NaOH, species of the same oxidation states as for the reaction of water with Cl_2 in part **A1** would be formed, but these species would exist as the anions instead of the acids. Write a balanced net ionic equation for the disproportionation of Cl_2 in basic solution.

   b. If a basic solution to which Cl_2 has been added is heated, further disproportionation of the ion with chlorine in the positive oxidation state occurs, to form ClO_3^- and Cl^-. Write a balanced net ionic equation for this reaction, and indicate the various oxidation states for chlorine that are involved.

3. Bromine, like chlorine, disproportionates in basic solution but to a much smaller extent. Similarly, further disproportionation occurs in hot basic solution, just as for chlorine. Write balanced net ionic equations for these two reactions which are analogous to those in POST-LABORATORY QUESTION 2.

4. Iodine also disproportionates in basic solution, but to a much smaller extent than either chlorine or bromine. Similarly, further disproportionation occurs in hot basic solution, just as for chlorine and bromine. Write balanced net ionic equations for these two reactions which are analogous to those in POST-LABORATORY QUESTION 2.

5. On the basis of what you have learned about the physical and chemical properties and reactions of Cl_2, Br_2, and I_2 and/or of HF, HCl, HBr, and HI, write an essay describing clearly what physical and chemical properties and reactions you would expect for elemental astatine (atomic number 85) and/or its acid, HAt (as designated by your instructor).

Chemistry of Copper and of Bromine

PURPOSE OF EXPERIMENT: Study some chemical reactions of some copper compounds and of bromine.

In this experiment you will study some of the chemistry of copper relating to its multiple oxidation states, and you will study some of the physical and chemical properties of bromine. In addition, you will utilize a variety of simple experimental techniques and will observe several types of chemical reactions. For example, you will prepare and use a liquid nitrogen-organic solvent slush bath for the purpose of maintaining a temperature well below that achievable with an ice bath or a salt-ice bath. Moreover, you will study a thermal decomposition reaction involving oxidation and reduction, a reduction of a salt to a metal by an active metal reducing agent, an oxidation of an active metal by a strong acid to produce hydrogen, and several oxidations by bromine. Finally, questions in this experiment will illustrate the use of thermodynamic calculations for predicting the feasibility of chemical reactions.

In this experiment a weighed sample of copper(II) bromide, $CuBr_2$, will be decomposed by gentle heating to copper(I) bromide, $CuBr$, and bromine, Br_2. The Br_2 will be removed under partial vacuum during the reaction and will be collected as a solid at low temperature. The behavior of aqueous Br_2 solutions toward a variety of reactants will be studied. After the $CuBr$ is weighed, it will be reduced with excess zinc metal, Zn, to produce copper metal, Cu, and zinc bromide, $ZnBr_2$. The excess Zn will be oxidized by a solution of hydrochloric acid, HCl. Since $ZnBr_2$ is soluble in HCl solution, both Zn and $ZnBr_2$ can then be separated from the unreactive Cu which will be dried and weighed. The various products can be differentiated qualitatively by their physical appearance (phase and color), and the stoichiometry of the two primary reactions can be determined quantitatively from the mass data. You will be graded on your quantitative results, so work very carefully.

REFERENCES

(1) Kotz, J. C., and Purcell, K. F., _Chemistry and Chemical Reactivity_, Saunders College Publishing, Philadelphia, 1987, sections 4.1, 13.6, 18.6, 22.3, 25.1, and 25.2.

(2) Masterton, W. L., Slowinski, E. J., and Stanitski, C. L., _Chemical Principles_, 6th ed., Saunders College Publishing, Philadelphia, 1985, sections 3.2, 3.3, 4.2, 8.6, 14.1-14.3, 25.2, and 26.3.

EXPERIMENTAL PROCEDURE

(Study this section and the PRE-LABORATORY QUESTIONS before coming to the laboratory. **Wear safety goggles when performing this experiment.**)

A. Thermal Decomposition of Copper(II) Bromide, $CuBr_2$

Weigh a clean, dry, 20 x 150-mm test tube on an analytical balance (Laboratory Methods C), and record the mass in TABLE 40.1A. Insert a rolled 4" x 6" piece of paper almost to the bottom of the weighed test tube to allow later transfer of $CuBr_2$ without any adhering to the walls.

Since $CuBr_2$ picks up moisture easily, it is important that you minimize exposure to air during the following transfers, handling, and weighing. Weigh approximately 1.1 g of solid copper(II) bromide, $CuBr_2$, onto weighing paper using a beam balance. Transfer it quickly to a mortar, and grind the $CuBr_2$ carefully with a pestle. Scrape the powdered $CuBr_2$ back onto the weighing paper, and then pour as much as possible of the powdered $CuBr_2$ through the rolled paper to the bottom of the test tube so that slightly more than 1 g of $CuBr_2$ is added. Tap the test tube <u>gently</u> on a clean desktop so that all the $CuBr_2$ falls to the bottom of the test tube; then remove the rolled paper. Weigh the test tube and contents on the analytical balance, and record the mass in TABLE 40.1A.

Describe in TABLE 40.2A the physical appearance of the $CuBr_2$.

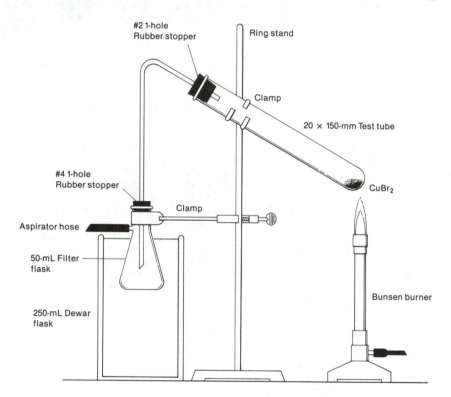

FIGURE 40.1. Apparatus for thermal decomposition of copper(II) bromide.

Part of the apparatus in FIGURE 40.1 may have already been made in Experiment 1.

(1) 50-cm glass tube with a 60°-bend 15 cm from one end

If so, obtain the two required 1-hole rubber stoppers (#2 and #4) from the equipment bench or the stockroom, and skip to the next paragraph. If not, the glass tubing will already be fitted in the appropriate 1-hole stoppers, and you can skip the next paragraph.

Following rigorously the procedures for inserting glass tubing into rubber stoppers that are given in Laboratory Methods F, carefully insert the 35-cm end of the 50-cm glass tube into a #4 1-hole rubber stopper so that the glass tube projects 5 cm below the rubber stopper. Likewise, carefully insert the 15-cm end of the 50-cm glass tube into a #2 1-hole rubber stopper so that the glass tube just projects through the stopper. Wipe any excess glycerol from the ends of the glass tube.

Using the procedure described in Laboratory Methods O, prepare carefully in a hood a slush bath composed of 200. mL of 1-hexanol, $CH_3(CH_2)_5OH$, and sufficient liquid nitrogen, contained in a 250-mL Dewar flask that is wrapped with tape or plastic mesh. This slush bath is a solid-liquid equilibrium mixture of 1-hexanol and maintains a temperature of about -52°C as long as both phases are present. If you make a thick slush, the equilibrium will be maintained long enough for use in this experiment. CAUTION: Liquid nitrogen is very cold (-196°C) and can cause severe burns if spilled on your hands; wear the gloves that are provided while preparing the slush bath. CAUTION: Liquid 1-hexanol is handled in a hood because its vapor is a moderate local irritant to mucous membranes of the eyes and the upper respiratory tract. At its melting point (-52°C) the vapor pressure of 1-hexanol is so low that the slush bath can be utilized at your laboratory desk.

Assemble on the desktop in the following sequence the apparatus shown in FIGURE 40.1. Attach an aspirator hose to a 50-mL filter flask. Clamp the filter flask to the left side from a ring stand in such a way that the ring stand base is toward you and the jaws of the clamp both encircle the neck of the flask and impinge on the aspirator tube on the sidearm. Adjust the height of the filter flask so that the bottom 1 cm of the filter flask is immersed in the slush bath contained in the Dewar flask; then remove the filter flask from the slush bath by lifting the ring stand. Insert the #2 stopper on the 15-cm end of the previously bent glass tube into the test tube containing $CuBr_2$. Slant the test tube at about a 30°-angle from horizontal, and clamp it near its mouth so that heat from a Bunsen burner flame will not melt the rubber of the clamp onto the test tube and change its mass. Then raise or lower the test tube clamp so that you can insert the other stopper into the filter flask. You may also have to adjust the filter flask horizontally, but do not move the filter flask vertically. CAUTION: The glass tube should project well below the sidearm on the filter flask, but should still be approximately 2 cm above the bottom of the filter flask; otherwise, solid Br_2 will later condense in the tip of the tube and plug it. CAUTION: Don't make the clamps too tight; you are stronger than the glassware!

If your slush bath is no longer very thick, add a bit more liquid nitrogen with stirring. Turn on the aspirator to its maximum position. Then lift the ring stand and attached apparatus, and lower the filter flask into the slush bath again. You are now ready to begin the thermal decomposition. Ask your instructor to inspect your apparatus.

Heat the $CuBr_2$ gently for a few minutes by slowly moving a very cool, luminous, yellow Bunsen burner flame back and forth beneath the base of the test tube. Do not melt the $CuBr_2$. CAUTION: Do not bring your Bunsen burner flame near the 1-hexanol slush bath because 1-hexanol, at least at room temperature, is a moderate fire hazard when exposed to a flame.

Describe in TABLE 40.2A what you observe in the test tube upon heating the $CuBr_2$. CAUTION: If an intensely colored brown gas forms throughout your test tube and glass tubing, stop heating and consult your instructor because your tubing is plugged, you have a leak, and/or your aspirator is not functioning properly. Also describe in TABLE 40.2A what you observe in the cooled filter flask.

CAUTION: The Br_2 that forms is highly irritating to mucous membranes of the eyes and upper respiratory tract as a vapor. As a liquid, Br_2 is very corrosive, causing severe burns that are slow to heal if spilled on the skin, and reacts vigorously with reducing materials. Wear rubber gloves when handling any containers of liquid bromine. However, it is collected in this experiment as a relatively harmless nonvolatile solid. Continue heating the test tube gently until there is no evidence of evolution of Br_2 or of changing color of the solid. The total time required for heating will probably be about 10. minutes.

Disassemble the apparatus as follows. Loosen the filter flask clamp slightly while still keeping the filter flask in the slush bath. Remove the aspirator hose from the sidearm, place a dropper bulb over the sidearm, turn off the aspirator, and tighten the filter flask clamp again. Loosen the test tube clamp, carefully remove the test tube from the rubber stopper, and clamp the test tube on the ring stand to cool to room temperature. Remove the rubber stopper and glass tube from the filter flask, and place a #4 solid rubber stopper loosely in the filter flask. Raise the filter flask out of the slush bath, wipe it with several Kimwipes, place a label on it with your name, and place the filter flask in the hood until you are ready to begin part C. Note the rubber stoppers at opposite ends of your bent glass tubing. One is eroded due to attack by bromine gas, whereas the other remains in good condition.

Describe in TABLE 40.2A what phase(s) of Br_2 is (are) present in the filter flask, and record the color of each phase.

After the test tube has cooled to room temperature, weigh the test tube and contents on the analytical balance, and record the mass in TABLE 40.1A.

B. Reduction of Copper(I) Bromide, CuBr

Weigh 0.25 g of powdered zinc, Zn, onto a piece of filter paper on a beam balance. Insert a rolled 4" x 6" piece of paper almost to the CuBr sample in the test tube used in part A, and add the powdered Zn to the test tube. Tap the test tube gently on the desktop so that all the Zn falls to the bottom of the test tube; then remove the rolled paper. Swirl the test tube to mix the two powders, but not so vigorously as to get powder up the sides of the test tube. Clamp the test tube to a ring stand at the same angle and the same height as shown in FIGURE 40.1.

Heat the CuBr and Zn powders gently for several minutes by slowing moving a very cool, luminous, yellow Bunsen burner flame back and forth beneath the base of the test tube. The reaction is initiated with a small puff of white "gas" at the bottom of the test tube.

Describe anything else that you observe in TABLE 40.2B.

Turn the test tube occasionally to mix the powder, and continue heating the test tube, first gently, then somewhat more strongly, for about 10. minutes, or longer if there is still evidence of reaction. Then allow the test tube to cool until it is warm to the touch.

Add enough 6 $\underline{M}$ hydrochloric acid, HCl, solution to the test tube to cover the solids. Swirl the test tube carefully. CAUTION: HCl of this concentration causes burns on your skin or holes in your clothing. If any spills or spatters occur onto your skin or clothing, rinse the affected area thoroughly with water.

Describe in TABLE 40.2B what you observe in the test tube.

When reaction has nearly ceased, pour the HCl solution and whatever solid comes with it into a 50-mL beaker. Again add enough 6 $\underline{M}$ HCl solution to the test tube to cover the remaining solids, swirl the test tube until reaction ceases, and pour the contents into the 50-mL beaker. Repeat, if necessary, until all solid has been transferred from the test tube to the beaker. Crush any solid in the beaker with a stirring rod, and stir the contents to make certain that the reactions have gone to completion.

Place a glass fiber filter paper circle (if required) into a clean, dry, Gooch filter crucible. Weigh the crucible on an analytical balance, and record the mass in TABLE 40.1B. Insert the Gooch crucible into a filter adapter and the filter adapter into a 500-mL filter flask. Connect an aspirator hose to the filter flask, turn on the aspirator, and decant the solution in the 50-mL beaker through the Gooch crucible (Laboratory Methods I). Wash the remaining solid in the 50-mL beaker twice with 5-mL portions of distilled water, decanting each portion through the Gooch crucible. Then pour the solid into the Gooch crucible, rinsing the last bits of solid into the Gooch crucible with distilled water.

Remove the aspirator hose, take apart the Gooch filter assembly, place the Gooch crucible on a watch glass labeled with your name, and dry the Gooch crucible and contents in the oven for about 1 hour at 110°C. Go on to part C while you are waiting for the Gooch crucible and contents to dry.

Remove the Gooch crucible and contents from the oven, and let them cool to room temperature. Weigh the Gooch crucible and contents on the analytical balance, and record the mass in TABLE 40.1B. Place the solid in a bottle on the reagent bench marked USED Cu.

C. Physical and Chemical Properties of Bromine, Br_2

Working at the hood and wearing rubber gloves, remove the rubber stopper from your 50-mL filter flask containing Br_2, and pour some of the Br_2 vapor into a dry 16 x 150-mm test tube just as you would pour water, **being careful to avoid transfer of any Br_2 liquid**. Replace the rubber stopper.

Describe in TABLE 40.2C whether the Br_2 vapor is more dense or less dense than air and why you reached this conclusion.

Remove the rubber stopper from the filter flask, add 5 mL of water to the Br_2 remaining in the filter flask, replace the rubber stopper, and swirl the contents to dissolve the Br_2. Remaining experiments may be done at your desk.

Using a medicine dropper, add 1 drop of the Br_2 solution to each of the following contained in the concave dip of a watch glass (Laboratory Methods P). Note carefully your observations in each case, and record them in TABLE 40.2C.

A strip of blue litmus paper

A strip of dyed cloth followed by rinsing with water

A strip of starch-potassium iodide test paper

Using a medicine dropper, add a few drops of Br_2 solution to a few drops of 0.1 $\underline{M}$ potassium chloride, KCl, solution contained in a 10 x 75-mm test tube. Describe what you observe in TABLE 40.2C. Add a few drops of cyclohexane, C_6H_{12}. Swirl the contents, and let the tube stand for a minute. Describe what you observe in TABLE 40.2C.

Using a medicine dropper, add a few drops of Br_2 solution to a few drops of 0.1 $\underline{M}$ potassium iodide, KI, solution contained in a 10 x 75-mm test tube. Describe what you observe in TABLE 40.2C. Add a few drops of cyclohexane, C_6H_{12}. Swirl the contents, and let the tube stand for a minute. Describe what you observe in TABLE 40.2C.

Using a medicine dropper, add a few drops of 0.25 $\underline{M}$ tin(II) chloride, $SnCl_2$, solution to a few drops of Br_2 solution contained in a 10 x 75-mm test tube. Swirl the contents. Describe the colors of the solution before and after swirling in TABLE 40.2C.

Perform the calculations in TABLE 40.3 before handing in your report.

Student ID number _____

Section _____ Date _____

TABLE 40.1. Mass data for thermal decomposition of $CuBr_2$ and reduction of CuBr.

A	Mass of test tube (g)	
	Mass of test tube and $CuBr_2$ (g)	
	Mass of test tube and CuBr (g)	
B	Mass of Gooch crucible and paper (g)	
	Mass of Gooch crucible, paper, and Cu (g)	

TABLE 40.2. Experimental observations and related questions.

(Answer all QUESTIONS before handing in your report.)

A Observation Physical appearance of $CuBr_2$

Observation Behavior in the test tube upon heating $CuBr_2$

Observation Behavior in the filter flask during heating of $CuBr_2$

QUESTION Write a balanced complete equation describing the reaction that occurred in the test tube.

Observation Phase(s) and color(s) of Br_2 in the filter flask

TABLE 40.2   (continued)

B   **Observation**   Behavior of CuBr and Zn upon heating

    QUESTION   Write a balanced complete equation describing the reaction that occurred in the test tube.

    Observation   Behavior of solids from reaction of CuBr and Zn toward 6 $\underline{M}$ HCl

    QUESTION   Write balanced net ionic equations for the reactions that describe what happened to the Zn and the $ZnBr_2$.

C   **Observation**   Density of Br_2 compared to air, and your reasoning

    Observation   Behavior of Br_2 solution toward blue litmus paper

    QUESTION   Write a balanced complete equation describing why a Br_2 solution should show your observed behavior toward blue litmus paper.

TABLE 40.2   (continued)

Observation   Behavior of Br_2 solution toward dyed cloth

Observation   Behavior of Br_2 solution toward starch-KI paper

QUESTION   Write a balanced net ionic equation describing the behavior of a Br_2 solution toward starch-KI paper.

Observation   Behavior of Br_2 solution toward a KCl solution

Observation   Behavior of Br_2 solution and KCl solution with cyclohexane

QUESTION   When dissolved in C_6H_{12}, Cl_2 gives no color, Br_2 gives an orange color, and I_2 gives a violet color. What do you conclude about the reaction of a Br_2 solution with a KCl solution? Why?

Observation   Behavior of Br_2 solution toward a KI solution

Observation   Behavior of Br_2 solution and KI solution with cyclohexane

TABLE 40.2   (continued)

QUESTION   Write a balanced net ionic equation describing the redox reaction that occurred.

QUESTION   From your results, arrange the elements, Cl_2, Br_2, and I_2, in order of increasing oxidizing strength. Indicate clearly your reasoning.

Observation   Colors of solution before and after swirling $SnCl_2$ solution with Br_2 solution

QUESTION   Write a balanced net ionic equation for the redox reaction that explains the color change.

Instructor's initials _____

TABLE 40.3.   Yield of products.

A	Mass of $CuBr_2$ used (g)	
	Theoretical yield of CuBr expected (g)	
	Actual yield of CuBr recovered (g)	
	Percent yield of CuBr (%)	
B	Theoretical yield of Cu expected based on $CuBr_2$ (g)	
	Actual yield of Cu recovered (g)	
	Percent yield of Cu (%)	

Name _____

Student ID number _____

Section _____ Date _____

Instructor _____

1. In scientific experiments, and even in daily life, people often miss important occurrences simply because they are not primed to look for them. You have a better chance both for making significant observations in an experiment and for understanding the meaning of such observations if your eye is trained and your mind is prepared. Look up in your textbook and in the Chemical Rubber Company Handbook of Chemistry and Physics the physical properties of each of the following elements or compounds, and record them below. You should at least be familiar with physical appearance, melting point and boiling point, and solubility in water, if each is available in the handbook. Note that the properties of the salts may depend upon the degree of hydration.

$CuBr_2$

$CuBr$

Br_2

Cu

Zn

$ZnBr_2$

H_2

$ZnCl_2$

2. Why is it not so important quantitatively to minimize the exposure of $CuBr_2$ to moist air once the test tube containing $CuBr_2$ has been weighed?

3. Why is it important quantitatively that any excess glycerol be wiped off of the end of the glass tubing in FIGURE 40.1 that projects into the 20 x 150-mm test tube?

4. Calculate the theoretical yield of Cu that could be produced from 1.104 g of $CuBr_2$ in this experiment and the percent yield of Cu if 0.312 g of Cu is recovered. Indicate clearly your calculations and reasoning.

Student ID number _____

Section _____ Date _____

Instructor _____

POST-LABORATORY QUESTIONS

1. a. The following thermodynamic functions have been measured: ΔH_f^o's for $CuBr_2(s)$, $CuBr(s)$, and $Br_2(g)$ are -136.8, -105, and +30.7 kJ/mol, respectively; S^o's for $CuBr_2(s)$ and $Br_2(g)$ are +134 and +174.9 J/K mol, respectively. In addition, S^o for $CuBr(s)$ is estimated to be about +65 J/K mol. Using this data, and assuming that ΔH and ΔS are constant over the <u>long</u> temperature range, calculate the minimum temperature (in ^oC) at which $CuBr_2(s)$ should decompose thermally under standard conditions. Indicate clearly your calculations and reasoning.

   b. Note that your calculated temperature is above the melting point for $CuBr_2(s)$. However, you did not require such a high temperature to decompose $CuBr_2(s)$ thermally. Why not?

2. One student recovered considerably more than the theoretical yield of CuBr in part **A** of this experiment, but then recovered exactly the theoretical yield of Cu expected in part **B**. Is this behavior reasonable? How can you account for it?

3. a. The standard free energies of formation, ΔG_f^o, for CuBr(s) and ZnBr$_2$(s) are -99.6 and -312.1 kJ/mol, respectively. Using this data determine whether the redox reaction involving the reduction of CuBr(s) by Zn(s) should proceed spontaneously at room temperature under standard conditions. Indicate clearly your reasoning.

    b. Why did you have to heat the CuBr(s) and Zn(s) to cause the reduction of CuBr(s) to proceed?

4. Write a balanced net ionic equation for the reaction expected, if any, under each of the following conditions.

    a. An aqueous solution of Br$_2$ is mixed with Hg$_2$$^{2+}$(aq).

    b. Concentrated nitric acid is mixed with Br$^-$(aq).

Multiple Oxidation States
of Transition Elements

PURPOSE OF EXPERIMENT: Study properties of aqueous ions representing multiple oxidation states of several transition elements, and determine the identity of two vanadium ions by redox titrations with MnO_4^-.

Transition elements make up the lower left-central portion of the periodic table of the elements, bridging the s-block elements at the left and the p-block elements at the right. Five properties are commonly attributed to transition elements: (1) they are all metals, (2) many of them form compounds involving a variety of oxidation states, (3) many of them form compounds and aqueous solutions that are colored, (4) many of them form complex ions, and (5) many of them form compounds exhibiting paramagnetic behavior. This experiment involves multiple oxidation states of several transition elements a variety of colored solutions, and several complex ions involving a transition metal cation and oxygen.

Three transition elements will be studied: vanadium, manganese, and iron. As elements each of these are metals, but the metals can be dissolved in water by reaction with appropriate chemical oxidizing agents, a process that you may have accomplished already using acids in Experiments 3, 12, or 36. Vanadium commonly forms the +2, +3, +4, and +5 oxidation states in aqueous solution. In acidic solutions these oxidation states take the form of V^{2+}, V^{3+}, VO^{2+}, and VO_2^+, respectively. These species are distinctly colored and are easily recognized in aqueous solution. Manganese commonly forms the +2, +6, and +7 oxidation states as soluble aqueous species and forms insoluble $MnO_2(s)$ in the +4 state. Each of these species is also distinctly colored and easily recognizable except for Mn^{2+} which is very pale pink or essentially colorless and resembles water itself. If MnO_4^- is used as an oxidizing agent in acidic solutions, only Mn^{2+} is formed as a product. Iron commonly forms the +2 and +3 oxidation states. The simple hydrated ions, Fe^{2+} and Fe^{3+}, occur in acidic solutions. Fe^{2+} is distinctly colored, but Fe^{3+} is essentially colorless and resembles water itself.

In this experiment you will observe and try to identify qualitatively these ions and their characteristic colors. Your identification is made somewhat more difficult by the fact that mixtures of these ions will produce a great variety of intermediate colors that will not be characteristic of a particular ion. Therefore, you must observe all color changes carefully and take detailed notes so that you can organize and explain your observations meaningfully by the end of the experiment. In addition, you will determine quantitatively the identity of two oxidation states of vanadium by means of redox titrations with MnO_4^-.

REFERENCES

(1) Hentz, Jr., F. C., and Long, G. G., J. Chem. Educ., **1978**, <u>55</u>, 5.

(2) Kotz, J. C., and Purcell, K. F., <u>Chemistry and Chemical Reactivity</u>, Saunders College Publishing, Philadelphia, 1987, sections 3.4, 4.4, 25.1, 25.3, and 25.6.

(3) Masterton, W. L., Slowinski, E. J., and Stanitski, C. L., <u>Chemical Principles</u>, 6th ed., Saunders College Publishing, Philadelphia, 1985, sections 2.6, 3.7, 21.3, 23.2, and 25.1-25.3.

EXPERIMENTAL PROCEDURE

(Study this section and the PRE-LABORATORY QUESTIONS before coming to the laboratory. **Wear safety goggles when performing this experiment.**)

A. Preparation of Zinc Amalgam

Zinc, Zn, will be used in this experiment to reduce vanadium in the +5 oxidation state to lower states. An active metal like Zn normally liberates hydrogen gas, H_2, from acidic solutions such as those used in this experiment. Thus, large quantities of Zn would be consumed without causing the reactions that we would like to observe to form the several ions of vanadium. This problem is avoided by forming an amalgam of Zn. An amalgam is produced by mixing mercury and another metal to coat the surface of the other metal or to form an alloy with the other metal. Specifically, when granular Zn is treated with a dilute acidic solution of mercury(II) chloride, $HgCl_2$, Hg(II) is reduced to metallic mercury which coats the surface of the Zn. This coating diminishes the reactivity of the Zn to the point where it will reduce several vanadium species but will be too weak a reducing agent to liberate H_2 because of a high overvoltage for H_2 on mercury.

Zinc amalgam is prepared according to the following procedure. Weigh 5.0 g of 20-mesh granular zinc metal on a beam balance (Laboratory Methods **C**), and place it in a 50-mL Erlenmeyer flask. Add 20. mL of 0.1 $\underline{M}$ mercury(II) chloride, $HgCl_2$, solution that has already been dissolved in 1 $\underline{M}$ hydrochloric acid, HCl, solution. **CAUTION:** <u>HCl of this concentration causes minor burns on your skin or holes in your clothing. If any spills or spatters occur onto your skin or clothing, rinse the affected area thoroughly with water</u>. Swirl the contents of the flask gently for about 5 minutes, being very careful to keep the zinc covered by water and thus protected from direct contact with air. Describe in TABLE 41.1A what you observe with respect to the appearance of the granular zinc. The granular zinc is said to be <u>amalgamated</u>.

Decant most of the $HgCl_2$ solution into a bottle marked USED $HgCl_2$ (Laboratory Methods **H**), making certain, however, that the amalgam remains covered with solution in order to protect it from contact with air. If the amalgam is exposed to air even for short periods, its effectiveness in part **B** is greatly decreased. Wash the amalgam several times by nearly filling the Erlenmeyer flask with distilled water, swirling the flask gently, and decanting most of the wash water, again keeping the amalgam covered with solution. This amalgam is ready for use as the reducing agent in part **B**.

B. Reduction of VO_2^+ by Zinc Amalgam

To the 50-mL Erlenmeyer flask containing the zinc amalgam, add 10. mL of 1 $\underline{M}$ sulfuric acid, H_2SO_4, solution. **CAUTION:** <u>H_2SO_4 of this concentration is corrosive and causes burns on your skin or holes in your clothing. If any spills or spatters occur onto your skin or clothing, rinse the affected area thoroughly with water</u>. Then add 10.00 mL of standard VO_2^+ stock solution, using a 10-mL pipet and <u>being certain to follow the procedure in Laboratory Methods B very carefully in order to draw solution into the pipet using a safety pipet filler</u>. Describe the color of the VO_2^+ solution in TABLE 41.1B. Record the exact concentration of the VO_2^+ stock solution in TABLE 41.2B. Stopper the flask with a #2 solid rubber stopper, and swirl it gently; then set it aside. While you are waiting for the reduction of the VO_2^+ stock solution to proceed, continue with parts **C** and **D** of this experiment, pausing <u>regularly</u> to swirl the 50-mL Erlenmeyer flask and to note the color of its contents. Describe carefully in TABLE 41.1B the sequence of color changes that occurs, including the final stable color. You can assume that the reduction is complete when no color change occurs after swirling intermittently for 15 minutes. This solution designated vanadium species $\underline{B}$ is now ready for use in part **E**.

C. Reduction of VO_2^+ by Fe^{2+}

Place a few crystals of iron(II) sulfate, $FeSO_4 \cdot 7H_2O$, in a 10 x 75-mm test tube. Then add several drops of water, and stir the solution with a glass stirring rod to dissolve the $FeSO_4 \cdot 7H_2O$ and produce a solution containing Fe^{2+} ions (Laboratory Methods K). Describe the color of the Fe^{2+} solution in TABLE 41.1C.

To a 250-mL Erlenmeyer flask, add 25.00 mL of standard VO_2^+ stock solution, using a 25-mL pipet and <u>being certain to follow the procedure in Laboratory Methods B very carefully in order to draw solution into the pipet using a safety pipet filler</u>. Then add 10. mL of 6 M sulfuric acid, H_2SO_4, solution and 5 mL of 15 M orthophosphoric acid, H_3PO_4, solution. **CAUTION: <u>Especially the H_2SO_4 solution of this concentration is corrosive and causes burns on your skin or holes in your clothing. If any spills or spatters occur onto your skin or clothing, rinse the affected area thoroughly with water</u>.** Finally, weigh 1 g of iron(II) sulfate, $FeSO_4 \cdot 7H_2O$, on a beam balance, add it to the flask, and swirl the solution in the flask to dissolve the solid $FeSO_4 \cdot 7H_2O$. Let the flask stand for 3 minutes. Describe carefully in TABLE 41.1C what color changes you observe during that time span. This solution designated vanadium species C is now ready for use in part D.

D. Oxidation of Vanadium Species C by MnO_4^-

To the 250-mL Erlenmeyer flask containing vanadium species C, add 1 g of ammonium peroxodisulfate, $(NH_4)_2S_2O_8$, weighed on a beam balance, and stir the solution vigorously for about 2 minutes. Describe what you observe in TABLE 41.1D.

The peroxodisulfate anion, $S_2O_8^{2-}$, is a powerful oxidizing agent that is capable of oxidizing excess Fe^{2+} rapidly, but is unable to oxidize vanadium species C for kinetic reasons. Furthermore, unlike the peroxides studied in Experiment 35 ($NaBO_3$ and H_2O_2) which acted as reducing agents toward MnO_4^-, $S_2O_8^{2-}$ cannot reduce MnO_4^- for kinetic reasons even though it is a parallel peroxo species.

The following MnO_4^- titration must be carried out at room temperature. At much below room temperature, the oxidation of vanadium species C is much too slow. At higher temperatures, the excess $S_2O_8^{2-}$ from the previous step oxidizes product Mn^{2+} back to MnO_4^-.

Obtain 35 mL of standard MnO_4^- solution from the reagent bench, and pour it into a clean buret. Record the exact concentration of the standard MnO_4^- solution in TABLE 41.2D. Then titrate vanadium species C in the 250-mL Erlenmeyer flask with the MnO_4^- solution, recording your initial and final volumes of MnO_4^- solution in TABLE 41.2D and observing carefully and recording in TABLE 41.1D any color changes that occur during the titration (Laboratory Methods B). The endpoint is indicated by the first permanent darkening of the solution (for at least 30. seconds) after the appearance of the bright lemon yellow color of VO_2^+ (compare with the VO_2^+ stock solution). One additional drop of MnO_4^- solution beyond the endpoint produces a peach color resulting from excess purple MnO_4^- dominating over lemon yellow VO_2^+. Doubling the volume of your solution by adding distilled water once the lemon yellow color appears decreases the intensity of the yellow color and allows you to see the endpoint more clearly.

E. Oxidation of Vanadium Species B by Oxygen in Air

Part of the apparatus required for this oxidation may have already been made in Experiment 1.

  (1) 17-cm straight tube, drawn to a glass jet at one end

  (2) 20-cm tube with a 90°-bend in the middle

If so, obtain the required #6 2-hole rubber stopper from the equipment bench or the stockroom, and skip to the next paragraph. If not, the glass tubing will already be fitted in the 2-hole rubber stopper, and you can skip the next paragraph.

<u>Following rigorously the procedures for inserting glass tubing into rubber stoppers that are given in Laboratory Methods F</u>, carefully insert the two pieces of glass tubing into the #6 2-hole rubber stopper. First, take the 17-cm straight tube, drawn to a glass jet at one end, and insert the 6-mm diameter end into the <u>bottom</u> side of the #6 2-hole rubber stopper just far enough that the glass jet almost reaches the bottom of a 250-mL Erlenmeyer flask when the stopper is fitted snugly into the mouth of the flask. Then remove the stopper from the flask, and insert the 20-cm tube with a 90°-bend in the middle into the <u>top</u> side of the other hole in the #6 2-hole rubber stopper so that the tube just projects through the stopper.

Pour the solution of vanadium species <u>B</u> from the 50-mL Erlenmeyer flask into a 250-mL Erlenmeyer flask, **being careful not to transfer any of the amalgam.** Wash the amalgam and flask with three 15-mL portions of distilled water, pouring each washing carefully into the 250-mL Erlenmeyer flask **without transferring any amalgam.** Then stopper the 250-mL Erlenmeyer flask with the #6 2-hole rubber stopper and tubes assembled previously. Transfer your amalgam to a bottle marked USED AMALGAM.

Clamp the 250-mL Erlenmeyer flask carefully, connect an aspirator hose to the right-angle glass tube, and turn on the aspirator slowly so as to pull air through the glass jet and into the solution of reduced vanadium species <u>B</u>. Make certain that the aspirator is <u>not</u> turned on so far that bubbling solution is sucked out through the aspirator hose. Describe carefully in TABLE 41.1E what color changes you observe in the solution.

After pulling air through the solution for about 5 minutes, remove the aspirator hose, and turn off the aspirator. Remove the rubber stopper and tubes, and wash the glass jet both inside and out with distilled water, making certain that the washings go into the 250-mL Erlenmeyer flask. This solution designated vanadium species <u>E</u> is now ready for use in part **F**.

F. Oxidation of Vanadium Species $\underline{E}$ by MnO_4^-

To the solution of vanadium species <u>E</u>, add 10. mL of 6 <u>M</u> sulfuric acid, H_2SO_4, solution. **CAUTION: <u>H_2SO_4 of this concentration is corrosive and causes burns on your skin or holes in your clothing. If any spills or spatters occur onto your skin or clothing, rinse the affected area thoroughly with water</u>.** Heat the solution just to boiling (Laboratory Methods D). Obtain 25 mL of standard MnO_4^- solution from the reagent bench, and pour it into your buret. Record the exact concentration of the standard MnO_4^- solution in TABLE 41.2D if you have not already done so while performing part D. Then titrate the hot solution of vanadium species <u>E</u> with the MnO_4^- solution, recording your initial and final volumes of MnO_4^- solution in TABLE 41.2F and observing carefully and recording in TABLE 41.1F any color changes that occur during the titration. The endpoint is indicated by the first permanent darkening of the solution (for at least 30. seconds) after the appearance of the bright lemon yellow color of VO_2^+ (compare with the VO_2^+ stock solution). One additional drop of MnO_4^- solution beyond the endpoint produces a peach color resulting from excess purple MnO_4^- dominating over lemon yellow VO_2^+. Doubling the volume of your solution by adding distilled water once the lemon yellow color appears decreases the intensity of the yellow color and allows you to see the endpoint more clearly.

Perform the calculations in TABLE 41.3 including sample calculations for one part on the back of TABLE 41.3. Also write the related equations requested under TABLE 41.3.

TABLE 41.1. Experimental observations and related questions.

(Answer all QUESTIONS before handing in your report.)

A Observation Behavior of granular Zn treated with 0.1 $\underline{M}$ $HgCl_2$

QUESTION Write a balanced net ionic equation describing the redox reaction that occurred, treating zinc amalgam as Zn.

B Observation Color of VO_2^+ solution

Observation Color changes during the reduction of VO_2^+ by zinc amalgam

---

C Observation Color of Fe^{2+} solution

Observation Color changes during the reduction of VO_2^+ by Fe^{2+}

QUESTION What oxidation state of vanadium was formed? Indicate clearly your reasoning. Hint: You may not be able to answer this question until after completing calculations for part D.

TABLE 41.1 (continued)

D **Observation** Behavior of vanadium species $\underline{C}$ toward $S_2O_8^{2-}$

QUESTION Is there any evidence for a change of oxidation state for vanadium species $\underline{C}$? Why?

Observation Color changes during titration of vanadium species $\underline{C}$ by MnO_4^-

E **Observation** Color changes during bubbling of air through vanadium species $\underline{B}$

QUESTION What oxidation state of vanadium is present after completion of the bubbling? Indicate clearly your reasoning. Hint: You may not be able to answer this question until after completing calculations for part **F**.

F **Observation** Color changes during titration of vanadium species $\underline{E}$ by MnO_4^-

TABLE 41.2. Titration data.

B	Concentration of VO_2^+ solution ($\underline{M}$)	
D	Concentration of MnO_4^- solution ($\underline{M}$)	
	Initial volume of MnO_4^- solution (mL)	
	Final volume of MnO_4^- solution (mL)	
F	Initial volume of MnO_4^- solution (mL)	
	Final volume of MnO_4^- solution (mL)	

Instructor's initials _____

Student ID number _____
Section _____ Date _____
Instructor _____   CALCULATIONS

TABLE 41.3. Moles of reactants and electrons gained or lost.

	Part D	Part F
Volume of MnO_4^- solution used (mL)		
Moles of MnO_4^- used (mol)		
Number of electrons gained per MnO_4^-		
Moles of VO_2^+ used (mol)		
Number of electrons lost per vanadium species __	C	E
Final oxidation state of vanadium		
Initial oxidation state of vanadium species __	C	E

Based on your results, write a balanced net ionic equation describing each of the following.

a. The oxidation of vanadium species B by oxygen from the air.

b. The oxidation of vanadium species C by MnO_4^-.

c. The oxidation of vanadium species E by MnO_4^-.

Experiment 41

Sample calculations for part ___

    Volume of MnO_4^- solution used

    Moles of MnO_4^- used

    Moles of VO_2^+ used

    Number of electrons lost per vanadium species ___

Student ID number _____

Section _____ Date _____

Instructor _____   **PRE-LABORATORY QUESTIONS**

1. An active metal like zinc normally liberates hydrogen gas from solutions of strong acids such as those used in part **A** of this experiment. Write a balanced net ionic equation describing this reaction.

2. Peroxodisulfate anion, $S_2O_8^{2-}$, oxidizes excess Fe^{2+} in part **D**. Write a balanced net ionic equation describing this redox reaction. To predict the correct sulfur-containing product, you must consider both the pH of the solution and that sulfur undergoes no change of oxidation state.

3. Why is it important for excess Fe^{2+} to be oxidized prior to the MnO_4^- titration in part **D**? Write a balanced net ionic equation to support your reasoning.

4. Although $S_2O_8^{2-}$ can oxidize Fe^{2+} rapidly, it is unable to oxidize vanadium species $\underline{C}$ for kinetic reasons. Look up and use appropriate data to show that oxidations of $\underline{any}$ reduced vanadium species by $S_2O_8^{2-}$ are thermodynamically favored (and therefore this one must be kinetically slow since it does not proceed).

5. a. If in a redox reaction, 10.24 mL of 0.0250 $\underline{M}$ $Cr_2O_7^{2-}$ solution is required to oxidize the Fe^{2+} in 20.53 mL of freshly prepared 0.0752 $\underline{M}$ Fe^{2+} solution completely to Fe^{3+}, calculate each of the following.

   (1) Moles of $Cr_2O_7^{2-}$ used.

   (2) Moles of Fe^{2+} used.

   (3) Number of electrons gained by $Cr_2O_7^{2-}$, recognizing that the total number of electrons lost equals the total number of electrons gained in the redox reaction.

   (4) The final oxidation state of chromium.

   b. Write a balanced net ionic equation describing the redox reaction that occurred in $\underline{a}$.

   c. Why was it important that the Fe^{2+} solution was freshly prepared, just as is done in part C of this experiment?

1. Having now studied experimentally all of the ions in the following
    table, describe clearly the color for each one in acidic water solu-
    tions, and indicate in which part(s) of this experiment you observed
    each ion.

Ion	Color	Part
V^{2+}		
V^{3+}		
VO^{2+}		
VO_2^+		
Mn^{2+}		
MnO_4^-		
Fe^{2+}		
Fe^{3+}		

2. You recorded a sequence of color changes for vanadium species in part B
    when you reduced a VO_2^+ solution with zinc amalgam. Write down in a
    column each color that you observed. Assume that no more than two ions
    were present at any given time, and identify the ion or ions responsible
    for each color based on your answer to POST-LABORATORY QUESTION 1.
    Remember that mixtures of ions produce a great variety of intermediate
    colors that are not characteristic of a particular ion, though one ion
    may dominate over another.

3. The MnO_4^- titration in part **D** must not be carried out much above room temperature because excess $S_2O_8^{2-}$ oxidizes product Mn^{2+} back to MnO_4^- at higher temperatures.

    a. Write a balanced net ionic equation describing the redox reaction that occurs at higher tempeatures. To predict the correct sulfur-containing product, you must consider both the pH of the solution and that sulfur undergoes no change of oxidation state.

    b. Would the reaction in <u>a</u>. cause you to use more MnO_4^- than should be required to oxidize vanadium species <u>C</u> or less? Indicate clearly your reasoning.

    c. Would the reaction in <u>a</u>. cause your calculated oxidation state for vanadium species <u>C</u> to be too high, too low, or unchanged? Indicate clearly your reasoning.

4. In neutral or weakly basic solution MnO_4^- is reduced to $MnO_2(s)$ rather than Mn^{2+}. If your pH were too high in part **F** and some MnO_4^- was converted to $MnO_2(s)$, would this cause your calculated oxidation state for vanadium species <u>E</u> to be too high, too low, or unchanged? Indicate clearly your reasoning.

Identification of Inorganic Solids

PURPOSE OF EXPERIMENT: Perform tests to identify an element present as a metal or nonmetal, or the ions present in an inorganic salt.

You may have performed several experiments that helped you to understand solubilities of inorganic salts, in particular, fractional crystallizations in Experiments 5 and 19, and the variation of solubility with temperature in Experiment 22. You may also have synthesized a number of inorganic solids in Experiments 13, 15, 16, 17, 18, 19, 33, 35, 37, 38, and 40. Finally, you may have studied the properties and reactions of inorganic solids in Experiments 5, 13, 15, 16, 17, 18, 19, 33, 34, 35, 36, 37, 38, 39, 40, and 41. This experiment combines all of these aspects and more while involving you in the identification of unknown inorganic solids. However, having performed any or all of the mentioned experiments, while certainly beneficial, is not a prerequisite for this experiment because it is designed to stand by itself. You will simply have to use data tables more frequently and carefully rather than depending on previously acquired knowledge.

This experiment is designed to involve you in the identification of inorganic salts on a research basis. You will become aware of some of the kinds of problems that chemists face, some of the kinds of techniques that they use to obtain data that will solve certain specific problems, and, most important, some of the kinds of thought processes that they use to interpret and explain their findings. You will quickly realize that simply performing an experiment may not bring an answer that you desire. First, you must devise experiments that provide appropriate evidence for the problem at hand. Then, even having performed the experiments and having obtained the results, you must learn to reason systematically in fitting the pieces of evidence together like the pieces of a crossword puzzle and in creating meaningful conclusions from seemingly unrelated pieces of evidence.

The number of inorganic species you will study will be limited, but the methods of attack open to you will be essentially unlimited. This experiment will teach you to use simple reactions and tests, many of which are already familiar, and then to think about the results. No single experiment will identify a species, but the results should provide hints that will direct you toward possibilities which you can then confirm. This whole process is a good illustration of the scientific method. You will gain an appreciation of the physical and chemical properties characteristic of given materials, and you will become aware of the relationship of these properties to the nature of the materials, their structure, and their reactivity.

You will examine one or more known or unknown metals, nonmetals, or salts by dissolving the samples, performing a series of preliminary tests, and identifying cations and anions through correlation of your results. Proper use of the procedures given and study of the results you obtain will certainly help you to identify your samples more readily, make you confident of your laboratory technique, and provide you with a considerable knowledge of inorganic chemistry. However, **thoughtless and injudicious use of these procedures will not meet the above objectives and will also consume much time.** You must, at all times, think about what you are doing, why you are doing it, what results you expect, and what such results will mean. Be very careful not to report cations or anions that you have added as reagents and that gave positive tests later. Use extensively the tables in this experiment and the literature available for your course to discover the meaning of your results. Don't perform unnecessary steps. For example, results from your first or second preliminary test may narrow your choices so far that common sense would dictate that you move immediately to specific tests not in the order provided in the experiment.

Use the designated DATA pages in this experiment to keep careful notes of all observations. Leave space liberally so that you can add conclusions and explanations or balanced chemical equations before turning in your report.

The possible elements, cations, and anions that you may study in this experiment are given in TABLE 42.1. Your instructor may delete some of these possibilities at his or her discretion. You will not be given a mixture of solids unless you are specifically told that by your instructor.

TABLE 42.1. Possible elements, cations, and anions.

Elements	Cations	Anions
Aluminum, Al	Aluminum(III), Al^{3+}	Bromide, Br^-
Bismuth, Bi	Ammonium, NH_4^+	Carbonate, CO_3^{2-}
Boron, B	Barium(II), Ba^{2+}	Chloride, Cl^-
Cadmium, Cd	Bismuth(III), Bi^{3+}	Hydroxide, OH^-
Chromium, Cr	Cadmium(II), Cd^{2+}	Iodide, I^-
Cobalt, Co	Calcium(II), Ca^{2+}	Nitrate, NO_3^-
Copper, Cu	Chromium(III), Cr^{3+}	Phosphate, PO_4^{3-}
Iodine, I_2	Cobalt(II), Co^{2+}	Sulfate, SO_4^{2-}
Iron, Fe	Copper(II), Cu^{2+}	Sulfide, S^{2-}
Lead, Pb	Iron(II), Fe^{2+}	
Magnesium, Mg	Lead(II), Pb^{2+}	
Manganese, Mn	Magnesium(II), Mg^{2+}	
Nickel, Ni	Manganese(II), Mn^{2+}	
Phosphorus, P	Nickel(II), Ni^{2+}	
Silicon, Si	Sodium(I), Na^+	
Sulfur, S	Tin(II), Sn^{2+}	
Tin, Sn	Zinc(II), Zn^{2+}	
Zinc, Zn		

REFERENCES

(1) Kotz, J. C., and Purcell, K. F., Chemistry and Chemical Reactivity, Saunders College Publishing, Philadelphia, 1987, sections 2.11, 3.3b, 3.4, 15.1, 15.8, 17.5, 20.3, 20.4, 21.1, 21.2, 22.1-22.3, 25.1, and 25.6.

(2) Masterton, W. L., Slowinski, E. J., and Stanitski, C. L., Chemical Principles, 6th ed., Saunders College Publishing, Philadelphia, 1985, sections 3.4, 3.6, 8.4-8.6, 13.1, 18.1, 18.2, 19.5, 22.1-22.3, 25.1, 25.2, 26.2, 26.5, and 26.6.

EXPERIMENTAL PROCEDURE

(Study this section and the PRE-LABORATORY QUESTIONS before coming to the laboratory. Wear safety goggles when performing this experiment. It is not wise for safety reasons to mix large quantities of unknown solutions; therefore, all solutions in this experiment should be added dropwise to a 10 x 75-mm test tube, or whatever is the smallest test tube in your desk set, unless otherwise stated in the directions.)

A. Physical Appearance

Describe the color, crystalline appearance, and any other unusual or distinctive physical characteristics of your sample. Examination under a magnifying glass, if available, may help you define the crystalline appearance.   If your sample is in the form of crystals, **cautiously** grind a small portion of it to a powder using a mortar and pestle, and note any changes in its color and appearance. If possible, use finely powdered samples for all tests that follow.

The color of a solid can sometimes vary depending upon how it was formed, how much water it contains, and whether it is crystalline or powdered.   Colors of some common powdered solids are given in TABLE 42.2. Since cations are more frequently responsible for color in solids than anions, the information is presented in terms of cations. This is not a complete list and includes only solids possible in this experiment.   Note that colorless crystals yield white powders upon grinding. In addition, some colors such as brown and black are very difficult to distinguish, and both should be considered when interpreting your evidence.

TABLE 42.2. Colors of some common solids.

White: Al(III) salts except yellow Al_2S_3 and brown AlI_3; NH_4^+ salts; Ba(II) salts; Bi(III) salts except yellow $BiBr_3$, red BiI_3, and brown Bi_2S_3; Cd(II) salts except yellow CdS and brown CdI_2; Ca(II) salts; Pb(II) salts except yellow PbI_2 and black PbS; Mg(II) salts except pale red-brown MgS; Na(I) salts; Sn(II) salts except orange SnI_2, yellow $SnBr_2$, and brown SnS; and Zn(II) salts except yellow ZnI_2

Red: BiI_3, elemental phosphorus, some hydrated Co(II) salts, and some hydrated Mn(II) salts and MnS (pink)

Orange: SnI_2

Yellow: Al_2S_3, $BiBr_3$, CdS, PbI_2, elemental sulfur, $SnBr_2$, ZnI_2, and some anhydrous Ni(II) salts

Green: $CoBr_2$; $Fe(OH)_2$ (pale green to white); $Ni(OH)_2$; some Cr(III) salts such as $CrBr_3$, $Cr(OH)_3$ (gray-green), and Cr_2S_3; some Cu(II) salts; and some hydrated Fe(II) salts and Ni(II) salts

Blue: Some anhydrous Co(II) salts and some hydrated Cu(II) salts

Violet: Many Cr(III) salts and elemental iodine (violet to black)

Brown: AlI_3, Bi_2S_3, elemental boron, CdI_2, Cu, $CuCl_2$, $FeBr_2$, and SnS

Gray: Elemental silicon, most metals except Cu, and FeI_2

Black: CrI_3, CoI_2, CoS, $CuBr_2$, CuS, FeS, NiI_2, NiS, and PbS

B. Behavior toward Water

Place about 0.1 g of your powdered sample (this is approximately the mass you pick up when you load sample on 5 mm of the tip of a wooden splint or the tip of a metal spatula) in 2 mL of water in a 16 x 150-mm test tube, and stir thoroughly. Does the sample dissolve (Laboratory Methods **K**)?   If not, warm the mixture in a boiling water bath in a beaker (Laboratory Methods **D**). Does the sample dissolve? If not, gradually increase the amount of water up to 10. mL, and repeat the heating. Does the sample dissolve? During these tests, watch for color changes of both the sample and the solution, for the evolution of gases, and for any other distinctive behavior. Centrifuge the solution (Laboratory Methods **J**) if you are in doubt about whether solid remains. If none of the sample dissolves in water, skip to part **C**.

Experiment 42

1. **Solubilities.** Solubilities of various powdered solids in water are given in TABLE 42.3. Note that solubilities sometimes decrease with aging of a sample; therefore, be careful when interpreting evidence for samples that don't seem to dissolve easily.

TABLE 42.3. Solubilities of some common solids in water.

a. Nitrates are generally soluble.

b. Chlorides, bromides, and iodides are generally soluble except for the halides of Pb(II).

c. Sulfates are generally soluble except those of Ba(II), Ca(II), and Pb(II).

d. Salts of NH_4^+ and Na(I) are generally soluble.

e. Carbonates, hydroxides, phosphates, and sulfides are generally insoluble. Common exceptions are covered in <u>d</u>. as well as hydroxides of Ca(II) and Ba(II) and sulfides of Mg(II), Ca(II), Ba(II), Cr(III), and Al(III). All of these salts are more soluble in weak or strong acids than in water.

f. All the elements in this experiment are insoluble except that I_2 is sparingly soluble.

2. **Colors.** The colors of ions in aqueous solution can be dependent upon concentrations, and some colors can combine with each other to produce a new color. The colors indicated in TABLE 42.4 are frequently exhibited in <u>acidic</u> aqueous solution by the designated ions. Other simple cations and anions in this experiment generally exhibit colorless solutions.

TABLE 42.4. Colors of ions in acidic solution.

Pink:	Co(II), Mn(II) almost colorless
Green:	Fe(II), Ni(II)
Blue:	Cu(II)
Violet:	Cr(III)

3. **pH.** Test the behavior of your water solution toward wide range pH paper (Laboratory Methods **P**), and record the approximate pH. An acidic reaction may result from hydrolysis of highly charged cations, but is particularly apparent for Al(III), Bi(III), Cr(III), and Sn(II). A basic reaction to litmus may result from OH^- or from hydrolysis of anions that are conjugate bases of weak acids, such as CO_3^{2-}, PO_4^{3-}, and S^{2-}.

C. Behavior toward Acids

1. **Hydrochloric acid, HCl.** Place about 0.1 g of your powdered sample in a 16 x 150-mm test tube, add dropwise up to 1 mL of 6 <u>M</u> HCl solution, and stir thoroughly. **CAUTION: <u>HCl of this concentration causes burns on your skin or holes in your clothing. If any spills or spatters occur onto your skin or clothing, rinse the affected area thoroughly with water</u>.** Note the solubility of the sample. Centrifuge the solution if you are in doubt about whether solid remains. Look for a new precipitate, a color change, or the evolution of a gas. Perform tests given in part **C1a** to identify any evolved gases. If no reaction with HCl occurs, warm the test tube in a hot water bath. Note any changes. Most samples that are insoluble in water but soluble in acid are salts that contain anions that are conjugate bases of weak acids and thus readily accept protons, or are metals that can reduce H^+ liberating hydrogen gas. Examples are CO_3^{2-}, OH^-, PO_4^{3-}, and S^{2-} or Al, Cr, Co, Mg, Mn, Ni, and Zn. However, HCl does not dissolve many water-insoluble sulfides that are dissolved by HNO_3.

a. Gas tests. The following tests for commonly evolved gases may prove useful because they provide evidence for the species shown in parentheses at each gas test. You must be ready to perform these tests rapidly and efficiently. If testing gases evolved from acid solution, have available <u>before</u> starting the test a watch glass, the underside of which has a piece of moist blue litmus paper (test for HCl, HBr, and HI), a piece of moist lead acetate paper (for H_2S), and a piece of starch-potassium iodide paper (for Cl_2), as well as a dropper of $Ba(OH)_2$ solution (for CO_2). As soon as a gas is evolved, note its color, and **very, very cautiously** note its odor; then test in sequence for HCl, HBr, HI, H_2S, Cl_2, CO_2, and H_2. If testing a gas evolved from basic solution, have available <u>before</u> starting the test a watch glass, the underside of which has a piece of moist red litmus paper (for NH_3). As soon as a gas is evolved, note its color, and **very, very cautiously** note its odor; then test in sequence for NH_3 and H_2. NO_2 may be evolved only when heating a sample to try to dissolve it or when heating a nitric acid, HNO_3, solution.

<u>Hydrogen Chloride</u> (Cl^-), <u>Hydrogen Bromide</u> (Br^-), and <u>Hydrogen Iodide</u> (I^-). These gases will turn blue litmus red. For further confirmation place a piece of filter paper, moistened with 6 <u>M</u> aqueous ammonia, NH_3, solution, on the underside of a watch glass that is covering the test tube from which gas is evolving. All three of these gases will cause the filter paper to fume, owing to the formation of NH_4Cl, NH_4Br, and NH_4I, respectively. Note that if HCl is used as the acid in a gas evolution test, HCl vapor may be evolved in addition to another gas. If in doubt, cause the gas evolution by using HNO_3 or H_2SO_4.

<u>Hydrogen Sulfide</u> (S^{2-}). The characteristic odor of rotten eggs will be readily detected. A brown or black color of lead acetate paper, due to the formation of PbS, indicates the presence of H_2S.

<u>Chlorine</u> (Cl^-). A dark blue color of starch-potassium iodide paper, due to the formation of a starch-iodine complex, indicates the presence of Cl_2. Be careful: reddish-brown NO_2 gas and Br_2 gas may cause a faint test.

<u>Carbon Dioxide</u> (CO_3^{2-}). Form a drop of saturated barium hydroxide, $Ba(OH)_2$, solution on the tip of a medicine dropper. Hold this clear drop in the mouth of the test tube from which the gas is evolving. (Do not allow the drop to fall into the tube.) A turbidity or white precipitate of $BaCO_3$ indicates the presence of CO_2.

<u>Hydrogen</u> (some metals and Si). Invert a 10 x 75-mm test tube over the tube evolving a gas in order to collect a small sample of the gas by downward displacement of air. Place the mouth of this inverted tube near a small flame. A mild pop will be heard, or the gas in the tube will burn with an almost colorless or light blue flame, if the evolved gas is hydrogen.

<u>Ammonia</u> (NH_4^+). The characteristic odor of NH_3 will be detected. NH_3 will turn red litmus blue.

<u>Nitrogen Dioxide</u> (NO_3^-). NO_2 is a reddish-brown gas and has a characteristic pungent odor. Be careful: reddish-brown Br_2 gas or violet I_2 gas may be obtained on heating and may make the identification of NO_2 doubtful.

Use great care when interpreting evidence from gas tests. Many of them are quite insensitive. Thus, a positive gas test should be taken very seriously, but a negative gas test may simply mean that you did not evolve enough gas to get a positive test.

b. Flame tests. If any of your sample was soluble in HCl, perform a flame test on the HCl solution. Clean a platinum, Pt, wire loop by plunging the loop into a test tube containing 12 $\underline{M}$ HCl and then placing the loop in a nonluminous flame. **CAUTION:** <u>Concentrated HCl causes burns on your skin or holes in your clothing. If any spills or spatters occur onto your skin or clothing, rinse the affected area thoroughly with water</u>. This method of cleaning may have to be repeated several times before no color is observed in the flame. Dip your **clean** Pt wire loop into the HCl solution containing your sample, immediately place the loop in a nonluminous flame, and note the color of the flame. Clean your Pt wire loop after each test.

If your sample was insoluble in HCl, perform a flame test as follows. Moisten a few crystals of your sample in a 10 x 75-mm test tube with 2 drops of 12 $\underline{M}$ HCl, dip your **clean** Pt wire loop into the resulting slurry or solution, immediately place the loop in a nonluminous flame, and note the color of the flame.

Many ions impart characteristic colors to a nonluminous flame. The colors indicated in TABLE 42.5 relate to possible cations in this experiment. Note that a Na(I) flame is very intense and frequently shows up from an impurity.

TABLE 42.5. Characteristic flame colors for common cations.

Cation	Color
Ba(II)	yellow-green
Ca(II)	brick red
Cu(II)	emerald green
Pb(II)	pale blue
Na(I)	persistent yellow

You may not need to study behavior toward additional acids if your sample was soluble in water and/or HCl solution. However, useful information may be gained.

2. Nitric acid, HNO_3. Repeat the tests described in part **C1** using 6 $\underline{M}$ HNO_3 solution in place of 6 $\underline{M}$ HCl, except that you should not repeat the flame test. **CAUTION:** <u>HNO_3 of this concentration causes burns on your skin or holes in your clothing. If any spills or spatters occur onto your skin or clothing, rinse the affected area thoroughly with water</u>. Elemental boron is insoluble in water and HCl, but slowly dissolves upon heating in HNO_3. In addition, HNO_3 dissolves a number of water-insoluble sulfides that are not soluble in HCl.

3. Sulfuric acid, H_2SO_4. Repeat the tests described in part **C1** using 6 $\underline{M}$ H_2SO_4 solution in place of 6 $\underline{M}$ HCl, except that you should not repeat the flame test. **CAUTION:** <u>H_2SO_4 of this concentration is corrosive and causes burns on your skin or holes in your clothing. If any spills or spatters occur onto your skin or clothing, rinse the affected area thoroughly with water</u>. Remember that an insoluble sulfate may form from reaction of H_2SO_4 with your original solid.

D. Behavior toward Bases

1. Sodium hydroxide, NaOH. If your sample was soluble in water or acids, place 10. drops of that solution in a 10 x 75-mm test tube. Down the side of the tilted test tube, add carefully dropwise, stirring between drops, 6 M NaOH solution until the solution is basic to litmus with 3 drops in excess. **CAUTION:** <u>NaOH of this concentration is corrosive and causes burns on your skin or holes in your clothing. If any spills or spatters occur onto your skin or clothing, rinse the affected area thoroughly with water</u>. Does any precipitate form? What color is it? Centrifuge if you are in doubt about whether a precipitate formed.

If a precipitate formed, centrifuge the mixture, and decant the supernatant liquid. Add 10. drops of 6 M NaOH to the precipitate, and stir the mixture thoroughly while heating the test tube in a hot water bath. Does the precipitate redissolve in excess 6 M NaOH? The hydroxides of Al(III), Cr(III), Sn(II), and Zn(II) are amphoteric and dissolve in excess 6 M NaOH.

If your sample was not soluble in water or acids, repeat the tests described in part **C1** using 6 M sodium hydroxide, NaOH, solution in place of 6 M HCl, except that you should not repeat the flame test. Elemental Al and Si are unique among elements that don't react with water in that they liberate hydrogen gas upon reaction with 6 M NaOH.

2. Aqueous ammonia, NH$_3$. The following test should be performed only on an aqueous solution of your sample, either in water or acid. Place 10. drops of that solution in a 10 x 75-mm test tube. Down the side of the tilted test tube, add carefully dropwise, stirring between drops, 6 M NH$_3$ solution until the solution is basic to litmus paper with 5 drops in excess. Does any precipitate form? What color is it? Centrifuge if you are in doubt about whether a precipitate formed.

If a precipitate formed, centrifuge the mixture, and decant the supernatant liquid. Add 4 drops of 2 M ammonium chloride, NH$_4$Cl, solution and 10. drops of 15 M NH$_3$ to the precipitate. Stir the mixture thoroughly while heating the test tube in a hot water bath. Does the precipitate redissolve in the presence of NH$_4^+$ and excess 15 M NH$_3$? The hydroxides of Cd(II), Co(II), Cu(II), Ni(II), and Zn(II) may dissolve under these conditions to form ammine complexes.

Only phosphorus and sulfur and a couple phosphate salts, sulfate salts, or sulfide salts may not dissolve easily under any of the conditions in parts **B**, **C**, and **D**.

E. Behavior toward Other Reagents

If you are still in doubt about your cation, the following tests might be helpful. To 10. drops of an aqueous solution of your sample (not an HCl solution) in a 10 x 75-mm test tube, add 1 drop of 6 M hydrochloric acid, HCl, solution. Does any precipitate form? What color is it? If so, it is an insoluble chloride salt of your cation. If no precipitate forms, add 5 drops of 1 M sodium sulfide, Na$_2$S, solution. Does any precipitate form? What color is it? If so, it is an insoluble sulfide salt of your cation. If no precipitate forms, add 6 M aqueous ammonia, NH$_3$, solution dropwise until the solution is basic to litmus paper with 3 drops in excess. Does any precipitate form? What color is it? If so, it is an insoluble sulfide salt of your cation or an insoluble hydroxide of Al(III) or Cr(III).

Most of the information that you have obtained thus far except from gas tests relates to cations. The following test provides information primarily about anions and should be performed only on an aqueous solution of your sample from part B. To 10. drops of an aqueous solution in a 10 x 75-mm test tube, add 3 drops of 0.1 $\underline{M}$ silver nitrate, $AgNO_3$, solution. **CAUTION: $AgNO_3$ <u>stains your skin by reacting with protein and is toxic. Wash your hands thoroughly after use or if there is any spill</u>**. Does a precipitate form? What color is it? TABLE 42.6 gives the colors of various insoluble silver salts. Each color is then an indication of the boldfaced anion being present. Remaining uncertainties can be removed through the use of infrared spectra in part F.

TABLE 42.6. Colors of some insoluble silver salts.

Formula	Color
$AgBr$	pale yellow
Ag_2CO_3	yellow (trace, sparingly soluble)
$AgCl$	white
Ag_2O (from OH^-)	brown-black
AgI	yellow
Ag_3PO_4	yellow
Ag_2S	black

F. Infrared Spectra

Polyatomic ions and molecules have several atoms bonded to each other with at least partial covalent character and can be represented as balls connected by springs. Just as the ball and spring model stretches and bends when the model is perturbed by plucking one of the atoms and thus extending one of the springs, so does the polyatomic ion or molecule exhibit stretching and bending vibrations when the ion or molecule absorbs and is excited by electromagnetic radiation in the infrared region of the spectrum. The infrared region extends from the lower energy end or the red end of the visible region to the microwave or radar region of the electromagnetic spectrum.

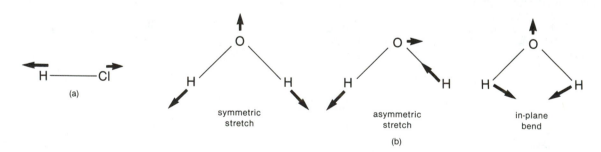

FIGURE 42.1. Fundamental modes of vibration for HCl and H_2O.

Each polyatomic ion or molecule has its own specific set of vibrational frequencies, and different polyatomic ions or molecules have different sets of vibrations. The number of absorptions depends on the number of atoms in the polyatomic ion or molecule and on the structure or specific arrangement of the atoms. The intensity of these absorptions depends on the kinds of atoms. For a diatomic molecule such as hydrogen chloride, HCl, only one simple vibrational pattern called a <u>fundamental mode</u> is possible. This involves the stretching and compression of the bond between the two atoms as shown in FIGURE 42.1a. For molecules with a greater number of atoms, the vibrational motion appears more complex, but is still comprised of a relatively small number of fundamental modes. For example, water has a bent planar structure and exhibits the two kinds of stretching vibrations and one in-plane bending vibration illustrated in FIGURE 42.1b. Both O-H bonds stretch and compress simultaneously in the symmetric stretching vibration whereas one O-H bond stretches while the other compresses in the asymmetric stretching vibration. The bending vibration involves a change of the H-O-H bond angle.

Just as an atom possesses a ground state and excited states for the energy of a particular electron, so does a polyatomic ion or molecule possess a ground state and excited states for the vibrational energy of each unique fundamental mode. Since these states are quantized, so is the energy that must be absorbed to cause a jump between states. For many molecules infrared radiation has the necessary energy to cause such excitations. For example, the symmetric stretching vibration for water is excited by energy at about 3220 cm^{-1}, the asymmetric stretching vibration is excited by energy at about 3445 cm^{-1}, and the bending vibration is excited by energy at about 1630 cm^{-1}.

An infrared spectrum represents a continuous record of the extent of absorption of infrared radiation as a function of frequency. At frequencies where there is no vibrational energy excited, there is no absorption, and the spectrum is a flat baseline representing 100% transmittance. Where there is excitation of vibrational energy, absorption occurs, and a dip or peak occurs in the infrared spectrum showing considerably less than 100% transmittance. An instrument that records such a spectrum is called an infrared spectrophotometer. An infrared spectrum of water is shown in FIGURE 42.2. Note that the spectrum is not as simple as our explanation might suggest. Since the stretching vibrations were described above to be quite close in energy, only one broad band including the two overlapping stretching vibrations appears in the infrared spectrum. A weaker band just above 1600 cm^{-1} arises from absorption by the bending vibration. Another broad band involving rocking, wagging, and twisting (not a purely vibrational mode of a single H_2O molecule) is centered at about 700 cm^{-1}. Another complexity, not a problem in the H_2O spectrum, is that peaks for totally symmetric vibrations normally do not appear in an infrared spectrum.

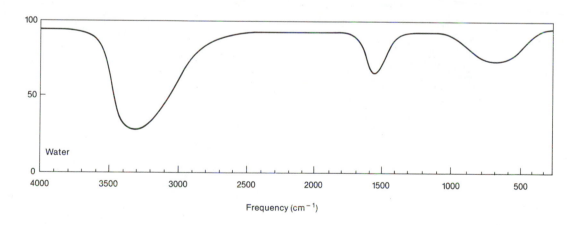

FIGURE 42.2. Infrared spectrum of water.

A working knowledge of the theory of infrared spectroscopy is not required in this experiment since you will simply use infrared spectra for identification of unknown samples in the same fashion that the Federal Bureau of Investigation uses fingerprints of unknown persons. Just as the FBI tries to match fingerprints of unknown persons against known fingerprints in their files, so will you try to match infrared spectra (or infrared fingerprints) of your unknown polyatomic ions against known spectra from various literature sources. Identification then becomes a matter of sorting and matching. The sorting in this experiment is very easy because the number of possible polyatomic species is very limited just as sorting for the FBI is easy if they have already narrowed their search to a few persons.

Your instructor will provide you with an infrared spectrum for each unknown sample and will either post known spectra for comparison or will urge you to use appropriate sources in the chemical literature. Particularly useful literature sources for matching spectra are Nakamoto, _Infrared and Raman Spectra of Inorganic and Coordination Compounds_, Wiley, 1978, especially pages 109, 127, 129, 134, 135, and 142, and also _Sadtler Standard Infrared Spectra_ that provide actual spectra for many compounds. You can even determine that a particular salt is at least partially hydrated (contains water in its crystal lattice) by noting the presence of water absorption bands in the infrared spectrum of the salt.

Your primary use of infrared spectra will be for the confirmation of NH_4^+, CO_3^{2-}, OH^-, NO_3^-, PO_4^{3-}, and SO_4^{2-} ions. Note that monatomic ions and most elements do not give rise to characteristic infrared spectra; however, monatomic ions may cause minor perturbations of the infrared spectra of polyatomic ions, significant enough on occasion to provide information concerning the monatomic ion. For example, the infrared spectra of Na_2CO_3 and $ZnCO_3$ can be differentiated easily.

Name ______________________________ **Experiment 42**

Student ID number ________________

Section __________ Date __________

Instructor ____________________ **DATA**

Unknown number __

Specify briefly in outline form each test used, your results, your conclusion, and an explanation (including a balanced chemical equation where appropriate, after you have identified your unknown sample). It is often helpful to tell what is eliminated or what is confirmed by each test.

A Physical Appearance

B Behavior toward Water

C Behavior toward Acids (specify acid or reagent for gas test)

D Behavior toward Bases (specify base)

E Behavior toward Other Reagents (specify reagent)

F Infrared Spectrum (attach your spectrum to this report)

Name of unknown sample _____

Formula of unknown sample _____

Instructor's initials _____

Name ______________________ **Experiment 42**

Student ID number ______________

Section __________ Date __________

Instructor ______________________ **DATA**

Unknown number __

Specify briefly in outline form each test used, your results, your conclusion, and an explanation (including a balanced chemical equation where appropriate, after you have identified your unknown sample). It is often helpful to tell what is eliminated or what is confirmed by each test.

A Physical Appearance

B Behavior toward Water

C Behavior toward Acids (specify acid or reagent for gas test)

Experiment 42

D Behavior toward Bases (specify base)

E Behavior toward Other Reagents (specify reagent)

F Infrared Spectrum (attach your spectrum to this report)

Name of unknown sample _____

Formula of unknown sample _____

Instructor's initials _____

Name _____   **Experiment 42**

Student ID number _____

Section _____ Date _____

Instructor _____   **D A T A**

Unknown number __

Specify briefly in outline form each test used, your results, your conclusion, and an explanation (including a balanced chemical equation where appropriate, after you have identified your unknown sample). It is often helpful to tell what is eliminated or what is confirmed by each test.

A   Physical Appearance

B   Behavior toward Water

C   Behavior toward Acids (specify acid or reagent for gas test)

Experiment 42

D Behavior toward Bases (specify base)

E Behavior toward Other Reagents (specify reagent)

F Infrared Spectrum (attach your spectrum to this report)

Name of unknown sample _____

Formula of unknown sample _____

Instructor's initials _____

Name ______________________________ **Experiment 42**

Student ID number ______________

Section __________ Date __________

Instructor _______________________ **DATA**

Unknown number __

Specify briefly in outline form each test used, your results, your conclusion, and an explanation (including a balanced chemical equation where appropriate, after you have identified your unknown sample). It is often helpful to tell what is eliminated or what is confirmed by each test.

A Physical Appearance

B Behavior toward Water

C Behavior toward Acids (specify acid or reagent for gas test)

Experiment 42

D Behavior toward Bases (specify base)

E Behavior toward Other Reagents (specify reagent)

F Infrared Spectrum (attach your spectrum to this report)

Name of unknown sample _____

Formula of unknown sample _____

Instructor's initials _____

PRE-LABORATORY QUESTIONS

1. Look in appropriate tables of the <u>Handbook of Chemistry and Physics</u>, Chemical Rubber Publishing Company, and find examples or data illustrating each of the following statements in this experiment.

   a. The color of a solid can sometimes vary depending upon how much water it contains.

   b. Nitrates are generally soluble in water.

   c. HNO_3 dissolves a number of water-insoluble sulfides that are not soluble in HCl.

2. Describe clearly the procedures for using a centrifuge, noting particularly the necessary safety precautions.

3. Explain clearly the reasons for each of the following.

   a. It is advantageous to use finely powdered samples for tests on solids in this experiment.

  b. Samples are frequently heated during solubility tests.

  c. Test papers are never placed directly into solutions to be tested.

  d. You must be fully prepared to test for gases <u>before</u> the gas is evolved from a reaction.

  e. Drops of 6 <u>M</u> hydrochloric acid, HCl, 6 <u>M</u> nitric acid, HNO$_3$, 6 <u>M</u> sulfuric acid, H$_2$SO$_4$, and 6 <u>M</u> sodium hydroxide, NaOH, solutions onto the skin are more dangerous than drops of 6 <u>M</u> aqueous ammonia, NH$_3$, solution.

POST-LABORATORY QUESTIONS

1. A student had a shiny gray solid that was insoluble in water, but dissolved rapidly with fizzing in 6 $\underline{M}$ hydrochloric acid, HCl, solution. He later reported the presence of Cl^- because of the formation of a white precipitate when 0.1 $\underline{M}$ silver nitrate, $AgNO_3$, solution was added to an aqueous solution of his unknown.

    a. What was the major flaw in his interpretation of his evidence?

    b. What test could he perform to convince himself of that flaw?

2. When you return from Spring vacation in the Bahamas, you find that you are being pressed into service as a laboratory instructor in a work-study position. Two labels (to be designated by your laboratory instructor) have fallen off reagent bottles over the Spring break, and one of your first tasks is to determine what samples are contained in the two bottles from which these labels came. The two labels are

    _____ and _____.
    This is, of course, very important to your students, and you must be absolutely certain of your conclusions. Therefore, describe clearly at least four simple tests and observations that you can make in the laboratory to distinguish which salt is which. Be specific as to the details of your tests and observations for each of the two samples.

3. Tests and observations are provided for an unknown salt containing a cation and an anion among the possible cations and anions for this experiment. You are to draw conclusions and give explanations (including balanced chemical equations wherever appropriate) for each of the lettered items below and ultimately to identify the unknown salt. Indicate clearly your reasoning.

   a. The unknown is rose colored as large crystals and pale pink when powdered.

   b. The pale pink powder in a. is only very slightly soluble in water.

   c. The pale pink powder in a. is readily soluble in dilute HCl, dilute HNO_3, and dilute H_2SO_4. An apparently colorless gas is liberated in all three cases.

   d. A flame test on the aqueous HCl solution from c. shows no emission of color.

   e. An infrared spectrum of the pale pink powder in a. shows a strong broad band centered at 1445 cm^{-1}, but having shoulders at 1460 cm^{-1} and 1380 cm^{-1}; a medium intensity band at 870 cm^{-1}; a weaker doublet at 730 cm^{-1} and 720 cm^{-1}; and possibly a very weak band at 1080 cm^{-1}.

   f. Heating the aqueous HCl solution in c. with 1 $\underline{M}$ Na_2S solution produces a pink precipitate only when the solution is made basic with 6 $\underline{M}$ NH_3.

   g. Making the aqueous HCl solution in c. basic by addition of 6 $\underline{M}$ NaOH produces a white precipitate that turns brown on standing and does not dissolve in excess 6 $\underline{M}$ NaOH.

Separation and Identification of Cations

PURPOSE OF EXPERIMENT: Separate and identify aqueous cations from group 1, group 2, or from groups 1 and 2 combined of one conventional qualitative analysis scheme.

An area of chemistry called qualitative analysis involves the study and use of techniques by which one can determine the nature of, but not the amount of, species in a mixture, a significant problem in the detection of impurities in metals, alloys, drugs, and other products as well as in environmental pollution. The mixture can be homogeneous or heterogeneous, and the species can be elements or groups of elements, or single-atom or multi-atom ions. However, this experiment is restricted to cations in aqueous solutions. A variety of chemical and physical techniques will be applied. Indeed, some measure of quantitative information is sometimes obtained simply because you may know the lower limits of detection of your techniques. You will perform qualitative analysis on a semi-micro level in which the sample solutions are about 0.1 $\underline{M}$ in each cation and contain an average of 10. mg of solute in 1 mL of solution. Your tests for the most part will be ineffective if you have less than 1-2 mg of solute present. Thus, procedures must be carried out carefully to avoid losing the major portion of any component somewhere during the analysis. Sloppy, careless work will often produce precipitates where you should get solutions, precipitates of the wrong color, or solutions where you should get precipitates, and generally will make the analyses more difficult and less pleasant than they ought to be.

Cations often interfere with each other in the final tests designed to detect the presence of specific cations. Therefore, cations must first be separated before identification can be accomplished. In fact, as with many chemical mixtures, separation of cations may be considerably more difficult than identification. Careful work is again very important; if the separations are not clean, results in identification tests may be masked by interfering cations. Separation of a complex mixture of cations is by no means simple and is generally broken down into several parts. Each part involves a fairly small group of cations which can be isolated from the mixture on the basis of some property which is common to the ions in the group and then studied as a separate set. After isolation, the cations within a group are further resolved by means of a series of chemical reactions into soluble and insoluble fractions which are sufficient to allow identification of each cation by one or more tests specific to that ion once interferences have been removed. Various types of chemical reactions will be used for separations and identifications in this experiment: precipitation reactions, acid-base reactions, complex ion formations, and oxidation-reduction reactions.

You will study two groups of cations in this experiment. **Group 1 cations form chloride salts that are insoluble in water.** Be careful not to confuse these cations with the alkali metal cations from Group I of the periodic table which all form soluble salts. Group 1 includes Ag^+, Hg_2^{2+} or Hg(I), and Pb^{2+}. They are separated from a mixture of cations because they precipitate white insoluble chlorides, for example, silver chloride, AgCl, when a small excess of Cl^- is added.

$$Ag^+(aq) + Cl^-(aq) \longrightarrow AgCl(s)$$

Care must be taken to ensure that only a small excess of Cl^- is added because a large excess of Cl^- causes the formation of soluble complexes, for example, $[AgCl_2]^-$,

$$AgCl(s) + Cl^-(aq) \longrightarrow [AgCl_2]^-(aq)$$

that prevent isolation of the group.

Once isolated as a group, Ag^+, Hg_2^{2+}, and Pb^{2+} are separated from each other and identified as follows. Lead(II) chloride, $PbCl_2$, dissolves in hot water while AgCl and mercury(I) chloride, Hg_2Cl_2, remain insoluble. The hot solution of Pb^{2+} is separated from the solids, and Pb^{2+} is identified by the formation of yellow, insoluble lead(II) chromate, $PbCrO_4$.

$$Pb^{2+}(aq) + CrO_4^{2-}(aq) \longrightarrow PbCrO_4(s)$$

AgCl forms a soluble complex, $[Ag(NH_3)_2]^+$, with aqueous ammonia, NH_3,

$$AgCl(s) + 2 NH_3(aq) \longrightarrow [Ag(NH_3)_2]^+(aq) + Cl^-(aq)$$

while Hg_2Cl_2 forms a black, insoluble mixture of mercury and mercury(II) amidochloride.

$$Hg_2Cl_2(s) + 2 NH_3(aq) \longrightarrow Hg(l) + HgNH_2Cl(s) + NH_4^+(aq) + Cl^-(aq)$$

White AgCl can be reformed from $[Ag(NH_3)_2]^+$ upon the addition of HNO_3.

$$[Ag(NH_3)_2]^+(aq) + 2 H^+(aq) + Cl^-(aq) \longrightarrow AgCl(s) + 2 NH_4^+(aq)$$

The following scheme illustrates the separation and identification of group 1 cations, and you will want to refer to it as you implement the EXPERIMENTAL PROCEDURES relating to group 1.

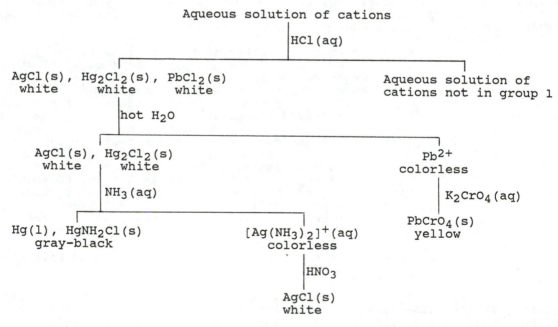

Group 2 cations form sulfide salts which are insoluble in dilute hydrochloric acid, HCl, solution. Be careful not to confuse these cations with the alkaline earth metal cations from group II in the periodic table which all form sulfide salts that are soluble even in water. We shall consider only Bi^{3+}, Cd^{2+}, Cu^{2+}, Pb^{2+}, and Hg^{2+}, though additional cations also form insoluble sulfide salts in dilute HCl solution. Assuming that Ag^+ and Hg_2^{2+}, and much of the Pb^{2+}, have already been isolated (since they also exhibit this behavior), the group 2 cations are isolated from a mixture of cations because they precipitate black (except for yellow CdS), insoluble sulfides, for example, bismuth(III) sulfide, Bi_2S_3, when sodium sulfide, Na_2S, solution is added to the cation solution already at about pH 1.

$$2\ BiO^+(aq)\ +\ 3\ H_2S(aq)\ \text{--->}\ Bi_2S_3(s)\ +\ 2\ H_2O(l)\ +\ 2\ H^+(aq)$$

Once isolated as a group Bi^{3+}, Cd^{2+}, Cu^{2+}, Pb^{2+}, and Hg^{2+} are separated from each other and identified as follows. Mercury(II) sulfide, HgS, is separated from Bi_2S_3, CdS, CuS, and PbS because all of the latter, for example, cadmium sulfide, CdS, are soluble in dilute nitric acid, HNO_3.

$$3\ CdS(s)\ +\ 2\ NO_3^-(aq)\ +\ 8\ H^+(aq)\ \text{--->}$$
$$3\ Cd^{2+}(aq)\ +\ 3\ S(s)\ +\ 2\ NO(g)\ +\ 4\ H_2O(l)$$

The remaining solid HgS (present along with elemental sulfur) is confirmed by dissolving it in aqua regia (a mixture of concentrated HCl and HNO_3) to form a complex ion, $[HgCl_4]^{2-}$,

$$3\ HgS(s)\ +\ 2\ NO_3^-(aq)\ +\ 8\ H^+(aq)\ +\ 12\ Cl^-(aq)\ \text{--->}$$
$$3\ [HgCl_4]^{2-}(aq)\ +\ 3\ S(s)\ +\ 2\ NO(g)\ +\ 4\ H_2O(l)$$

which is reduced to black mercury by a solution of tin(II) chloride, $SnCl_2$.

$$[HgCl_4]^{2-}(aq)\ +\ Sn^{2+}(aq)\ \text{--->}\ Hg(l)\ +\ Sn^{4+}(aq)\ +\ 4\ Cl^-(aq)$$

From the portion that was originally soluble in dilute HNO_3, Pb^{2+} is separated as white, insoluble lead(II) sulfate, $PbSO_4$,

$$Pb^{2+}(aq)\ +\ SO_4^{2-}(aq)\ \text{--->}\ PbSO_4(s)$$

which is dissolved in ammonium acetate, $NH_4C_2H_3O_2$, solution

$$PbSO_4(s)\ +\ 4\ C_2H_3O_2^-(aq)\ \text{--->}\ [Pb(C_2H_3O_2)_4]^{2-}(aq)\ +\ SO_4^{2-}(aq)$$

and then converted to yellow, insoluble $PbCrO_4$ as in the group 1 tests. Bi^{3+} is separated from Cu^{2+} and Cd^{2+} by forming white, insoluble bismuth(III) hydroxide, $Bi(OH)_3$, with concentrated aqueous ammonia, NH_3,

$$Bi^{3+}(aq)\ +\ 3\ NH_3(aq)\ +\ 3\ H_2O(l)\ \text{--->}\ Bi(OH)_3(s)\ +\ 3\ NH_4^+(aq)$$

while Cu^{2+} and Cd^{2+} form soluble ammine complexes.

$$Cu^{2+}(aq)\ +\ 4\ NH_3(aq)\ \text{--->}\ [Cu(NH_3)_4]^{2+}(aq)$$

$Bi(OH)_3(s)$ produces black bismuth metal when a basic mixture is treated with $SnCl_2$ solution {$SnCl_2$ forms $[Sn(OH)_4]^{2-}$ in basic solution}.

$$2\ Bi(OH)_3(s)\ +\ 3\ [Sn(OH)_4]^{2-}(aq)\ \text{--->}\ 2\ Bi(s)\ +\ 3\ [Sn(OH)_6]^{2-}(aq)$$

Of the soluble ammine complexes above, $[Cu(NH_3)_4]^{2+}$ is deep blue whereas $[Cd(NH_3)_4]^{2+}$ is colorless. After half of the solution of ammine complexes is acidified, Cu^{2+} is identified by the formation of a red-brown precipitate when potassium hexacyanoferrate(II), $K_4[Fe(CN)_6]$, solution is added.

$$2\ [Cu(NH_3)_4]^{2+}(aq)\ +\ [Fe(CN)_6]^{4-}(aq)\ \text{--->}\ Cu_2[Fe(CN)_6](s)\ +\ 8\ NH_3(aq)$$

A brown precipitate of copper metal is formed when a sodium dithionite, $Na_2S_2O_4$, solution is added to the other half of the solution of ammine complexes.

$$[Cu(NH_3)_4]^{2+}(aq)\ +\ S_2O_4^{2-}(aq)\ +\ 4\ OH^-(aq)\ \text{--->}$$
$$Cu(s)\ +\ 2\ SO_3^{2-}(aq)\ +\ 4\ NH_3(aq)\ +\ 2H_2O(l)$$

Experiment 43

Treatment of the supernatant liquid containing $[Cd(NH_3)_4]^{2+}$ (which is not reduced by $S_2O_4{}^{2-}$) with Na_2S solution precipitates yellow, insoluble CdS to confirm the presence of Cd^{2+}.

$$[Cd(NH_3)_4]^{2+}(aq) + S^{2-}(aq) ---> CdS(s) + 4 NH_3(aq)$$

The following scheme illustrates the separation and identification of group 2 cations, and you will want to refer to it as you implement the EXPERIMENTAL PROCEDURES relating to group 2 cations.

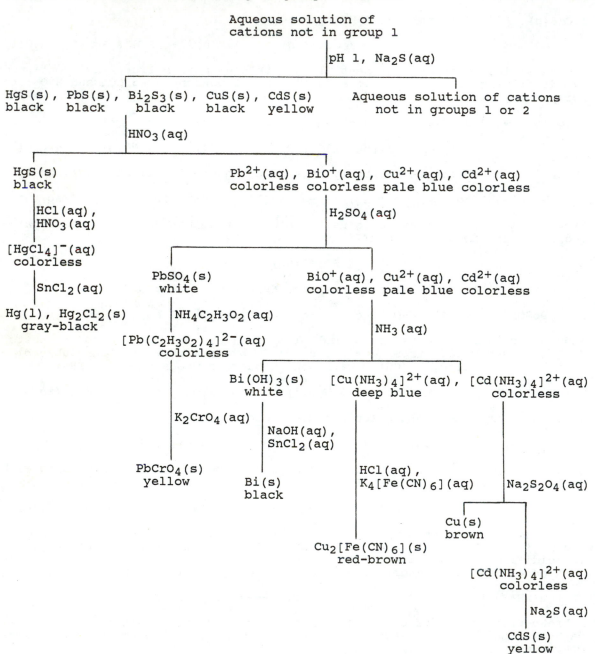

In this experiment you will separate and identify the cations in one or more unknown solutions of cations of groups 1 and/or 2 as designated by your instructor after having first tested the procedures on known solutions to see that you are achieving clean separations and know what the identification tests look like.

REFERENCES

(1) Kotz, J. C., and Purcell, K. F., _Chemistry and Chemical Reactivity_, Saunders College Publishing, Philadelphia, 1987, sections 1.3, 3.2, 3.3b, 3.4, 15.1, 15.6, 15.8, 17.1, 17.5, and 25.3.

(2) Masterton, W. L., Slowinski, E. J., and Stanitski, C. L., _Chemical Principles_, 6th ed., Saunders College Publishing, Philadelphia, 1985, sections 1.6, 3.4, 12.3, 18.1, 18.2, 19.6, 21.1, 22.1-22.4, 23.2, and 25.1-25.3.

EXPERIMENTAL PROCEDURE

(Study this section and the PRE-LABORATORY QUESTIONS before coming to the laboratory. **Wear safety goggles when performing this experiment.**)

Your instructor will designate how many weeks you will spend on this experiment and whether you will do part or all of the experiment. Normally you will test solutions containing known cations first; then you will receive and test one or more unknown solutions. Record the number of your unknown on the appropriate DATA page as soon as you receive your sample. Your instructor will describe clearly the ground rules as to what your unknown solution may contain. Record clearly and concisely on the appropriate DATA page the tests used, observations made, conclusions drawn, and any explanations or balanced equations required by your instructor for each unknown solution.

CAUTION: _Some of the cations used in this experiment are poisonous. In addition, several strong acids (HNO_3, HCl, and H_2SO_4) as well as a strong base (NaOH) are used frequently, fortunately in very small amounts. If there are any spills or spatters onto your hands or clothing, rinse the affected area thoroughly with water. Moreover, wash your hands carefully before leaving the laboratory._

A. **Analysis of Group 1 Cations** (for known or unknown)

Place 10. drops of a known solution or 20. drops of an unknown solution in a 10 x 75-mm or 13 x 100-mm test tube. Test the solution with litmus paper (Laboratory Methods P). If the solution is basic, add 6 $\underline{M}$ nitric acid, HNO_3, solution dropwise until the solution is acidic. One of the most important variables controlling chemical reactions is the pH of the solution. Frequently it is necessary to make a basic solution acidic or vice versa in order to make a desired reaction take place. Frequently, the solution changes its character at the neutral point; a precipitate may dissolve or form, or the color may change. In any event, add enough acid so that after mixing, blue litmus turns red when touched with the stirring rod.

Add 2 drops of concentrated (12 $\underline{M}$) hydrochloric acid, HCl, solution. Stir the solution for 5 minutes. It is hard to overstir. Some precipitates form slowly and require both time and thorough mixing. Then allow the precipitate to settle. A precipitate will form when the product of the concentration of the ions that react to form the precipitate just exceeds the solubility product. Furthermore, in any precipitation process, the insoluble compound will keep precipitating until the concentrations of its ions remaining in solution reach values at which their product just equals the solubility product. When this point is reached, the precipitate will be in equilibrium with its ions, which means that the rate at which the precipitate is forming is equal to the rate at which it is dissolving and dissociating into its ions. This is a true equilibrium, and the equilibrium point will be shifted by changes in the concentrations of the reacting ions in accordance with the mass law. It follows, therefore, that an excess of the group reagent ion, for example Cl^- in Group 1, is desirable since if the concentration of Cl^- is high, the concentration of Ag^+, Hg_2^{2+}, and Pb^{2+} remaining in solution will be low. An inordinately large excess of the reagent may however form soluble complex ions thus nullifying the usual advantage resulting from having an excess of reagent.

Add 1 drop of 6 $\underline{M}$ hydrochloric acid, HCl, solution to determine whether precipitation is complete. If precipitation is not complete, stir and add more drops of 6 $\underline{M}$ HCl until no more precipitate is formed.

Cool the test tube by dipping it into cold water. Then centrifuge (Laboratory Methods J) the mixture. If the sample is still suspended, centrifuge again. If this still does not work, you may find it helpful to heat the sample in the water bath for a few minutes, thereby promoting formation of larger crystals of solid, which tend to centrifuge more easily.

Carefully decant (Laboratory Methods H) the supernatant liquid into another test tube. If the solution is not decanted cleanly and some precipitate is present, centrifuging and decanting must be repeated.

Wash the precipitate with a few drops of distilled water. The solid remaining in the test tube has residual liquid around it. Since this liquid contains ions that may interfere with further tests on the solid, they must be removed. This is accomplished by diluting the liquid with a wash liquid, often water, which does not interfere with the analysis, and stirring thoroughly to disperse the solid in the wash liquid and to remove foreign ions adhering to the solid. Failure to wash precipitates thoroughly is one of the primary sources of error in qualitative analysis. Centrifuge, and decant the wash solution into the test tube containing the original supernatant liquid. If you are working on an unknown comprised of cations of both groups 1 and 2, save your supernatant liquid for the group 2 analysis in part B.

Add 10. drops of <u>hot</u> water to the precipitate, stir, and heat the test tube in a hot water bath (Laboratory Methods D) to near boiling. Then quickly centrifuge the hot mixture, and decant the hot supernatant liquid into another test tube. Wash the precipitate with 2-3 drops of hot water, centrifuge, and decant the wash solution into the test tube containing supernatant liquid. Save the precipitate for Ag^+ and Hg_2^{2+} tests. If $PbCl_2$ is not completely dissolved and separated from AgCl and Hg_2Cl_2 solids, Pb^{2+} will later react with aqueous NH_3 in the mercury(I) test. This will be particularly confusing if Hg_2^{2+} is not present because you will obtain white, insoluble Pb(OH)Cl rather than a gray-black solid.

Allow the supernatant liquid from the previous paragraph to cool. Then add a few drops of 1 $\underline{M}$ potassium chromate, K_2CrO_4, solution to several drops of the supernatant liquid. A yellow precipitate of lead chromate, $PbCrO_4$, indicates the presence of Pb^{2+}.

To the precipitate of silver chloride, AgCl, and/or mercury(I) chloride, Hg_2Cl_2, from the second paragraph above, add 10. drops of concentrated (15 $\underline{M}$) aqueous ammonia, NH_3. A gray-black precipitate of mercury, Hg, and mercury(II) amidochloride, $Hg(NH_2)Cl$, indicates the presence of Hg_2^{2+}. Centrifuge the mixture, and decant the supernatant liquid. To the supernatant liquid, add 6 $\underline{M}$ nitric acid, HNO_3, solution dropwise until the solution gives an acidic reaction to blue litmus paper. A white precipitate of AgCl indicates the presence of Ag^+.

B. Analysis of Group 2 Cations (for known or unknown)

Place 10. drops of the supernatant solution from the chloride precipitation of the group 1 cations in a 10 x 75-mm or 13 x 100-mm test tube, and dilute to about 1.0 mL. Measuring an approximate volume of this sort is easy and fast, if you know roughly the volume 1 mL occupies in a small test tube. This can be determined by putting about 5 mL of water in a small graduated cylinder and then pouring the water into the test tube 1 mL at a time. After you have done this a few times you should be able to judge the increase in level that corresponds to 1 mL.

Using wide-range indicator paper, adjust the pH of the solution to about 2 by adding dropwise either 6 $\underline{M}$ nitric acid, HNO_3, solution or 6 $\underline{M}$ aqueous ammonia, NH_3, solution. If the H^+ concentration is too large (pH too low) the concentration of S^{2-} from Na_2S will be too small to precipitate all the group 2 cations. Then add 5-7 drops of 1 $\underline{M}$ sodium sulfide, Na_2S, solution, stir for about 2 minutes, and allow the precipitate to settle. Centrifuge, and decant and discard the supernatant liquid.

To the precipitate, add 1.0 mL (about 20. drops) of 3 $\underline{M}$ nitric acid, HNO_3, solution, and stir the mixture. Only 3 $\underline{M}$ HNO_3 is used because if the HNO_3 is too concentrated, it may dissolve some of the HgS and may also oxidize PbS to white, insoluble lead sulfate, $PbSO_4$, which will remain with and alter the color of black HgS. Place the test tube in a boiling water bath until the contents bubble and most of the precipitate is dissolved. A remaining black HgS [or yellow $Hg(NO_3)_2 \cdot 2HgS$] residue indicates the presence of Hg^{2+}.

Centrifuge the mixture, decant the supernatant liquid into a clean casserole to save for later tests, and confirm the presence of Hg^{2+} by the following procedure. To the residue add 8 drops of concentrated (12 $\underline{M}$) hydrochloric acid, HCl, solution and 3 drops of concentrated (15 $\underline{M}$) nitric acid, HNO_3, solution. Heat the mixture in a boiling water bath for 2 minutes. Then add 10. drops of hot distilled water, and boil the mixture for about 1 minute. Cool the mixture, centrifuge, and decant the supernatant liquid into another test tube. Finally, add 5 drops of 0.2 $\underline{M}$ tin(II) chloride, $SnCl_2$, solution to the supernatant liquid. A black precipitate of Hg and Hg_2Cl_2 confirms the presence of Hg^{2+}. A white precipitate may form initially, but this will often turn black in a few minutes.

To the supernatant liquid in the casserole, add with extreme care 4 drops of concentrated (18 $\underline{M}$) sulfuric acid, H_2SO_4, solution. **CAUTION: <u>Concentrated sulfuric acid is very corrosive. Wash your hands immediately if there is any contact with H_2SO_4</u>. <u>Neutralize</u>** (with your instructor's guidance), **<u>and clean up any H_2SO_4 spills</u>. Working in a hood,** evaporate the solution over a very low flame (to avoid spattering) until dense white fumes of SO_3 are visible due to the thermal decomposition of H_2SO_4. **CAUTION: <u>Overheating can easily occur with such a small volume</u>.** Since HNO_3 has a lower boiling point than H_2SO_4, the appearance of white fumes confirms the fact that all HNO_3 has been evaporated, and it cannot dissolve the lead sulfate that was formed. Add 10. drops of water **cautiously**, and mix thoroughly. Transfer the mixture to a test tube by stirring the mixture and pouring as much as you can, using the liquid as a carrier. Since Pb^{2+} is largely removed as a group 1 cation, only very small amounts of it will ordinarily appear in group 2.

Centrifuge, and decant the supernatant liquid into another test tube. A white precipitate of $PbSO_4$ indicates the presence of Pb^{2+}; however, you should confirm Pb^{2+} by the following procedure even if there is only a trace of precipitate. Add enough 2 $\underline{M}$ ammonium acetate, $NH_4C_2H_3O_2$, solution (probably about 5 drops) to dissolve the precipitate. Warming the mixture, even directly over a flame, may be necessary, but do not boil the solution.

After the precipitate has dissolved, add a few drops of 1 $\underline{M}$ potassium chromate, K_2CrO_4, solution. Formation of a yellow precipitate confirms that Pb^{2+} is present.

To the supernatant liquid from the previous paragraph, add concentrated (15 $\underline{M}$) aqueous ammonia, NH_3, solution dropwise until the solution is basic to litmus paper; then add 5 more drops of 15 $\underline{M}$ NH_3, and stir for 2 minutes. Cu^{2+} and Cd^{2+} may initially form insoluble hydroxides, $Cu(OH)_2$ and $Cd(OH)_2$, which slowly dissolve to form ammine complexes. If the solution turns deep blue from the formation of $[Cu(NH_3)_4]^{2+}$, Cu^{2+} is present. Cd^{2+} forms a colorless ammine complex that is masked by the deep blue $[Cu(NH_3)_4]^{2+}$. Place the test tube in a water bath at about 60.^{0}C for 1 minute. The formation of a white precipitate of $Bi(OH)_3$ indicates that Bi^{3+} is present. Nevertheless, both Cu^{2+} and Bi^{3+} should be confirmed by the following procedures.

Centrifuge the mixture, and decant the supernatant liquid into another test tube. Wash the precipitate twice with 15-drop portions of hot water, centrifuge, and discard the wash solution. To the washed precipitate add 3 drops of 8 $\underline{M}$ sodium hydroxide, NaOH, solution and 2 drops of 0.2 $\underline{M}$ tin(II) chloride, $SnCl_2$, solution. The formation of jet black metallic bismuth confirms the presence of Bi^{3+}.

To about half of the supernatant liquid from the previous paragraph, add 6 $\underline{M}$ hydrochloric acid, HCl, solution until the solution is acidic to litmus paper. Add 3 drops of 0.2 $\underline{M}$ potassium hexacyanoferrate(II), $K_4[Fe(CN)_6]$, solution. A red-brown precipitate of $Cu_2[Fe(CN)_6]$ confirms the presence of Cu^{2+}. Cd^{2+} also forms white, insoluble $Cd_2[Fe(CN)_6]$; however, this is masked by the colored Cu^{2+} salt, and Cd^{2+} must be confirmed independently.

If the other half of the supernatant solution from the previous paragraph is colorless because Cu^{2+} is absent, skip directly to the next paragraph. If the supernatant liquid is blue, Cu^{2+} must first be removed. To the blue supernatant liquid, add a pinch of sodium dithionite, $Na_2S_2O_4$, solution in order to decolorize the blue solution and form brown copper metal. Add more $Na_2S_2O_4$, if necessary. Cd^{2+} is not reduced by $S_2O_4^{2-}$ and is now separated for independent testing. Heat the test tube at 60.^{0}C for 2 minutes. Centrifuge, decant the supernatant liquid into another test tube, and discard the residue.

To the supernatant liquid, add 2 drops of 1 $\underline{M}$ sodium sulfide, Na_2S, solution, stir the solution, and allow it to stand. Formation of a yellow precipitate of CdS indicates the presence of Cd^{2+}. If the precipitate is dirty yellow and contains some black precipitate, either Hg^{2+} or Pb^{2+} was not separated cleanly in earlier steps, and the separation of group 2 cations should be repeated.

Name _____

Student ID number _____

Section _____ Date _____

Instructor _____

DATA

TABLE 43.1.   Tests, observations, conclusions, and explanations.

Unknown number ___

TABLE 43.1 (continued)

Name _____   Experiment 43

Student ID number _____

Section _____ Date _____

Instructor _____   D A T A

TABLE 43.1.   Tests, observations, conclusions, and explanations.

Unknown number ___

TABLE 43.1 (continued)

Name _____

Student ID number _____

Section _____ Date _____

Instructor _____

TABLE 43.1. Tests, observations, conclusions, and explanations.

Unknown number ____

TABLE 43.1 (continued)

Name _____   **Experiment 43**

Student ID number _____

Section _____ Date _____

Instructor _____   D A T A

    TABLE 43.1. Tests, observations, conclusions, and explanations.

Unknown number ___

TABLE 43.1 (continued)

PRE-LABORATORY QUESTIONS

1. Define clearly what is meant by each of the following.

    a. Qualitative analysis.

    b. Separation versus identification.

    c. Group 1 cations.

    d. Group 2 cations.

2. Describe clearly and concisely how you would accomplish each of the following (on a semi-micro scale).

    a. Adjusting the pH of a 2 $\underline{M}$ HCl solution to pH 2.

    b. Centrifuging a mixture.

    c. Decanting a supernatant liquid.

    d. Washing a precipitate.

3. Examine carefully the reactions in the Introduction, and write each reaction in the appropriate category. (Note that some reactions fit into more than one category.)

    a. Acid-base reactions.

    b. Complex ion formations.

    c. Oxidation-reduction reactions.

    d. Precipitation reactions.

Name _____   **Experiment 43**

Student ID number _____

Section _____ Date _____

Instructor _____   **POST-LABORATORY QUESTIONS**

1. An unknown solution containing only group 1 cations produced a white precipitate **P1** upon treatment with HCl solution. **P1** was soluble in hot water to form solution **S1**, and a yellow precipitate formed when **S1** was treated with K_2CrO_4 solution. The remainder of **P1** dissolved completely in aqueous NH_3, and the resulting solution yielded a white precipitate when treated with HNO_3 solution. Which cation(s) is(are) definitely present, which is(are) definitely absent, and which is(are) uncertain based on the evidence given? Write balanced net ionic equations to illustrate your reasoning.

2. Copper can be identified in at least three different ways in this experiment. Write a balanced net ionic equation describing each of the ways for which Cd^{2+} would not interfere.

3. An unknown solution containing only group 2 cations is adjusted to pH 2. Addition of Na_2S solution gives a black precipitate **P1**. When **P1** is heated with HNO_3 solution, it partly dissolves, forming solution **S1**. The remainder of **P1** dissolves in a mixture of HCl and HNO_3 solutions, and the solution gives a black precipitate upon treatment with $SnCl_2$ solution. There is no reaction when **S1** is partially evaporated with H_2SO_4 solution. **S1** forms a white precipitate **P2** and a colorless solution **S2** upon treatment with aqueous NH_3. **P2** yields a black precipitate upon treatment with a basic solution of $SnCl_2$. **S2** was inadvertently discarded. Which cation(s) is(are) definitely present, which is(are) definitely absent, and which is(are) uncertain based on the evidence given? Write balanced net ionic equations to illustrate your reasoning.

Multi-Bottle Challenge

PURPOSE OF EXPERIMENT: Identify the chemical species in a number of aqueous solutions.

In previous experiments you used a large number of aqueous solutions for a variety of purposes. Several acids or bases were used simply to change the pH of a solution. For example, in Experiment 35 you may have added a drop of 6 $\underline{M}$ aqueous ammonia, NH_3, solution to distilled water containing a drop of chlorophenol red indicator so that you could examine the color of the basic form of the indicator. Many solutions were used as reactants to convert chemicals to desired products. For example, in Experiments 3 and 37 you may have reacted an aqueous sodium iodide, NaI, solution with an aqueous solution of a group IVA +2 cation to produce an iodide salt of the +2 cation. In this experiment you will apply what you have learned about the properties and reactions of aqueous solutions used in this course to identify the chemical species in aqueous solutions in a number of lettered bottles. The best way to prepare for this experiment is to study the chemical reactions of various aqueous solutions used in previous experiments in this course.

You will be provided with chemical names (in alphabetical order) and chemical formulas for the species in lettered bottles for your experimentation. You will also be told the approximate molar concentration of each solution or at least the concentration range for the solutions. This information should be entered in TABLE 44.1. However, you will not be given the chemical name corresponding to the letter identifying each solution.

Your goal is to determine which chemical species is solution A, which chemical species is solution B, and so forth, and to enter your conclusions in TABLE 44.1. You may also be asked to identify any precipitates produced, gases evolved, or weak electrolytes formed during your experimentation and to write balanced chemical equations to describe any reactions that you observe.

In order to interpret the results that you will observe upon mixing pairs of aqueous solutions you must recognize that reactions in aqueous solution that do <u>not</u> involve oxidation and reduction generally go to completion only if one or more of the products formed removes ions from aqueous solution. The reverse reaction cannot then occur to an appreciable extent, and the equilibrium is shifted toward the right. This can happen in one of three ways.

Formation of a sparingly soluble solid

$$Ca(NO_3)_2(aq) + (NH_4)_2CO_3(aq) \rightarrow \textbf{CaCO}_3\textbf{(s)} + 2\ NH_4NO_3(aq)$$

Formation of a sparingly soluble gas

$$CaCO_3(s) + 2\ HCl(aq) \rightarrow \textbf{CO}_2\textbf{(g)} + CaCl_2(aq) + H_2O(l)$$

Formation of a weak electrolyte

$$Ca(OH)_2(aq) + 2\ HI(aq) \rightarrow 2\ \textbf{H}_2\textbf{O(l)} + CaI_2(aq)$$

Therefore, you must know solubility rules for solids and gases, and you must know common weak acids and bases. You should have memorized the strong acids and bases and then can assume that other acids and bases are weak.

Experiment 44

In order to interpret the results that you will observe upon mixing pairs of aqueous solutions you must recognize and be able to predict possible oxidation-reduction reactions, and you must be able to balance redox equations. You may have already studied some trends in oxidizing and reducing strength: diatomic halogens decrease in oxidizing strength in the order $F_2 > Cl_2 > Br_2 > I_2$ whereas halide anions decrease in reducing strength in the order $I^- > Br^- > Cl^- > F^-$; species containing an element having a high oxidation number for that element such as +7 for Mn in MnO_4^- or +4 for Pb in PbO_2 tend to be good oxidizing agents; species containing an element having a low oxidation number for that element such as -1 for H in NaH or -2 for S in S^{2-} tend to be good reducing agents; and species containing an element having an intermediate oxidation number for that element such as -1 for O in H_2O_2 may be either oxidized or reduced depending on the other reagent. Thus, H_2O_2 will be oxidized by MnO_4^-, a strong oxidizing agent;

$$5\ H_2O_2(aq)\ +\ 2\ MnO_4^-(aq)\ +\ 6\ H^+(aq)\ \longrightarrow$$
$$2\ Mn^{2+}(aq)\ +\ 5\ O_2(g)\ +\ 8\ H_2O(l)$$

H_2O_2 will be reduced by I^-, a good reducing agent;

$$H_2O_2(aq)\ +\ 2\ I^-(aq)\ +\ 2\ H^+(aq)\ \longrightarrow\ I_2(aq)\ +\ 2\ H_2O(l)$$

and MnO_4^- and I^- will oxidize and reduce each other.

$$2\ MnO_4^-(aq)\ +\ 10\ I^-(aq)\ +\ 16\ H^+(aq)\ \longrightarrow$$
$$5\ I_2(aq)\ +\ 2\ Mn^{2+}(aq)\ +\ 8\ H_2O(l)$$

REFERENCES

(1) Kotz, J. C., and Purcell, K. F., _Chemistry and Chemical Reactivity_, Saunders College Publishing, Philadelphia, 1987, sections 3.3, 3.4, 12.2, and 15.1, and TABLES 2.7, 2.8, 16.4, 16.5, and 19.1.

(2) Masterton, W. L., Slowinski, E. J., and Stanitski, C. L., _Chemical Principles_, 6th ed., Saunders College Publishing, Philadelphia, 1985, sections 3.4-3.6, 12.3, 18.1, 18.2, 19.3-19.6, 23.2, and 24.1, and TABLES 18.1, 19.4-19.6, and 24.1.

EXPERIMENTAL PROCEDURE

(Study this section and the PRE-LABORATORY QUESTIONS before coming to the laboratory.) **Wear safety goggles when performing this experiment.)**

You may use any equipment in your desk for tests, but you may not use chemical reagents other than those in the lettered bottles unless one or more reagents are specifically designated by your instructor. Preliminary information for identifying the solution in each lettered bottle will come from examining the physical properties of each solution individually. Additional useful information will come from mixing pairs of solutions from lettered bottles with each other; observing any chemical reactions that occur; and interpreting any observations of color, odor, heat, gases evolved, or precipitates formed. **CAUTION: It is not wise for safety reasons to mix large quantities of unknown solutions; three drops of each solution in a 10 x 75-mm test tube, or whatever is the smallest test tube in your desk set, should be sufficient for these mixing tests. Please cooperate by returning a bottle to its designated place each time you finish using it.**

Record your experimental evidence in TABLE 44.2. This table will help you to organize your data and to think about your evidence so that you can reach conclusions as to the chemical identity of each lettered solution.

602

Name ____________________ **Experiment 44**

Student ID number ____________

Section ________ Date ________

Instructor ________________

DATA

TABLE 44.1. Aqueous solutions for the multi-bottle challenge.

Chemical name	Chemical formula	Concentration (M)	Letter

Experiment 44

TABLE 44.2. Observations of individual lettered solutions and from mixing pairs of lettered solutions.

	B	C	D	E	F	G	H	I	J	
										A
										B
										C
										D
										E
										F
										G
										H
										I
										J

Instructor's initials _____

Name _____

Student ID number _____

Section _____ Date _____

Instructor _____

Experiment 44

PRE-LABORATORY QUESTIONS

1. Decide whether aqueous solutions of each of the following salts will be acidic, basic, or neutral. Write net ionic equations to indicate clearly your reasoning.

   a. $KHSO_4$

   b. NaF

   c. $Ba(NO_3)_2$

   d. $AlCl_3$

2. Write balanced complete equations, complete ionic equations, and net ionic equations describing reactions between the following species. Circle the species that provides the driving force for the reaction.

   a. $KBr(aq)$ + $Pb(NO_3)_2(aq)$ --->

   b. $CaSO_3(s)$ + $HBr(aq)$ --->

   c. $H_2SO_4(aq)$ + $KOH(aq)$ --->

   d. $(NH_4)_2CO_3(aq)$ + $Ba(OH)_2(aq)$ --->

Experiment 44

3. Complete and balance equations describing the following oxidation-reduction reactions.

    a. $Cl_2(aq)$ + $S_2O_3{}^{2-}(aq)$ $\xrightarrow{\text{base}}$ $Cl^-(aq)$ + $SO_4{}^{2-}(aq)$

    b. $Br_2(aq)$ + $I^-(aq)$ $\xrightarrow{\text{neutral}}$

    c. $NO_3{}^-(aq)$ + $S^{2-}(aq)$ $\xrightarrow{\text{acid}}$ $S(s)$ + $NO(g)$

    d. $Cr_2O_7{}^{2-}(aq)$ + $Sn^{2+}(aq)$ $\xrightarrow{\text{acid}}$ $Cr^{3+}(aq)$ + $Sn^{4+}(aq)$

    e. $Cu(s)$ + Ag^+ $\xrightarrow{\text{neutral}}$

Name ______________________

Student ID number ______________

Section __________ Date __________

Instructor ____________________

Experiment 44

POST-LABORATORY QUESTIONS

1. Indicate the chemical identity (name and formula) of each precipitate that was produced and provided the driving force for reaction when designated pairs of lettered aqueous solutions were mixed.

2. Write a balanced complete equation for each chemical reaction in which a precipitate formed when designated pairs of lettered aqueous solutions were mixed.

3. Indicate the chemical identity (name and formula) of each gas that was evolved and provided the driving force for reaction when designated pairs of lettered aqueous solutions were mixed.

4. Write a balanced complete equation for each chemical reaction in which a gas was evolved when designated pairs of lettered aqueous solutions were mixed.

5. Indicate the chemical identity (name and formula) of each weak electrolyte that was formed and provided the driving force for reaction when designated pairs of lettered aqueous solutions were mixed.

6. Write a balanced complete equation for each chemical reaction in which a weak electrolyte was formed when designated pairs of lettered aqueous solutions were mixed.

7. Write a balanced net ionic equation for each case in which a redox reaction occurred when designated pairs of lettered aqueous solutions were mixed.

APPENDIX

A. Significant Figures

Some numbers are pure numbers and are not subject to the uncertainties of measurement. Such numbers are said to be <u>exact</u>. For example, the number of runs performed in an experiment is an exact number. If you performed three runs, there is no uncertainty as to whether you completed 2.8, 2.97, or 3.1 runs. You completed exactly 3 runs, and the number can be expressed as 3.00... to an infinite number of digits. Defined quantities such as 1000 mL/1 L are also exact numbers and have an infinite number of significant figures. You should learn to recognize exact numbers.

Most experimental measurements in the laboratory are inexact numbers that have particular uncertainties attached to them. The numbers represent approximations rather than exact quantities. The concept of significant figures is designed to indicate the reliability of a measurement or the uncertainty in that measurement and to provide the maximum amount of information and no misinformation.

By convention, **significant figures** in a number **include all digits known with certainty and one additional doubtful or estimated digit.** Calibrations on a device often give a clue as to the appropriate number of significant figures. An example using a Celsius thermometer is shown in FIGURE A.1.

FIGURE A.1. Reading a thermometer scale.

Note that the thermometer is calibrated to the nearest 0.5°C and that the end of the mercury column lies between 22.0°C and 22.5°C. Thus, there is no doubt about the first two digits, 22. Our only remaining task is to estimate the distance between 22.0 and 22.5 and to record one doubtful or estimated digit. The estimated digit is probably 3, and the temperature would be recorded as 22.3°C. This conveys to a reader that the uncertainty in the temperature is at least ±0.1 and maybe larger. Obviously, it would be foolish to record more digits to the right of the 3 because they would have no significance. Common sense sometimes dictates that there is greater uncertainty and fewer significant figures than calibrations might suggest. Thus, the temperature may be constantly changing because of the offsetting effects of lighted Bunsen burners around you and a cool breeze from a nearby window. Alternatively, later calibration of your thermometer may indicate that it reads about 0.7°C high at least at 0°C. In either of these instances, the second 2 of 22 becomes uncertain or doubtful, only two significant figures are justified, and 22°C should be recorded, implying that the uncertainty is at least ±1°C. When you are recording numerical data in the laboratory, make certain that you write down the proper number of significant figures.

A zero may or may not be significant. If zeroes are embedded <u>within</u> a string of nonzero numbers, the zeroes are <u>always</u> significant. Thus, the zeroes in 5004.3 are significant, and 5004.3 has five significant figures. If zeroes are to the <u>left</u> of a string of numbers where they are used solely to locate the decimal point, as in 0.0054, the zeroes are <u>not</u> significant. Thus, 0.0054 has only two significant figures, as shown clearly when written in exponential notation, 5.4×10^{-2}. If zeroes are to the <u>right</u> of a string of numbers, but also to the <u>right</u> of the decimal and thus not being used to locate the decimal, as in 612.80, the zeroes <u>are</u> significant. Thus, 612.80 has five significant figures, as shown clearly when written in exponential notation, 6.1280×10^2. If zeroes are to the <u>right</u> of a string of numbers that does not contain a decimal point, special problems arise that can easily lead to confusion. For example, how many significant figures are indicated by 400, one, two, or three? The answer is YES!, confusion reigns, and some conventions must be adopted. In this book, we will follow the convention that zeroes to the right of the last nonzero digit are significant only when the number contains a decimal point, whether written in exponential notation or not, as shown in the following examples. Note that two significant figures cannot be indicated for this number without using exponential notation.

400 or 4×10^3	1 significant figure	$\geq \pm 100$ uncertainty
4.0×10^3	2 significant figures	$\geq \pm 10$ uncertainty
400. or 4.00×10^3	3 significant figures	$\geq \pm 1$ uncertainty

A general rule for the use of significant figures in mathematical calculations is that **calculations cannot decrease the uncertainty of an experimental result.** Moreover, unlike maximum error calculations in APPENDIX B, the use of significant figures in calculations does not increase the uncertainty of a computed result either. Four specific rules covered below assure conformance with the above general rule. It is even more important to use these specific rules rigorously now that electronic calculators and microcomputers allow us to perform calculations easily with large numbers of digits, often giving us the false impression of much smaller uncertainties than are justified.

1. Addition and Subtraction. For addition and/or subtraction, the last digit retained in the result should correspond to the first doubtful decimal place in any of the added or subtracted numbers. Stated another way, the number of decimal places in the answer should be the same as the number of decimal places in the value with the fewest places. For example, a calculator gives

$$38.23 + 0.186 + 0.0021 = 38.4181$$

and

$$38.23 - 0.186 - 0.0021 = 38.0419$$

However, when these numbers are placed in a column and the uncertain digits are underlined, we see that each result should be rounded off to the hundredths place.

```
  38.23            38.23
+ 0.186          - 0.186
+ 0.0021         - 0.0021
---------        ---------
  38.4181 -> 38.42   38.0419 -> 38.04
```

2. Multiplication and Division. For multiplication and/or division, the result contains the same number of significant figures as the least number in the data used in the calculation. For example, a calculator gives

$$(38.23)(0.186)(0.0021) = 0.0149326 \rightarrow 0.015 \text{ or } 1.5 \times 10^{-2}$$

and

$$\frac{38.23}{(0.186)(0.0021)} = 97875.064 \rightarrow 98,000 \text{ or } 9.8 \times 10^4$$

However, since the least number of significant figures in the data is two in 0.0021, the product and quotient should be rounded to two significant figures.

3. Common Logarithms to the Base 10., log x. When a common logarithm is computed, the following approximate convention is used - the answer contains as many digits in the mantissa (the numbers after the decimal) as there are significant figures in x. For example, a calculator gives

$$\log 28.23 = 1.4507109 \to 1.4507$$
characteristic   mantissa

$$\log 0.186 = -0.7304871 \to -0.730$$

$$\log 0.0021 = -2.6777807 \to -2.68$$

The first one is rounded to four significant figures in the mantissa, the second one to three, and the third one to two. It follows that when an inverse logarithm (antilog) to the base 10. is required, the answer contains the same number of significant figures as there are in the original mantissa. For example, a calculator gives

$$\text{antilog } 1.4507 = 28.229293 \to 28.23$$

$$\text{antilog } -0.730 = 0.1862087 \to 0.186$$

$$\text{antilog } -2.68 = 0.0020893 \to 0.0021$$

Comparing to the mantissa, the first one is rounded to four significant figures, the second one to three, and the third one to two.

4. Natural Logarithms to the Base e, ln x. When a natural logarithm is computed, the following approximate convention is used - the answer contains the same number of significant figures as there are in the original value of x. For example, a calculator gives

$$\ln 28.23 = 3.3403852 \to 3.340$$

$$\ln 0.186 = -1.6820086 \to -1.68$$

$$\ln 0.0021 = -6.1658179 \to -6.2$$

The first one is rounded to four significant figures including both the characteristic and the mantissa, the second one to three, and the third one to two. It follows when an inverse logarithm (antiln) to the base e is required, the answer contains the same number of significant figures as there are in the original value. For example, a calculator gives

$$\text{antiln } 3.340 = 28.219127 \to 28.22$$

$$\text{antiln } -1.68 = 0.186374 \to 0.186$$

$$\text{antiln } -6.2 = 0.0020294 \to 0.0020$$

The fact that the final result is not always the same number as the original number above indicates that these rules are only approximate.

5. General Rules. Several conventions are available for reducing the number of significant figures by rounding. In this book, the last digit retained is increased by 1 whenever the first digit to be dropped is 5 or greater. In working multi-step problems on a pocket calculator or microcomputer, you should carry through the calculation all of the digits allowed by the calculator or microcomputer, and round only at the end of the problem. This reduces the minor errors introduced by multiple rounding, but in a more practical way, eliminates the need to clear one number and punch in a rounded one. Remember: the fewer punches on your calculator or microcomputer, the better. Calculators and microcomputers allow us to both do more calculations per unit time and also make more errors per unit time!

Appendix

B. Errors in Measurements

Every experimental measurement has an error associated with it. This error may be comprised of accidental errors and/or constant errors.

Accidental errors arise because you cannot make perfect measurements no matter how hard you try and how careful you are. For example, you may weigh a crucible five times and not obtain the same mass because of air currents, temperature fluctuations, vibrations of the balance table, or other reasons. However, the probability of overweighing and of underweighing should be comparable so that your average value may be quite accurate and may be very close to the average value obtained by another student weighing the same crucible on the same balance. You can reduce the consequences of accidental errors further by weighing the crucible an even greater number of times before averaging. Thus, the purpose of averaging results from multiple runs of the same experiment is to decrease the impact of accidental errors on your result.

Constant errors arise when you use equipment that is incorrectly calibrated or that functions incorrectly the same way each time. The constant error is not a result of your care or lack of it, but rather a faulty instrument. For example, a thermometer might be calibrated to read $1^{\circ}C$ too low at all temperatures. Unless you calibrate the thermometer and correct for this miscalibration, even temperatures measured with great care will be $1^{\circ}C$ too low. Constant errors also arise when you use procedures that create bias in your measurements. For example, if the volume of a liquid is measured in one container and the mass of the liquid is measured in another container after a transfer, the mass, and thus the density calculated from it, will usually be too small because not all of the liquid will be transferred when it is poured from one container to another. Such a constant error can be eliminated by changing the procedure to eliminate the transfer so that volume and mass can be measured in the same container.

1. **Accuracy.** Accuracy is a measure of how close an experimental value is to the true value of the property that is measured. This assumes that the true value is known from an independent measurement by someone else. For example, if the density of cyclohexane, C_6H_{12}, is known to be 0.777 g/mL at $25^{\circ}C$, a student's experimentally determined value of 0.778 g/mL is highly accurate whereas another student's value of 0.743 g/mL has poor accuracy. If the true value is not known, an experimental result having excellent precision is usually considered to be accurate. This statement clearly makes the assumption that there are no constant errors associated with the measurement, but only accidental errors. If constant errors were present, excellent precision and poor accuracy could be obtained simultaneously. For example, suppose you are cutting four identical bookshelves using a defective tape measure. All four shelves might be cut the same length (high precision), but all might be too short to fit the frame that was already constructed (poor accuracy).

2. **Precision.** Precision is a measure of how close experimental results from several identical runs are to each other. Precision is a measure of the reproducibility of a set of results from different runs. It does not matter how close the average of these runs is to the true value; thus, precision and accuracy are defined independently of each other. Accuracy is affected dramatically by the presence of constant errors in a measurement, whereas precision is more commonly affected by the magnitude of accidental errors. If there are either a large number of accidental errors or a small number with large magnitudes, the precision will be poor. On the other hand, if there are few accidental errors, all of small magnitude, the precision will be very good.

In this book we will use average deviation as a measure of precision. Average deviation is calculated as follows: (1) calculate the average value of an experimental result from a set of identical runs; (2) calculate an absolute deviation for each run, which is the absolute value of the difference between the experimental result for each run and the average value for all runs; (3) add the absolute deviations calculated in (2); and (4) divide the sum of the absolute deviations by the number of runs to obtain the average deviation. The calculation of average deviation is illustrated for melting point data for cyclohexane for two different students. A $\pm$ precedes the calculated average deviation because there is equal probability of being higher or lower than the average value.

Student A		Student B	
6.8^{o}C	0.1^{o}C	5.8^{o}C	0.4^{o}C
7.3^{o}C	0.6^{o}C	6.3^{o}C	0.9^{o}C
5.9^{o}C	0.8^{o}C	4.2^{o}C	1.2^{o}C
3)20.0^{o}C	3)1.5^{o}C	3)16.3^{o}C	3)2.5^{o}C
6.7^{o}C	0.5^{o}C	5.4^{o}C	0.8^{o}C
6.7 $\pm$ 0.5^{o}C		5.4 $\pm$ 0.8^{o}C	

Student A has the smaller average deviation ($\pm$0.5^{o}C) and has data with higher precision. Since the true melting point of cyclohexane is 6.5^{o}C, the average value for student A (6.7^{o}C) is also more accurate.

   3. **Maximum Error.** Maximum error, as the name implies, is the maximum error expected for an experimental result if all of the sources of error worked against you in the same direction and gave you the largest error possible. In the parlance of the risk analysis of our day, it is a "worst case scenario." This is in contrast to calculations involving significant figures in APPENDIX A, which are such that the uncertainty in a calculated result is of about the same magnitude as the uncertainty in the original data before the calculations. It necessarily follows then that the more steps in a calculation, the more the maximum error will increase until ultimately the maximum error will not even fall in the same decimal place as the last significant figure. This is alright because significant figures and maximum errors are based on two very different principles in their calculations.

   In order to calculate maximum error for an experimental result, an explicit _estimated error_ must be recorded for each piece of raw data. Just knowing which digit is the last significant figure (the estimated or doubtful one) is not enough. For example, an experimental melting point for cyclohexane, C_6H_{12}, of 6.8^{o}C shows that the doubtful digit is in the tenths place. However, you must estimate how doubtful this digit is if the data is to be used later in a maximum error calculation. Thus, you might record the melting point as 6.8 $\pm$ 0.6^{o}C where 0.6^{o}C is the estimated error. It takes much experience to make judgments about reasonable estimated errors. A word of caution at the beginning is not to think too highly of your abilities and to estimate your errors on the high side if you have any doubts. You will get a check on yourself in several experiments in which you can compare average deviation with maximum error. Common sense should tell you that, except where rather drastic procedural mistakes were made, your average deviation should be less than your maximum error, since the latter represents your worst case. If your average deviation is greater than your maximum error, it means that you were too kind to yourself or too optimistic and your estimated errors were too small.

   Four rules will be used in this course to calculate maximum errors so as to conform to the principles discussed previously.

   In computations requiring addition and/or subtraction, the maximum error in the sum or difference is obtained by adding absolute values of the

Appendix

individual estimated errors. For example, a Celsius temperature and its estimated error can be converted to a Kelvin temperature and its maximum error as follows.

$$
\begin{array}{r}
6.8 \pm 0.6\,^{\circ}C \\
+\ 273.15 \pm 0.02 \\
\hline
279.95 \pm 0.62\ K \ \rightarrow\ 280.0 \pm 0.6\ K
\end{array}
$$

Note that you pay attention to significant figures and round the answer where appropriate. Note also that only one of the estimated errors makes a contribution in this calculation because the other one is so small compared to it. The relative error actually decreased in this calculation (0.6/6.8 versus 0.6/280.0). As a second example, the mass of sample and its maximum error can be calculated as follows.

$$
\begin{array}{lr}
\text{Mass of beaker and sample} & 16.843 \pm 0.002\ g \\
\text{Mass of beaker} & -\ 16.318 \pm 0.002\ g \\
\hline
\text{Mass of sample} & 0.525 \pm 0.004\ g
\end{array}
$$

Note that the relative error increased markedly in this calculation (0.002/16.318 versus 0.004/0.525). This frequently happens when two comparable numbers are subtracted from each other.

In computations requiring multiplication and/or division, the maximum error in the product or quotient is obtained by adding the relative errors that arise from the estimated errors (the estimated error divided by the value itself). For example, the number of moles of acid titrated and its maximum error can be calculated as molarity times volume.

$$
(0.103 \pm \mathbf{0.001}\ \underline{M})(0.0313 \pm \mathbf{0.0002}\ L) =
$$

$$
(0.103 \pm 0.001/0.103\ \underline{M})(0.0313 \pm 0.0002/0.0313\ L) =
$$

$$
(0.103\ \underline{M})(0.0313\ L) \pm (0.001/0.103 + 0.0002/0.0313) =
$$

$$
0.00322 \pm (\mathbf{0.01_6})(0.00322)\ \text{mol}\ H^+ =
$$

$$
0.00322 \pm \mathbf{0.00005}\ \text{mol}\ H^+ \ \text{or}\ (3.22 \pm \mathbf{0.05}) \times 10^{-3}\ \text{mol}\ H^+
$$

Note that absolute values of estimated errors are converted to relative errors in the second step by dividing the absolute estimated errors by the original data. Moreover, an absolute value of the maximum error is obtained in the fourth step by multiplying the sum of the relative errors (with a subscripted nonsignificant figure) by the calculated result. For clarification, absolute errors are boldfaced, whereas relative errors are not. As a second example, the density of a sample and its maximum error can be calculated as mass divided by volume.

$$
\frac{6.85 \pm \mathbf{0.03}\ g}{10.0 \pm \mathbf{0.1}\ mL} = \frac{6.85 \pm 0.03/6.85\ g}{10.0 \pm 0.1/10.0\ mL} = \frac{6.85\ g}{10.0\ mL} \pm (0.03/6.85 + 0.1/10.0)
$$

$$
= 0.685 \pm (\mathbf{0.01_4})(0.685)\ g/mL
$$

$$
= 0.685 \pm \mathbf{0.01}\ g/mL
$$

Again absolute values of estimated errors are converted to relative errors in the second step by dividing absolute estimated errors by the original data, and an absolute value of the maximum error is obtained in the fourth step by multiplying the sum of the relative errors (with a subscripted nonsignificant figure) by the calculated result. Note also that we have reached the point where the last significant figure and maximum error occur in different decimal places even after only a one-step calculation. This illustrates the principle that the last significant figure and the <u>estimated error</u> in experimental data must occur in the same decimal place, but the last significant figure and the <u>maximum error</u> after one or more steps in a calculation may occur in different decimal places.

In computations requiring common logarithms to the base 10. and inverse logarithms to the base 10., the following formula is used.

$$\log (x \pm \Delta x) = \log x \pm \frac{\Delta x}{2.3x}$$

For example, the log of a vapor pressure and its maximum error is computed as follows.

$$\log (38 \pm 4) = \log 38 \pm \frac{4}{(2.3)(38)} = 1.58 \pm 0.05$$

When taking the inverse log, the process is reversed, and it is recognized that

$$\frac{\Delta x}{2.3x} = 0.05, \text{ and } \Delta x = (0.05)(2.3x) = (0.05)(2.3) \text{ antilog } 1.58$$

Therefore,

$$\text{antilog } (1.58 \pm 0.05) = \text{antilog } 1.58 \pm (0.05)(2.3) \text{ antilog } 1.58 = 38 \pm 4$$

In computations requiring natural logarithms to the base e and inverse logarithms to the base e, a similar formula is used.

$$\ln (x \pm \Delta x) = \ln x \pm \frac{\Delta x}{x}$$

For example, the ln of a vapor pressure and its maximum error is computed as follows.

$$\ln (38 \pm 4) = \ln 38 \pm \frac{4}{38} = 3.6 \pm 0.1$$

When taking the inverse ln, the process is reversed, and it is recognized that

$$\frac{\Delta x}{x} = 0.1, \text{ and } \Delta x = 0.1x = (0.1) \text{ antiln } 3.6$$

Therefore,

$$\text{antiln } (3.6 \pm 0.1) = \text{antiln } 3.6 \pm (0.1) \text{ antiln } 3.6 = 37 \pm 4$$

Appendix

C. Properties of Water

TABLE C.1. Density of water at various temperatures.

Temperature (^{O}C)	Density (g/mL)	Temperature (^{O}C)	Density (g/mL)
0.0	0.9999	26.0	0.9968
5.0	1.0000	27.0	0.9965
10.0	0.9997	28.0	0.9963
15.0	0.9991	29.0	0.9960
16.0	0.9990	30.0	0.9957
17.0	0.9988	35.0	0.9941
18.0	0.9986	40.0	0.9922
19.0	0.9984	45.0	0.9903
20.0	0.9982	50.0	0.9881
21.0	0.9980	60.0	0.9832
22.0	0.9978	70.0	0.9778
23.0	0.9976	80.0	0.9718
24.0	0.9973	90.0	0.9653
25.0	0.9971	100.0	0.9584

TABLE C.2. Vapor pressure of water at various temperatures.

Temperature (^{O}C)	Pressure (mmHg)	Temperature (^{O}C)	Pressure (mmHg)
0.0	4.6	26.0	25.2
5.0	6.5	27.0	26.7
10.0	9.2	28.0	28.3
15.0	12.8	29.0	30.0
16.0	13.6	30.0	31.8
17.0	14.5	35.0	42.2
18.0	15.5	40.0	55.3
19.0	16.5	45.0	71.9
20.0	17.5	50.0	92.5
21.0	18.7	60.0	149.4
22.0	19.8	70.0	233.7
23.0	21.1	80.0	355.1
24.0	22.4	90.0	525.8
25.0	23.8	100.0	760.0